Solutions to

Weatherburn's

Elementary Vector Analysis

With Applications to Geometry and Mechanics

Solutions to

Weatherburn's

Elementary Vector Analysis

With Applications to Geometry and Mechanics

Rajnikant Sinha MSc (Gold Medalist), PhD
Ex-Professor
Faculty Head, JRS Tutorials
Durgakund, Varanasi
UP, India

CBS

CBS Publishers & Distributors Pvt Ltd

New Delhi • Bengaluru • Pune • Kochi • Chennai

Solutions to

Weatherburn's

Elementary Vector Analysis

ISBN: 978-81-239-2006-1

First Edition: 2011

Published by Satish Kumar Jain and produced by Vinod K. Jain for
CBS Publishers & Distributors Pvt Ltd
4819/XI Prahlad Street, 24 Ansari Road, Daryaganj
New Delhi 110 002, India. Fax: 011-23243014 Website: www.cbspd.com
e-mail: delhi@cbspd.com
cbspubs@vsnl.com
cbspubs@airtelmail.in

Branches

- Bengaluru: Seema House 2975, 17th Cross, K.R. Road, Banasankari 2nd Stage, Bengaluru 560 070, Karnataka
  Ph: +91-80-26771678/79 Fax: +91-80-26771680 e-mail: bangalore@cbspd.com
- Pune: Bhuruk Prestige, Sr. No. 52/12/2+1+3/2 Narhe, Haveli (Near Katraj-Dehu Road Bypass), Pune 411 051, Maharashtra
  Ph: 020-64704058, 64704059, 32342277
  Fax: +91-020-24300160 e-mail: pune@cbspd.com
- Kochi: 36/14 Kalluvilakam, Lissie Hospital Road, Kochi 682 018, Kerala
  Ph: +91-484-4059061-65 Fax: +91-484-4059065 e-mail: cochin@cbspd.com
- Chennai: 20, West Park Road, Shenoy Nagar, Chennai 600 030, Tamil Nadu
  Ph: +91-44-26260666, 26208620 Fax: +91-44-45530020 email: chennai@cbspd.com

Printed at Somya Printer, Delhi 110 053

Preface

"The best way to spend life is to indulge in accomplishing a nearly impossible task." These words of Picasso worked well with me during the writing of this book.

The fascinating world of vectors and tremendous applications of vectors in geometry and mechanics incline one to believe that vector is the nature's favourite language in which he wrote most of the laws of the universe.

The problems given in CE Weatherburn's *Elementary Vector Analysis* are classics and quite thought-provoking. They are very popular questions in the sense as they are often asked in examinations.

Their detailed solutions are the present-day need for the students. If one wishes to solve these problems, this solution-book will be of tremendous help. The solutions have been presented in an explicit manner. Since I believe in natural development of the subject, this feature is reflected in each solution given in the book. I have left no gap in presenting these solutions so that precious time of students is utilized in genuine progress.

Each of the eight chapters contains some beautiful but challenging problems. Their solutions require deep insight and expert guidance. The last three chapters are related to the two pillars of Newtonian mechanics, namely, kinematics of particles and rigid body, and dynamics of particles and rigid body. It spans from the inverse square law to Euler's dynamical equations of rigid body. The last chapter on statics of rigid body contains some marvelous problems. Their full solutions are a kind of the gift for the readers.

In my endeavour, I am especially grateful to Mr Y N Arjuna, Senior Director, CBS Publishers & Distributors.

Rajnikant Sinha

In loving memory
of
Rita Sahay

Contents

1

ADDITION AND SUBTRACTION OF VECTORS
CENTROIDS

Examples (page 15): (2):

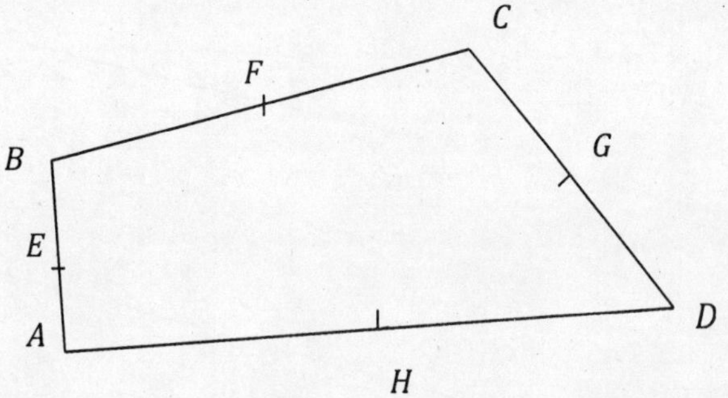

Let $ABCD$ be any skew quadrilateral, where $A(\vec{a}), B(\vec{b}), C(\vec{c}), D(\vec{d})$. Let E, F, G, H be mid-points of sides AB, BC, CD, DA respectively.

So, $E\left(\frac{\vec{a}+\vec{b}}{2}\right), F\left(\frac{\vec{b}+\vec{c}}{2}\right), G\left(\frac{\vec{c}+\vec{d}}{2}\right), H\left(\frac{\vec{d}+\vec{a}}{2}\right)$. So, mid-point of FH is $K_1\left(\frac{1}{2}\left(\frac{\vec{d}+\vec{a}}{2}+\frac{\vec{b}+\vec{c}}{2}\right)\right)$ or $K_1\left(\frac{1}{4}(\vec{a}+\vec{b}+\vec{c}+\vec{d})\right)$. Similarly $\vec{K}_2\left(\frac{1}{4}(\vec{a}+\vec{b}+\vec{c}+\vec{d})\right)$. Hence K_1 and K_2 are the same points. So, HF and EG bisect each other at their mid-point, whose position vector is $\frac{1}{4}(\vec{a}+\vec{b}+\vec{c}+\vec{d})$.

Also $\overrightarrow{HG} = \frac{\vec{c}+\vec{d}}{2} - \frac{\vec{a}+\vec{d}}{2} = \frac{1}{2}(\vec{c}-\vec{a})$ and $\overrightarrow{EF} = \frac{\vec{b}+\vec{c}}{2} - \frac{\vec{a}+\vec{b}}{2} = \frac{1}{2}(\vec{c}-\vec{a})$. So, $\overrightarrow{HG} = \overrightarrow{EF}$. Hence $EFGH$ is a parallelogram. Proved.

(5): Example 1. In this figure, P, Q are *harmonic conjugates* of A and B:

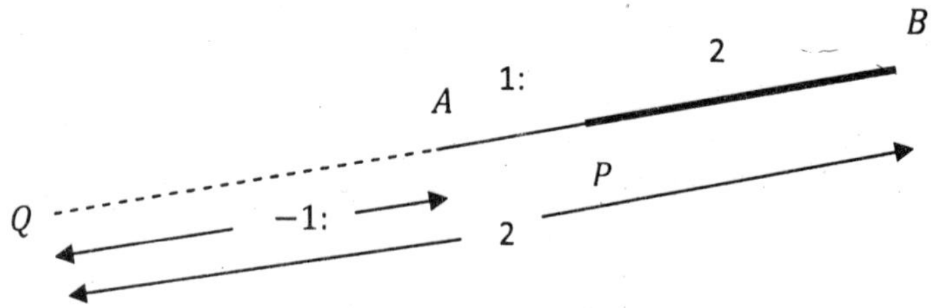

Example 2. In the following figure, P and Q are *harmonic conjugates* of A and B.

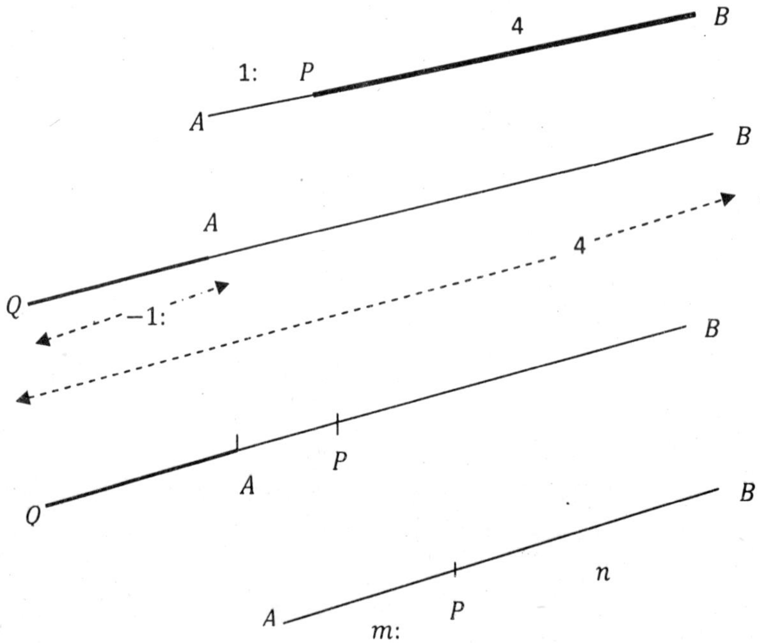

Here $\overrightarrow{AP}$ and $\overrightarrow{PB}$ have the same directions and $\frac{AP}{PB} = \frac{m}{n}$. So, $n(AP) = m(PB)$.

Also, $n(\overrightarrow{AP}) = m(\overrightarrow{PB})$ or, $n(\vec{p} - \vec{a}) = m(\vec{b} - \vec{p})$ or, $\vec{p} = \frac{m\vec{b}+n\vec{a}}{m+n}$. Ans. $\cdots(1)$

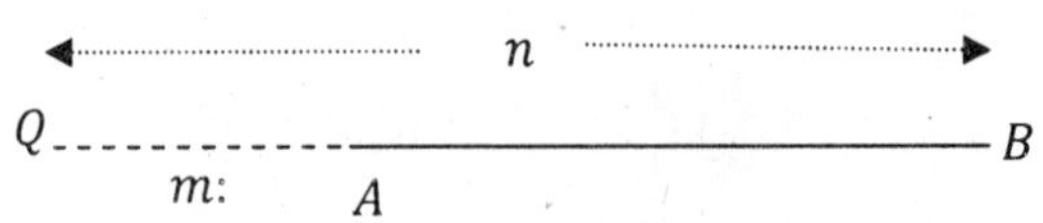

Here $\frac{QA}{QB} = \frac{m}{n}$, so $n(QA) = m(QB)$. From figure, $n(\overrightarrow{QA}) = m(\overrightarrow{QB})$ so,

$n(\vec{a} - \vec{q}) = m(\vec{b} - \vec{q})$ or, $\vec{q} = \frac{n\vec{a}-m\vec{b}}{n-m}$ $\cdots(2)$

From (1), $n\vec{a} + m\vec{b} = (m+n)\vec{p}$. From (2), $n\vec{a} - m\vec{b} = (n-m)\vec{q}$. On adding the last two equations, we get $2n\vec{a} = (m+n)\vec{p} + (n-m)\vec{q}$.

Hence $\vec{a} = \dfrac{(m+n)\vec{p} + (n-m)\vec{q}}{2n}$. Similarly $\vec{b} = \dfrac{(m+n)\vec{p} - (n-m)\vec{q}}{2m}$. Ans. Also, $\vec{a} = \frac{(n-m)\vec{q}+(n+m)\vec{p}}{(n-m)+(n+m)}$, and $\vec{b} = \frac{(-(n-m))\vec{q}+(n+m)\vec{p}}{(-(n-m))+(n+m)}$. So, A, B are harmonic conjugates of P, Q dividing PQ in the ratios $\pm(n-m):(n+m)$ respectively (see the figure given below). Proved

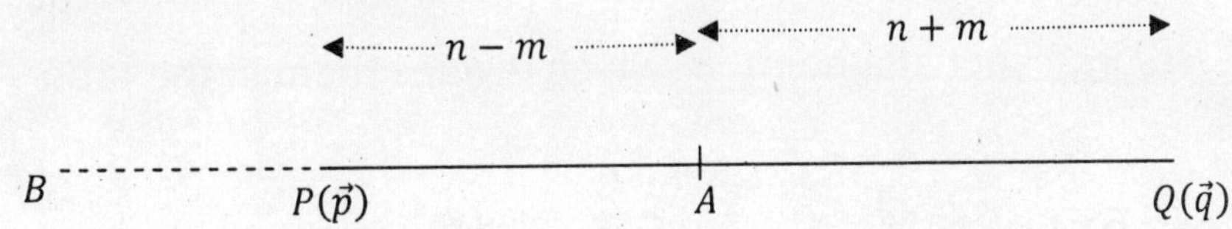

(6):

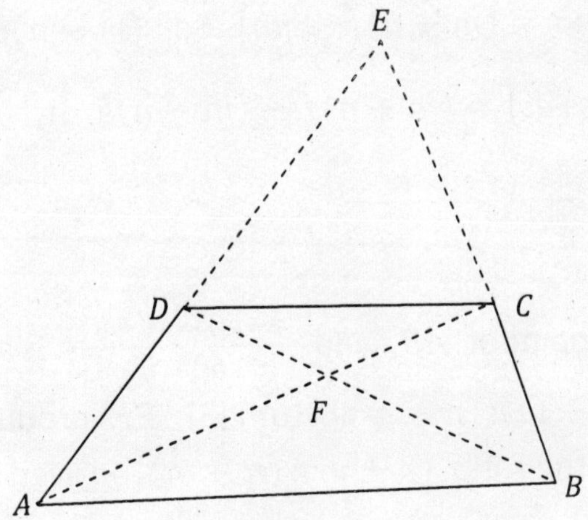

Trapezium

Given that $\frac{AB}{DC} = \frac{m}{n}$ or, $n(AB) = m(DC)$. So, from figure, $n\overrightarrow{AB} = m\overrightarrow{DC}$ or, $n(\vec{b} - \vec{a}) = m(\vec{c} - \vec{d})$ $\cdots(*)$ or, $m\vec{c} + n\vec{a} = n\vec{b} + m\vec{d}$ or, $\frac{m\vec{c}+n\vec{a}}{m+n}$

$$= \frac{m\vec{d}+n\vec{b}}{m+n} \qquad \cdots(1)$$

Here $\frac{m\vec{c}+n\vec{a}}{m+n}$ is the point on AC which divides it in the ratio $m:n$. Similarly $\frac{m\vec{d}+n\vec{b}}{m+n}$ is a point on BD which divides BD in the ratio $m:n$. So, from (1),

AC and BD has a common point. Since two straight line have only one common point, namely, their point of intersection, so AC and BD intersect at a common point F, which divides them in the ratio $m:n$. Proved.

From $(*)$, $m\vec{d} - n\vec{a} = m\vec{c} - n\vec{b}$ or $\frac{m\vec{d}-n\vec{a}}{m-n} = \frac{m\vec{c}-n\vec{b}}{m-n}$ $\cdots (2)$

Here $\frac{m\vec{d}-n\vec{a}}{m-n}$ is a point on AD, which divides it in the ratio $-m:n$. Similarly $\frac{m\vec{c}-n\vec{b}}{m-n}$ is a point on BC which divides it in the ratio $-m:n$. Hence, from (2), AD and BC intersect at a point E which divides them in the ratio $-m:n$. Proved.

Here $\vec{e} = \overrightarrow{OE} = \frac{m\vec{d}-n\vec{a}}{m-n}$ and $\vec{f} = \overrightarrow{OF} = \frac{m\vec{d}+n\vec{b}}{m+n}$ or $[m\vec{d} = (m-n)\vec{e}$ and $m\vec{d} = (m+n)\vec{f} - n\vec{b}]$. So, $(m-n)\vec{e} + n\vec{a} = (m+n)\vec{f} - n\vec{b}$ or $n(\vec{a}+\vec{b}) = (m+n)\vec{f} - (m-n)\vec{e}$ or,

$$\frac{\vec{a}+\vec{b}}{2} = \frac{(m+n)\vec{f}-(m-n)\vec{e}}{(m+n)-(m-n)} = \frac{(n+m)\vec{f}+(n-m)\vec{e}}{(n+m)+(n-m)} \qquad \cdots (3)$$

Here $\frac{\vec{a}+\vec{b}}{2}$ is the mid point of AB, and $\frac{(n+m)\vec{f}+(n-m)\vec{e}}{(n+m)+(n-m)}$ is a point on EF which divides it in the ratio $(n+m):(n-m)$. From (3), EF produced bisects AB while AB divides EF in the ratio $(n+m):(n-m)$. Proved

(7): Here $D\left(\frac{1\vec{c}+4\vec{b}}{1+4}\right), E\left(\frac{3\vec{a}+2\vec{c}}{3+2}\right), F\left(\frac{3\vec{b}+7\vec{a}}{3+7}\right)$.

So, $\overrightarrow{AD} + \overrightarrow{BE} + \overrightarrow{CF} = \left(\frac{\vec{c}+4\vec{b}}{5} - \vec{a}\right) + \left(\frac{3\vec{a}+2\vec{c}}{5} - \vec{b}\right) + \left(\frac{3\vec{b}+7\vec{a}}{10} - \vec{c}\right) = \frac{1}{10}(2\vec{c} + 8\vec{b} - 10\vec{a} + 6\vec{a} + 4\vec{c} - 10\vec{b} + 3\vec{b} + 7\vec{a} - 10\vec{c}) = \frac{1}{10}(3\vec{a} + \vec{b} - 4\vec{c})$ and $\overrightarrow{CK} = \frac{1\vec{b}+3\vec{a}}{1+3} - \vec{c} = \frac{1}{4}(3\vec{a} + \vec{b} - 4\vec{c})$. So, $4\overrightarrow{CK} = 10(\overrightarrow{AD} + \overrightarrow{BE} + \overrightarrow{CF}) \parallel \overrightarrow{CK}$ and their magnitudes are in the ratio 4:10 or 2:5. Ans.

Examples (page 20): (1): Here

$$AB = \sqrt{(4-2)^2 + (5-4)^2 + (1-(-1))^2} = \sqrt{4+1+4} = 3,$$

$BC = \sqrt{1^2+1^2+4^2} = 3\sqrt{2}$, $CA = \sqrt{1+4+4} = 3$. Hence $AB^2 + CA^2 = BC^2$. So, ABC is a right-angled triangle, where $\angle A = 90^\circ$. Here $\overrightarrow{AB} = (4,5,1) - (2,4,-1) = \left(4-2, 5-4, 1-(-1)\right) = (2,1,2)$ and $AB = |\overrightarrow{AB}| = |(2,1,2)| = \sqrt{2^2+1^2+2^2} = 3$. So, the unit vector along $\overrightarrow{AB}$ is $\left(\frac{1}{AB}\right)\overrightarrow{AB}$ or $\frac{1}{3}(2,1,2)$ or $\left(\frac{2}{3}, \frac{1}{3}, \frac{2}{3}\right)$. So required direction cosines of $\overrightarrow{AB}$ are $\frac{2}{3}, \frac{1}{3}, \frac{2}{3}$. Ans.

(2): Here $\overrightarrow{AB} = \overrightarrow{OB} - \overrightarrow{OA} = (4,2,3) - (2,-1,-3) = (2,3,6)$.
$\overrightarrow{AC} = (6,3,4) - (2,-1,-3) = (4,4,7)$. Also $AB = |\overrightarrow{AB}| = \sqrt{2^2+3^2+6^2} = \sqrt{49} = 7$ unit. $AC = \sqrt{4^2+4^2+7^2} = \sqrt{81} = 9$ unit. Ans.

Direction cosines of $\overrightarrow{AB}$ are $\frac{1}{7}(2,3,6)$ or $\frac{2}{7}, \frac{3}{7}, \frac{6}{7}$. Also the d.c. of $\overrightarrow{AC}$ are $\frac{1}{9}(4,4,7)$ or $\frac{4}{9}, \frac{4}{9}, \frac{7}{9}$. Ans.

(3): Here $P\left(\frac{3(6,3,4)+2(4,2,3)}{3+2}\right)$ or $P\left(\frac{1}{5}(26,13,18)\right)$ and $Q\left(\frac{-3(6,3,4)+2(4,2,3)}{-3+2}\right)$ or $Q\left(\frac{-3(6,3,4)+2(4,2,3)}{-3+2}\right)$ or, $Q\left(\frac{1}{-1}(-10,-5,-6)\right)$ or, $Q(10,5,6)$. Proved.

Here $\overrightarrow{AQ} = (10,5,6) - (2,-1,-3) = (8,6,9)$. So, direction cosines are proportional to 8,6,9. Ans.

(4): Let $\vec{a} \equiv (a_1, a_2, a_3)$ and $\vec{b} \equiv (b_1, b_2, b_3)$. Let $\vec{a}$ and $\vec{b}$ be parallel. If $\vec{a}$ and $\vec{b}$ have the same directions, then $\frac{1}{|\vec{a}|}\vec{a} = \frac{1}{|\vec{b}|}\vec{b}$.

So, $(a_1, a_2, a_3) = \vec{a} = \frac{|\vec{a}|}{|\vec{b}|}\vec{b} = \frac{|\vec{a}|}{|\vec{b}|}(b_1, b_2, b_3) = \left(\frac{|\vec{a}|}{|\vec{b}|}b_1, \frac{|\vec{a}|}{|\vec{b}|}b_2, \frac{|\vec{a}|}{|\vec{b}|}b_3\right)$.

So, $a_1 = \frac{|\vec{a}|}{|\vec{b}|}b_1$, $a_2 = \frac{|\vec{a}|}{|\vec{b}|}b_2$, $a_3 = \frac{|\vec{a}|}{|\vec{b}|}b_3$. So, components of $\vec{a}$ are proportional to the components of $\vec{b}$. If $\vec{a}$ and $\vec{b}$ have opposite directions, then $\frac{1}{|\vec{a}|}\vec{a} = -\frac{1}{|\vec{b}|}\vec{b}$. As above, we can see that the components of $\vec{a}$ are proportional to the components of $\vec{b}$. Here $\overrightarrow{AB} = (5,0,-2) - (1,-2,-8) = (4,2,6) = 2(2,1,3)$, and $\overrightarrow{BC} = (11,3,7) - (5,0,-2) = (6,3,9) = 3(2,1,3)$. So $\frac{1}{2}\overrightarrow{AB} = \frac{1}{3}\overrightarrow{BC}$. Hence $\overrightarrow{AB}$ and $\overrightarrow{BC}$ have the same directions. So, A, B, C are collinear points. Proved.

Also, $\frac{AB}{BC} = \frac{2}{3}$. Ans.

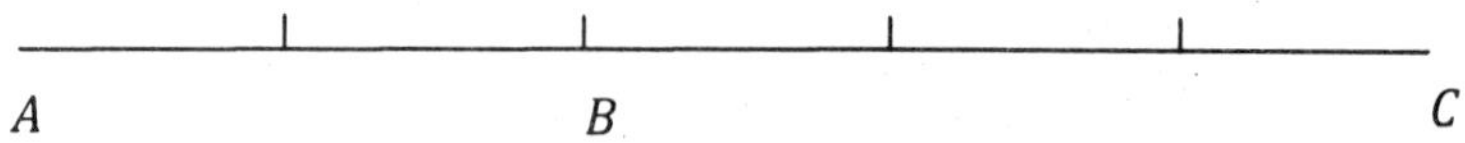

(5): Here D divide AC in the ratio $-2:3$. So, $D\left(\frac{(-2)(11,3,7)+3(1,-2,-8)}{(-2)+3}\right)$ or $D(-19,-12,-38)$. Ans.

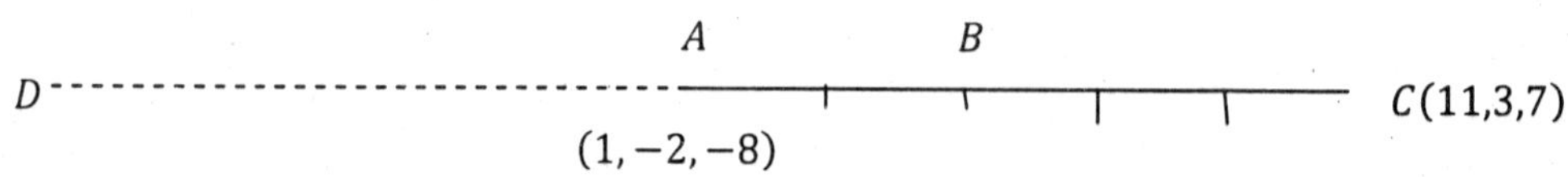

(6): Let $O(\vec{0}), B(\vec{b}), C(\vec{c})$ and $D(\vec{d})$.

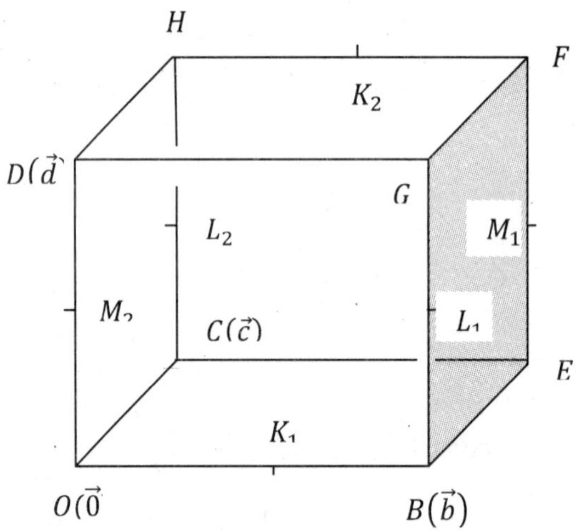

So, $\overrightarrow{OE} = \overrightarrow{OB} + \overrightarrow{BE} = \vec{b} + \overrightarrow{OC} = \vec{b} + \vec{c}$. So, $E(\vec{b} + \vec{c})$.

Here $\overrightarrow{OF} = \overrightarrow{OE} + \overrightarrow{EF} = (\vec{b} + \vec{c}) + \overrightarrow{EF} = (\vec{b} + \vec{c}) + \overrightarrow{OD} = \vec{b} + \vec{c} + \vec{d}$.

So, $F(\vec{b} + \vec{c} + \vec{d})$. Also, $\overrightarrow{OG} = \overrightarrow{OB} + \overrightarrow{BG} = \vec{b} + \overrightarrow{BG} = \vec{b} + \overrightarrow{OD} = \vec{b} + \vec{d}$.

So, $G(\vec{b}+\vec{d})$. Further

$\overrightarrow{OH}=\overrightarrow{OD}+\overrightarrow{DH}=\vec{d}+\overrightarrow{DH}=\vec{d}+\overrightarrow{OC}=\vec{d}+\vec{c}$. So, $H(\vec{c}+\vec{d})$. Here mid-point of edge OB is $K_1\left(\frac{\vec{b}}{2}\right)$ and mid-point of its opposite edge HF is $K_2\left(\frac{(\vec{c}+\vec{d})+(\vec{b}+\vec{c}+\vec{d})}{2}\right)$ or $K_2\left(\frac{\vec{b}}{2}+\vec{c}+\vec{d}\right)$. Here mid-point of edge BG is $L_1\left(\frac{\vec{b}+(\vec{b}+\vec{d})}{2}\right)$ or $L_1\left(\vec{b}+\frac{\vec{d}}{2}\right)$, and mid-point of opposite edge CH is $L_2\left(\frac{\vec{c}+(\vec{c}+\vec{d})}{2}\right)$ or $L_2\left(\vec{c}+\frac{\vec{d}}{2}\right)$. Here mid-point of edge EF is $M_1\left(\frac{(\vec{b}+\vec{c})+(\vec{b}+\vec{c}+\vec{d})}{2}\right)$ or $M_1\left(\vec{b}+\vec{c}+\frac{\vec{d}}{2}\right)$ and mid-point of its opposite edge OD is $M_2\left(\frac{\vec{d}}{2}\right)$. So, the mid-point of L_1L_2 is $L\left(\frac{\left(\vec{b}+\frac{\vec{d}}{2}\right)+\left(\vec{c}+\frac{\vec{d}}{2}\right)}{2}\right)$ or $L\left(\frac{\vec{b}+\vec{c}+\vec{d}}{2}\right)$. Also the mid-point of M_1M_2 is $M\left(\frac{\left(\vec{b}+\vec{c}+\frac{\vec{d}}{2}\right)+\frac{\vec{d}}{2}}{2}\right)$ or $M\left(\frac{\vec{b}+\vec{c}+\vec{d}}{2}\right)$. Finally, the mid-point of K_1K_2 is $K\left(\frac{\frac{\vec{b}}{2}+\left(\frac{\vec{b}}{2}+\vec{c}+\vec{d}\right)}{2}\right)$ or $K\left(\frac{\vec{b}+\vec{c}+\vec{d}}{2}\right)$. Hence K, L, M are the same points. This completes the proof.

<u>**Examples**</u> **(page 22): (1):** The mean centre of two points $A(\vec{a})$ and $B(\vec{b})$ is $\frac{\vec{a}+\vec{b}}{2}$, which is the mid-point of AB. <u>Proved.</u>

(2):

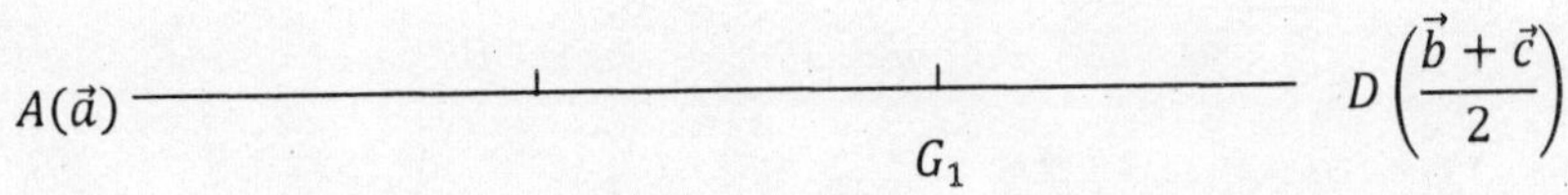

Let $A(\vec{a}), B(\vec{b}), C(\vec{c})$. Let D, E, F be the mid-points of BC, CA, AB respectively. So $F\left(\frac{\vec{a}+\vec{b}}{2}\right)$, $D\left(\frac{\vec{b}+\vec{c}}{2}\right)$, $E\left(\frac{\vec{c}+\vec{a}}{2}\right)$. Here the point of trisection of AD is $G_1\left(\frac{1\vec{a}+2\left(\frac{\vec{b}+\vec{c}}{2}\right)}{1+2}\right)$ or $G_1\left(\frac{\vec{a}+\vec{b}+\vec{c}}{3}\right)$, which is the mean centre of the vertices of $\Delta\, ABC$; etc. <u>Proved.</u>

(3): Here mean centre of vertices of triangle ABC is $G\left(\frac{\vec{a}+\vec{b}+\vec{c}}{3}\right)$. Let D, E, F divides each of BC, CA, AB in the ratio $\lambda: 1$ each, respectively. So, $D\left(\frac{\lambda\vec{c}+1\vec{b}}{\lambda+1}\right)$, $E\left(\frac{\lambda\vec{a}+1\vec{c}}{\lambda+1}\right)$, $F\left(\frac{\lambda\vec{b}+1\vec{a}}{\lambda+1}\right)$. So the mean centre of D, E, F is $H\left(\frac{1}{3}\left(\frac{\lambda\vec{c}+1\vec{b}}{\lambda+1}+\frac{\lambda\vec{a}+1\vec{c}}{\lambda+1}+\frac{\lambda\vec{b}+1\vec{a}}{\lambda+1}\right)\right)$ or $H\left(\frac{1}{3}\frac{1}{\lambda+1}\left(\lambda(\vec{a}+\vec{b}+\vec{c})+(\vec{a}+\vec{b}+\vec{c})\right)\right)$ or $H\left(\frac{1(\lambda+1)}{3(\lambda+1)}(\vec{a}+\vec{b}+\vec{c})\right)$ or $H\left(\frac{\vec{a}+\vec{b}+\vec{c}}{3}\right)$. So, H and G are the same points. Proved.

(4): Here mid-point of edge AB is $K_1\left(\frac{\vec{a}+\vec{b}}{2}\right)$ and mid-point of its opposite edge CD is $K_2\left(\frac{\vec{c}+\vec{d}}{2}\right)$. So mid-point of K_1K_2 is $K\left(\frac{1}{2}\left(\frac{\vec{a}+\vec{b}}{2}+\frac{\vec{c}+\vec{d}}{2}\right)\right)$ or $K\left(\frac{1}{4}(\vec{a}+\vec{b}+\vec{c}+\vec{d})\right)$. From symmetry, $L\left(\frac{1}{4}(\vec{a}+\vec{b}+\vec{c}+\vec{d})\right)$ and $M\left(\frac{1}{4}(\vec{a}+\vec{b}+\vec{c}+\vec{d})\right)$.

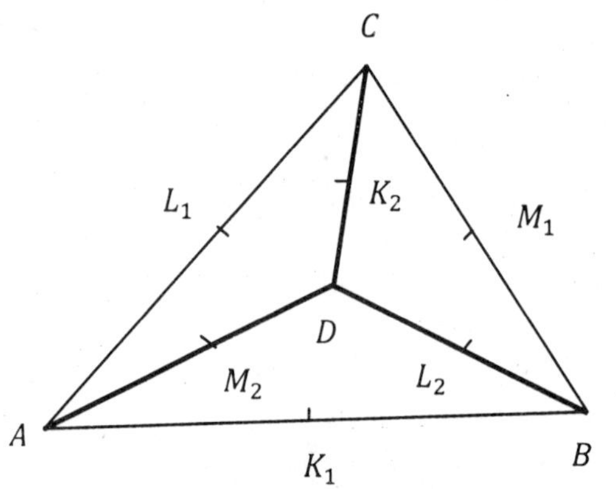

So, K, L, M are the same points, say H. Also the mean centre of tetrahedron is $G\left(\frac{\vec{a}+\vec{b}+\vec{c}+\vec{d}}{4}\right)$. So, G and H are also the same points. Proved.

Let G_1 be the mean centre of B, C, D. So, $G_1\left(\frac{\vec{b}+\vec{c}+\vec{d}}{3}\right)$. Also the point, which divides AG_1 in the ratio 3:1 is $T_1\left(\frac{3\left(\frac{\vec{b}+\vec{c}+\vec{d}}{3}\right)+1\vec{a}}{3+1}\right)$ or $T_1\left(\frac{\vec{a}+\vec{b}+\vec{c}+\vec{d}}{4}\right)$. So, T_1 coincides with G. Similarly T_2, T_3 coincide with G. Proved.

(5): Required centroid is $\frac{3(2,-3,3)+4(5,-3,-4)+5(2,-3,-1)}{3+4+5} = \frac{1}{12}(6+20+10,-9-12-15,9-16-5) = \frac{1}{12}(36,-36,-12) = (3,-3,-1)$. <u>Proved</u>.

(6): $LHS = \sum_i p_i \overrightarrow{A_iB_i} = \sum_i p_i(\vec{b_i} - \vec{a_i}) = \sum_i (p_i\vec{b_i} - p_i\vec{a_i}) = \sum_i p_i\vec{b_i} - \sum_i p_i\vec{a_i} = (\sum_i p_i)\overrightarrow{OG_2} - (\sum_i p_i)\overrightarrow{OG_1} = (N)\overrightarrow{OG_2} - (N)\overrightarrow{OG_1} = N(\overrightarrow{OG_2} - \overrightarrow{OG_1}) = (N)\overrightarrow{G_1G_2} = RHS$. <u>Proved</u>.

<u>Examples</u> (page 27): (1): This is solved in the text.

(2): Required resultant $= n\overrightarrow{OA} + m\overrightarrow{OB} = n(\overrightarrow{OR} + \overrightarrow{RA}) + m(\overrightarrow{OR} + \overrightarrow{RB}) = (m+n)\overrightarrow{OR} + m\overrightarrow{RB} + n\overrightarrow{RA}$ (since $\frac{AR}{RB} = \frac{m}{n}$, so, from figure, $n\overrightarrow{AR} = m\overrightarrow{RB}$ $\cdots(1)$

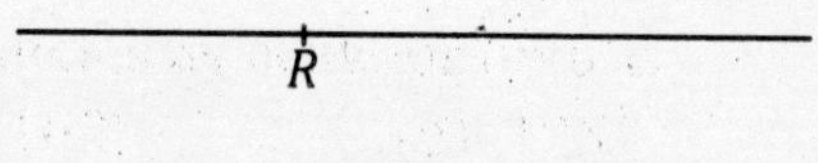

So required resultant $= (m+n)\overrightarrow{OR} + n\overrightarrow{AR} + n\overrightarrow{RA}$ (from (1))

$= (m+n)\overrightarrow{OR}$. <u>Proved</u>.

(3): This is solved in the text.

(4): This is solved in the text.

(5): This is solved in the text.

(6): This is solved in the text.

<u>Exercises on Chapter 1</u>

(1): This is solved in the text.

(2): This is solved in the text.

(3): This is solved in the text.

(4): Let $A_1, A_2, A_3, \cdots, A_n$ be n vertices of a regular polygon having centre O. Let A_i and A_j vertices have no mass and all other vertices have equal mass m. So, required c.m. G is given by $m\vec{a_1} + m\vec{a_2} + \cdots + m\overrightarrow{a_{i-1}} + m\overrightarrow{a_{i+1}} + \cdots +$

$m\overrightarrow{a_{j-1}} + m\overrightarrow{a_{j+1}} + \cdots + m\overrightarrow{a_n} = (n-2)m\overrightarrow{OG}$ or $\overrightarrow{OG} = \frac{(\overrightarrow{a_1}+\overrightarrow{a_2}+\cdots+\overrightarrow{a_n})+(\overrightarrow{a_i}+\overrightarrow{a_j})}{n-2} = \vec{0} + \frac{\overrightarrow{a_i}+\overrightarrow{a_j}}{n-2}$ (from question). Hence $\overrightarrow{OG} = \left(\frac{2}{n-2}\right)\left(\frac{\overrightarrow{a_i}+\overrightarrow{a_j}}{2}\right) = \left(\frac{2}{n-2}\right)\overrightarrow{OM}$ where M denotes the mid-point of the vacant vertices. So, required c.m. lies on the line OM such that $\overrightarrow{OG} = \left(\frac{2}{n-2}\right)\overrightarrow{OM}$. Ans.

(5): Here $\overrightarrow{CD} = \overrightarrow{OD} - \overrightarrow{OC} = \overrightarrow{BC} - \overrightarrow{AB} = \vec{b} - \vec{a}$ $\cdots(1)$. $\overrightarrow{DE} = -\overrightarrow{AB} = -\vec{a}$, $\overrightarrow{EF} = \overrightarrow{CB} = -\vec{b}$, $\overrightarrow{FA} = \overrightarrow{DC} = \vec{a} - \vec{b}$ (from (1). Ans.

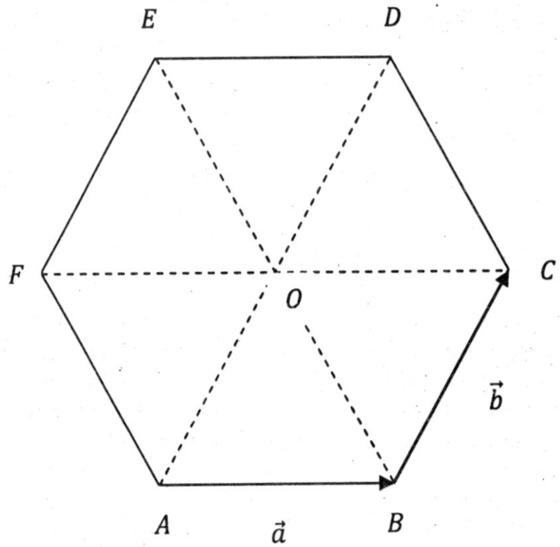

(6): Here $\omega = \frac{2\pi}{12}$rad/s. So, required position vectors are $\left(1\cos\left(\frac{2\pi}{12}\cdot 1\right), 1\sin\left(\frac{2\pi}{12}\cdot 1\right)\right)$ or $\left(\frac{\sqrt{3}}{2}, \frac{1}{2}\right)$ or $\frac{1}{2}(\sqrt{3}\vec{i} + \vec{j})$, $\left(1\cos\left(\frac{2\pi}{12}\cdot 3\right), 1\sin\left(\frac{2\pi}{12}\cdot 3\right)\right)$ or $(0,1)$ or $\vec{j}$, $\left(1\cos\left(\frac{2\pi}{12}\cdot 5\right), 1\sin\left(\frac{2\pi}{12}\cdot 5\right)\right)$ or $\left(\cos\left(\pi - \frac{\pi}{6}\right), \sin\left(\pi - \frac{\pi}{6}\right)\right)$ or $\left(-\cos\frac{\pi}{6}, \sin\frac{\pi}{6}\right)$ or $\left(-\frac{\sqrt{3}}{2}, \frac{1}{2}\right)$ or $\frac{1}{2}(\vec{j} - \sqrt{3}\vec{i})$; $\left(1\cos\left(\frac{2\pi}{12}\cdot 7\right), 1\sin\left(\frac{2\pi}{12}\cdot 7\right)\right)$ or $\left(\cos\left(\pi + \frac{\pi}{6}\right), \sin\left(\pi + \frac{\pi}{6}\right)\right)$ or $\left(-\cos\frac{\pi}{6}, -\sin\frac{\pi}{6}\right)$ or $\left(-\frac{\sqrt{3}}{2}, -\frac{1}{2}\right)$ or $\frac{-1}{2}(\vec{j} + \sqrt{3}\vec{i})$. Also, required position vectors are $\left(1\cos\left(\frac{2\pi}{12}\cdot\frac{3}{2}\right), 1\sin\left(\frac{2\pi}{12}\cdot\frac{3}{2}\right)\right)$ or

$\left(\cos\frac{\pi}{4}, \sin\frac{\pi}{4}\right)$ or $\left(\frac{1}{\sqrt{2}}, \frac{1}{\sqrt{2}}\right)$ or $\frac{1}{\sqrt{2}}(\vec{\imath}+\vec{\jmath})$; $\left(1\cos\left(\frac{2\pi}{12}\cdot\frac{9}{2}\right), 1\sin\left(\frac{2\pi}{12}\cdot\frac{9}{2}\right)\right)$ or $\left(\cos\frac{3\pi}{4}, \sin\frac{3\pi}{4}\right)$ or $\left(\cos\left(\pi-\frac{\pi}{4}\right), \sin\left(\pi-\frac{\pi}{4}\right)\right)$ or $\left(-\cos\frac{\pi}{4}, \sin\frac{\pi}{4}\right)$ or $\left(\frac{-1}{\sqrt{2}}, \frac{1}{\sqrt{2}}\right)$ or $\frac{1}{\sqrt{2}}(\vec{\jmath}-\vec{\imath})$. Ans.

(7): Required velocity vector

$\vec{v} = \frac{d}{dt}(r\cos\omega t, r\sin\omega t) = (-r\omega\sin\omega t, r\omega\cos\omega t) = r\omega(-\sin\omega t, \cos\omega t)$

where $r = 1$, and $\omega = \frac{2\pi}{12} = \frac{\pi}{6}$. So, $\vec{v} = \frac{\pi}{6}\left(-\sin\frac{\pi}{6}t, \cos\frac{\pi}{6}t\right)$.

Now $[\vec{v}]_{t=1.5s} = \frac{\pi}{6}\left(-\sin(\frac{\pi}{6}\cdot\frac{3}{2}), \cos(\frac{\pi}{6}\cdot\frac{3}{2})\right) = \frac{\pi}{6}\left(-\sin\frac{\pi}{4}, \cos\frac{\pi}{4}\right) = \frac{\pi}{6}\left(-\frac{1}{\sqrt{2}}, \frac{1}{\sqrt{2}}\right) = \frac{\pi}{6\sqrt{2}}(\vec{\jmath}-\vec{\imath})$. Ans.

$[\vec{v}]_{t=3s} = \frac{\pi}{6}\left(-\sin\left(\frac{\pi}{6}\cdot 3\right), \cos\left(\frac{\pi}{6}\cdot 3\right)\right) = \frac{\pi}{6}(-1,0) = -\frac{\pi}{6}\vec{\imath}$. Ans.

$[\vec{v}]_{t=7s} = \frac{\pi}{6}\left(-\sin\left(\frac{\pi}{6}\cdot 7\right), \cos\left(\frac{\pi}{6}\cdot 7\right)\right) = \frac{\pi}{6}\left(-\sin\left(\pi+\frac{\pi}{6}\right), \cos\left(\pi+\frac{\pi}{6}\right)\right) = \frac{\pi}{6}\left(\sin\frac{\pi}{6}, -\cos\frac{\pi}{6}\right) = \frac{\pi}{6}\left(\frac{1}{2}, \frac{-\sqrt{3}}{2}\right) = \frac{\pi}{12}(\vec{\imath}-3\vec{\jmath})$. Ans.

(8): Here $3\vec{\imath}+4\vec{\jmath} = \vec{v}_{boat} - \vec{v}_{water} = \vec{v}_{boat} - (\vec{\imath}-3\vec{\jmath})$. So, $\vec{v}_{boat} = 4\vec{\imath}+\vec{\jmath}$. Hence $\vec{v}_{boat} = \sqrt{4^2+1^2} = 16$ km/h. Here $\vec{v}_{boat}$

makes an angle $\tan^{-1}\frac{1}{4}$ North of East

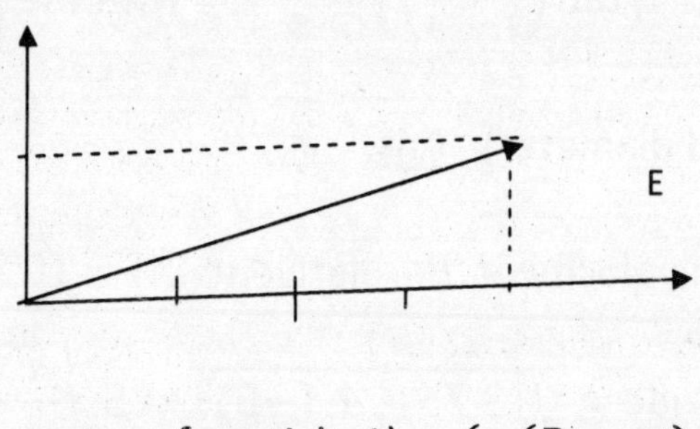

(9): Here $\vec{r_1} \equiv$ (position vector of particle 1) $= (-(R-vt), 0)$ and $\vec{r_2} \equiv$ (position vector of particle 2) $= \left(R\cos\left(\frac{v}{R}t\right), R\sin\left(\frac{v}{R}t\right)\right)$. So, required velocity

$= \dot{\vec{r_1}} - \dot{\vec{r_2}} = (v,0) — \frac{v}{R}R\sin\frac{vt}{R}, \frac{v}{R}R\cos\frac{vt}{R} = (v,0) — v\sin\frac{vt}{R}, v\cos\frac{vt}{R} =$
$\left(v + v\sin\frac{vt}{R}, -v\cos\frac{vt}{R}\right).$

So, magnitude of the required velocity.

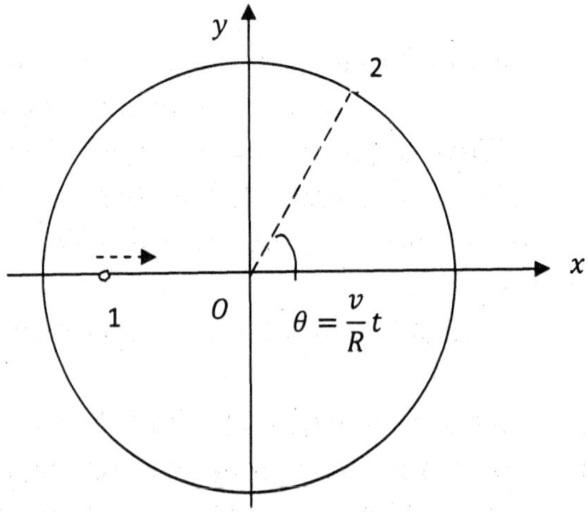

$$= \sqrt{\left\{v\left(1+\sin\frac{vt}{R}\right)\right\}^2 + \left(-v\cos\frac{vt}{R}\right)^2} = v\sqrt{2\left(1+\sin\frac{vt}{R}\right)} = v\sqrt{2(1+\sin\theta)}.$$

This velocity makes an angle

$$\tan^{-1}\frac{-v\cos\frac{vt}{R}}{v+v\sin\frac{vt}{R}} = \tan^{-1}\left(\frac{-\cos\theta}{1+\sin\theta}\right) = \tan^{-1}\frac{-\left(\cos^2\frac{\theta}{2}-\sin^2\frac{\theta}{2}\right)}{\left(\cos\frac{\theta}{2}+\sin\frac{\theta}{2}\right)^2} = \tan^{-1}\frac{\sin\frac{\theta}{2}-\cos\frac{\theta}{2}}{\sin\frac{\theta}{2}+\cos\frac{\theta}{2}} =$$

$\tan^{-1}\left(\frac{\tan\frac{\theta}{2}-1}{\tan\frac{\theta}{2}+1}\right) = \tan^{-1}\left(\tan\left(\frac{\theta}{2}-\frac{\pi}{4}\right)\right) = \frac{\theta}{2}-\frac{\pi}{4}$ with the positive direction of x-axis (which is a fixed diameter). Ans.

(10): Here (required velocity of B relative to A) $= (0,3.75) - (5,0) = (-5,3.75)$. Its magnitude $= \sqrt{(3.75)^2 + (-5)^2} = 5\sqrt{\frac{3^2}{4^2}+1} = 6.25$ m/s.

Here required shortest distance is given by $D = 15\sin\theta = 15\left(\frac{3.75}{\sqrt{(3.75)^2+(-5)^2}}\right) =$

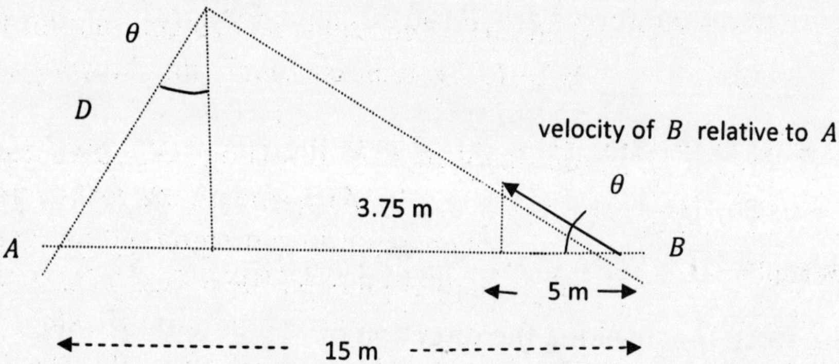

$15\left(\frac{15}{4}\right)\frac{1}{\frac{25}{4}} = 9$ m. Ans. Required duration $= \frac{15-D\sin\theta}{5} = \frac{15-9\left(\frac{15/4}{25/4}\right)}{5} = \frac{48}{25} = 1\frac{23}{25}$ s. Ans.

(11):

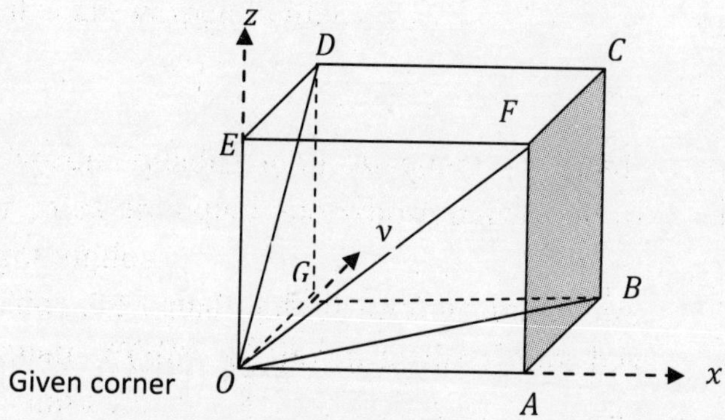

Let a be the length of each side of the cube. Here required sum $= \overrightarrow{OB} + \overrightarrow{OF} + \overrightarrow{OD} = (a,a,0) + (a,0,a) + (0,a,a) = (2a,2a,2a) = 2(a,a,a) = 2(\overrightarrow{OC}) =$ 2(diagonal of the cube from the given corner). Ans.

(12): Using the above figure of question (11), resultant force is $(3 \text{ lb. wt.})\frac{1}{|(1,1,0)|}(1,1,0) + (1 \text{ lb. wt.})\frac{1}{|(1,0,1)|}(1,0,1) + (2 \text{ lb. wt.})\frac{1}{|(0,1,1)|}(0,1,1) = \frac{1}{\sqrt{2}}(4,5,3)$ lb. wt. Its magnitude is $\frac{1}{\sqrt{2}}\sqrt{9+16+25} = \frac{5\sqrt{2}}{\sqrt{2}} = 5$ lb. wt. and it acts along $4\vec{\imath} + 5\vec{\jmath} + 3\vec{k}$. Ans.

(13):

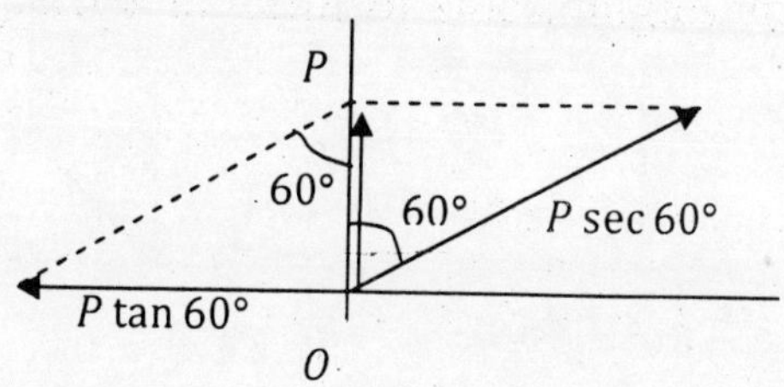

From figure, required forces are $P\tan 60°$ and $P\sec 60°$ or ($P\sqrt{3}$ and $2P$). Ans.

(14): If $|\vec{P}+\vec{Q}| = |\vec{P}|$ and $(\vec{P}+\vec{Q})\cdot\vec{P} = 0$ then $P^2+Q^2+2\vec{P}\cdot\vec{Q} = P^2$ and $P^2+\vec{P}\cdot\vec{Q} = 0$. So, $Q^2+2(\vec{P}.\vec{Q}) = 0$ and $\vec{P}.\vec{Q} = -P^2$. So, $Q^2-2P^2 = 0$ or $Q = \sqrt{2}P$. Also, $\vec{P}\cdot\vec{Q} = -P^2$ gives $P(\sqrt{2}P)\cos\theta = -P^2$. So, $\cos\theta = -\frac{1}{\sqrt{2}} = \cos 135°$. So, force $\vec{Q}$ acts along the direction of $135°$ with $\vec{P}$. Ans.

(15): Here required resultant $= \overrightarrow{AB}+\overrightarrow{AD}+\overrightarrow{CB}+\overrightarrow{CD} = (\overrightarrow{AP}+\overrightarrow{PQ}+\overrightarrow{QB}) + (\overrightarrow{AP}+\overrightarrow{PQ}+\overrightarrow{QD}) + (\overrightarrow{CP}+\overrightarrow{PQ}+\overrightarrow{QB}) + (\overrightarrow{CP}+\overrightarrow{PQ}+\overrightarrow{QD}) = 4(\overrightarrow{PQ})$. Proved.

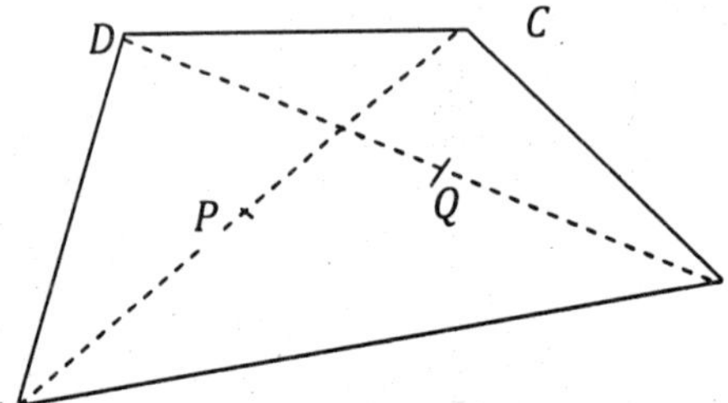

(16): Here $a = 1$. Now $\overrightarrow{r_{cm}} = \frac{1}{1+2+3+4+5+6+7+8}(1(0,0,0)+2(a,0,0)+3(a,a,0)+4(0,a,0)+5(0,0,a)+6(a,0,a)+7(a,a,a)+8(0,a,a)) = \frac{1}{4\cdot 9}(18a, 22a, 26a) = a\left(\frac{1}{2},\frac{11}{18},\frac{13}{18}\right) = \left(\frac{1}{2},\frac{11}{18},\frac{13}{18}\right)$. So, c.m. is situated at distance $\frac{1}{2}$ from the face $ADD'A'$, and $\frac{13}{18}$ from $ABCD$. Ans.

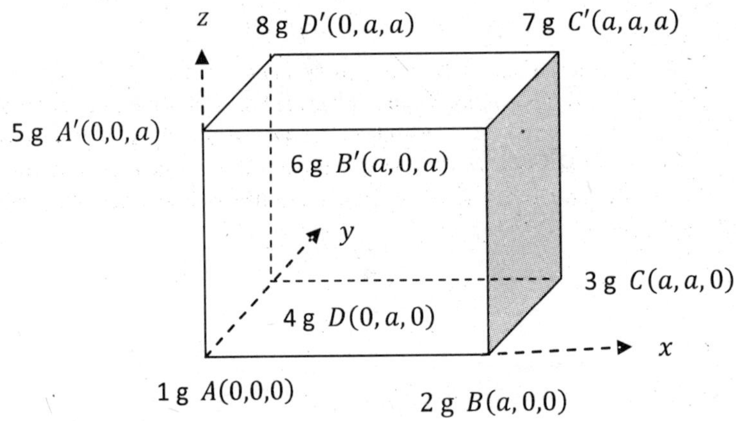

(17): Required centroid $\overrightarrow{r_{cm}} = \frac{1}{3n}\{(\vec{\imath}+2\vec{\imath}+3\vec{\imath}+\cdots+n\vec{\imath})+(\vec{\jmath}+2\vec{\jmath}+3\vec{\jmath}+\cdots+n\vec{\jmath})+(\vec{k}+2\vec{k}+3\vec{k}+\cdots+n\vec{k})\} = \frac{n(n+1)}{2}(\vec{\imath}+\vec{\jmath}+\vec{k})\left(\frac{1}{3n}\right) = \frac{1}{6}(n+1)(\vec{\imath}+\vec{\jmath}+\vec{k})$. Ans.

(18):

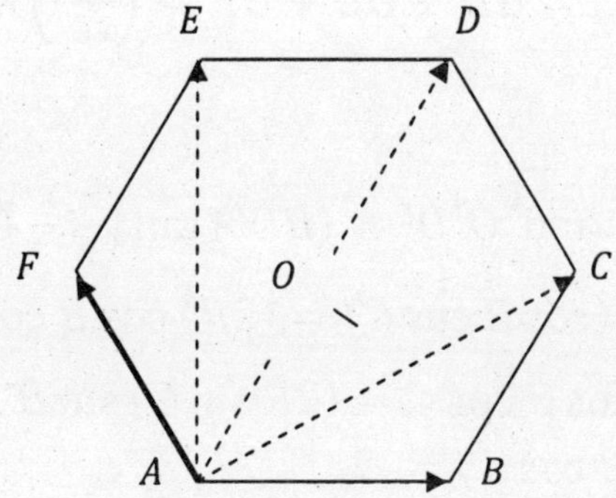

Required resultant $= k\overrightarrow{AB} + kAC + k\overrightarrow{AD} + k\overline{A}E + k\overrightarrow{AF}$ (where k is the constant of proportionality) $= k(\overrightarrow{OB} - \overrightarrow{OA} + \overrightarrow{OC} - \overrightarrow{OA} + \overrightarrow{OD} - \overrightarrow{OA} + \overrightarrow{OE} - \overrightarrow{OA} + \overrightarrow{OF} - \overrightarrow{OA}) = k(\overrightarrow{OD} + 5\overrightarrow{AO}) = k\left\{\frac{1}{2}\overrightarrow{AD} + 5\left(\frac{1}{2}\overrightarrow{AD}\right)\right\} = 3k\overrightarrow{AD} = 6k\overrightarrow{AO}$.

So required resultant force is proportional to $6\overrightarrow{AO}$ where O is the centre of the regular hexagon. Ans.

(19):

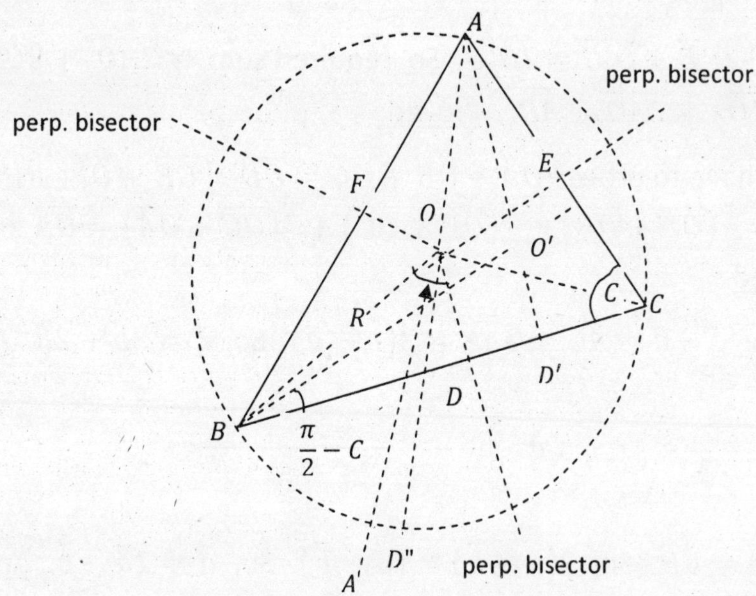

Here $\overrightarrow{OA}+\overrightarrow{OB}+\overrightarrow{OC}=\left(\overrightarrow{OO'}+\overrightarrow{O'A}\right)+\left(\overrightarrow{OO'}+\overrightarrow{O'B}\right)+\left(\overrightarrow{OO'}+\overrightarrow{O'C}\right)=3\overrightarrow{OO'}+\left(\overrightarrow{O'A}+\overrightarrow{O'B}+\overrightarrow{O'C}\right)$ $\cdots(1)$

Now $\overrightarrow{OA}+\overrightarrow{OB}+\overrightarrow{OC}=\frac{1}{2}\left\{\left(\overrightarrow{OA}+\overrightarrow{OB}\right)+\left(\overrightarrow{OB}+\overrightarrow{OC}\right)+\left(\overrightarrow{OC}+\overrightarrow{OA}\right)\right\}=\frac{1}{2}\left(2\overrightarrow{OF}+2\overrightarrow{OD}+2\overrightarrow{OE}\right)=\overrightarrow{OD}+\overrightarrow{OE}+\overrightarrow{OF}=\left(\frac{OD}{AO'}\right)\overrightarrow{AO'}+\left(\frac{OE}{BO'}\right)\overrightarrow{BO'}+\left(\frac{OF}{CO'}\right)\overrightarrow{CO'}$ $\cdots(2)$

Clearly $OD=R\cos A$ and $O'D'=(BD')\tan\left(\frac{\pi}{2}-C\right)=(BD')\cot C=(c\cos B)\cot C=\frac{c}{\sin C}(\cos B\cos C)=(2R)\cos B\cos C$. So, $AO'=AD'-O'D'=c\sin B-2R\cos B\cos C=(2R\sin C)\sin B-2R\cos B\cos C=2R(\sin B\sin C-\cos B\cos C)=$

$-2R\cos(B+C)=-2R\cos(\pi-A)$ $\therefore AO'=2R\cos A$. Hence $\frac{OD}{AO'}=\frac{1}{2}$, etc. $\cdots(*)$. From $(*)$ and (2), $\overrightarrow{OA}+\overrightarrow{OB}+\overrightarrow{OC}=\frac{1}{2}\overrightarrow{AO'}+\frac{1}{2}\overrightarrow{BO'}+\frac{1}{2}\overrightarrow{CO'}$.

So, $2\left(\overrightarrow{OA}+\overrightarrow{OB}+\overrightarrow{OC}\right)=-\left(\overrightarrow{O'A}+\overrightarrow{O'B}+\overrightarrow{O'C}\right)$. Hence, from (1), $\overrightarrow{OA}+\overrightarrow{OB}+\overrightarrow{OC}=3\overrightarrow{OO'}-2\left(\overrightarrow{OA}+\overrightarrow{OB}+\overrightarrow{OC}\right)$ or, $\overrightarrow{OA}+\overrightarrow{OB}+\overrightarrow{OC}=\overrightarrow{OO'}$ and $\overrightarrow{O'A}+\overrightarrow{O'B}+\overrightarrow{O'C}=2\overrightarrow{O'O}$. Proved.

(20): Here required sum $=\overrightarrow{AO'}+\overrightarrow{O'B}+\overrightarrow{O'C}=\overrightarrow{AO'}+\left(2\overrightarrow{O'O}-\overrightarrow{O'A}\right)$,

since $\overrightarrow{O'A}+\overrightarrow{O'B}+\overrightarrow{O'C}=2\overrightarrow{O'O}$. So, required sum $=2\overrightarrow{AO'}+2\overrightarrow{O'O}=2\left(\overrightarrow{AO'}+\overrightarrow{O'O}\right)=2\overrightarrow{AO}=\overrightarrow{AD}''$. Proved.

(21): We have to prove: $\overrightarrow{OA}+\overrightarrow{OB}+\overrightarrow{OC}=\overrightarrow{OD}+\overrightarrow{OE}+\overrightarrow{OF}$. $RHS=\overrightarrow{OD}+\overrightarrow{OE}+\overrightarrow{OF}=\frac{1}{2}\left(\overrightarrow{OA}+\overrightarrow{OB}\right)+\frac{1}{2}\left(\overrightarrow{OB}+\overrightarrow{OC}\right)+\frac{1}{2}\left(\overrightarrow{OC}+\overrightarrow{OA}\right)=\overrightarrow{OA}+\overrightarrow{OB}+\overrightarrow{OC}=LHS$. Proved.

(22): Here $\vec{c}-\vec{a}=\overrightarrow{AC}=3\overrightarrow{AB}=3\left(\vec{b}-\vec{a}\right)$. So, $\vec{c}=3\vec{b}-2\vec{a}$. Ans.

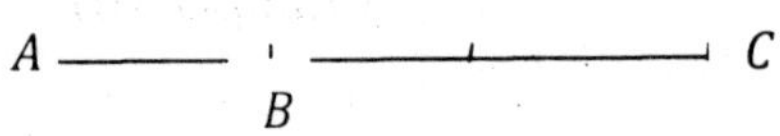

Given: $2\overrightarrow{BA}=\overrightarrow{BD}$. So, $2\left(\vec{a}-\vec{b}\right)=\left(\vec{d}-\vec{b}\right)$. So, $\vec{d}=2\vec{a}-\vec{b}$. Ans.

(23): Here resultant force $= \overrightarrow{PA} + \overrightarrow{PB}$. Let $P(x, y), C(c_1, c_2), B(c, 0)$ (see the above figure). Here $P(x, y), C(c_1, c_2), D\left(\frac{c}{2}, 0\right)$ are collinear, so $\frac{y-c_2}{x-c_1} = \frac{c_2-0}{c_1-\frac{c}{2}}$. So,

$y - c_2 = \frac{2c_2}{2c_1-c}(x - c_1)$, which is a straight line passing through C and the mid-point of AB. Ans.

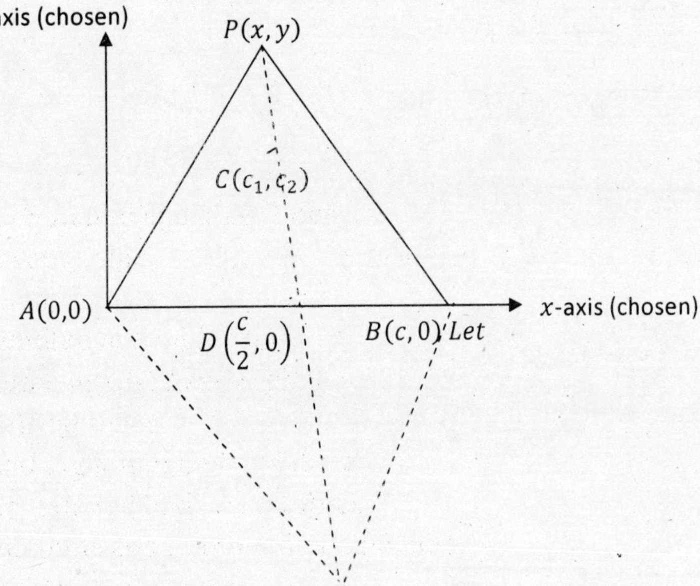

(24): Here $\overrightarrow{PQ} = \overrightarrow{AP} + \overrightarrow{PB} + \overrightarrow{PC} = \overrightarrow{AB} + \overrightarrow{PC}$. So, $\overrightarrow{CP} + \overrightarrow{PQ} = \overrightarrow{AB}$ and hence $\overrightarrow{CQ} = \overrightarrow{AB}$. So, $ABQC$ is a parallelogram for every position of P on BC. So, for

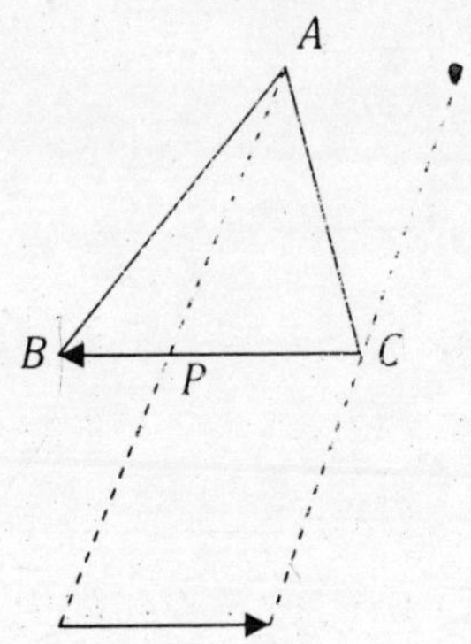

every position of P, resultant passes through the fixed point Q. Proved.

(25):

Let $\vec{a}, \vec{b}, \vec{c}, \vec{d}$ be any 4 non-coplanar vectors. Here $\vec{d} = \overrightarrow{OD} = \overrightarrow{OA} + \overrightarrow{AB} + \overrightarrow{BD} = \left(\frac{OA}{a}\right)\vec{a} + \left(\frac{AB}{b}\right)\vec{b} + \left(\frac{BD}{c}\right)\vec{c}$. So, $\vec{d} = \lambda\vec{a} + \mu\vec{b} + \nu\vec{c}$ where $\lambda \equiv \frac{OA}{a}$, $\mu \equiv \frac{AB}{b}$, $\nu \equiv \frac{BD}{c}$. So, $\vec{d}$ is a linear combination of $\vec{a}, \vec{b}, \vec{c}$. Next

let $\vec{d} = \lambda\vec{a} + \mu\vec{b} + \nu\vec{c}$ $\cdots(1)$

and $\vec{d} = \lambda'\vec{a} + \mu'\vec{b} + \nu'\vec{c}$ $\cdots(2)$

To prove: $\lambda = \lambda', \mu = \mu', \nu = \nu'$. (1) – (2): $(\lambda - \lambda')\vec{a} + (\mu - \mu')\vec{b} = (\nu' - \nu)\vec{c}$.
If $\nu \neq \nu'$ then $\vec{c} = \left(\frac{\lambda-\lambda'}{\nu'-\nu}\right)\vec{a} + \left(\frac{\mu-\mu'}{\nu'-\nu}\right)\vec{b}$ $\cdots(3)$

Since $\left(\frac{\lambda-\lambda'}{\nu'-\nu}\right)\vec{a} + \left(\frac{\mu-\mu'}{\nu'-\nu}\right)\vec{b}$ is coplanar with $\vec{a}$ and $\vec{b}$, so $\vec{c}, \vec{a}, \vec{b}$ are coplanar vectors, which is a contradiction. So, $\nu = \nu'$. Similarly $\mu = \mu'$ and $\lambda = \lambda'$. Proved

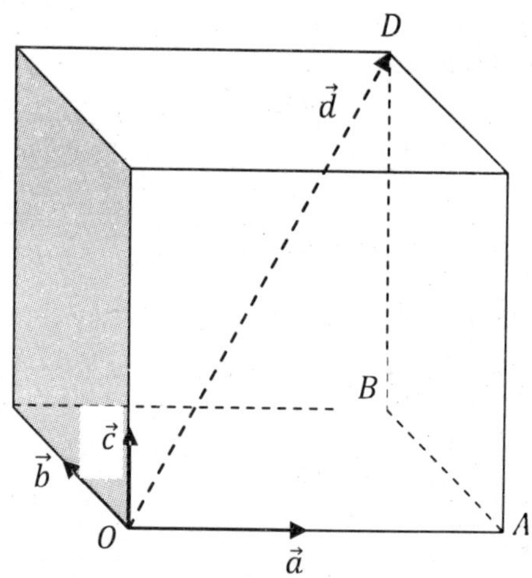

2

FURTHER GEOMETRICAL APPLICATIONS

<u>Examples</u> **(page 36): (1):** This is solved in the text.

(2): This is solved in the text.

(3): This is solved in the text.

<u>Examples</u> **(page 39): (1):** This is solved in the text.

(2): This is solved in the text.

(3): This is solved in the text.

(4): This is solved in the text.

Exercises on Chapter 2

(1): This is solved in the text.

(2): This is solved in the text.

(3): Equation of the straight line is $\vec{r} = (1,-2,1) + t\big((0,-2,3) - (1,-2,1)\big)$ or, $\vec{r} = (1,-2,1) + t(-1,0,2)$ <u>Ans</u>. Equation of the plane is $\vec{r} - (0,0,0) = \lambda\{(2,4,1) - (0,0,0)\} + \mu\{(4,0,2) - (0,0,0)\}$ or, $\vec{r} = \lambda(2,4,1) + \mu(4,0,2)$. <u>Ans</u>.

Let us find λ, μ, t such that $2\lambda + 4\mu = 1 - t$ $\cdots(1)$

$4\mu = -2$ $\cdots(2)$

and $\lambda + 2\mu = 2t + 1$. $\therefore$ $\lambda = -\frac{1}{2}$, $t = -\frac{1}{5}$, $\mu = \frac{11}{20}$. So, required point is $(1,-2,1) + \frac{-1}{5}(-1,0,2) = \left(1 + \frac{1}{5}, -2, 1 - \frac{2}{5}\right) = \left(\frac{6}{5}, -2, \frac{3}{5}\right)$. <u>Ans</u>.

(4): Required equation of line AB is $(x,y,z) = (2,-3,-1) + t\big((8,-1,2) - (2,-3,-1)\big)$ or, $(x-2, y+3, z+1) = t(6,2,3)$. So, $\frac{x-2}{6} = \frac{y+3}{2} = \frac{z+1}{3}$.

So required equation of line is $\frac{x-2}{6} = \frac{y+3}{2} = \frac{z+1}{3}$. Proved. Let $P(6t+2, 2t-3, 3t-1)$ be any point on the given line. If $AP = 14$, then $(6t+2-2)^2 + (2t-3+3)^2 + (3t-1+1)^2 = (14)^2$ or, $t = \pm 2$. So, required points are $(2 \pm 2\cdot 6, -3 \pm 2\cdot 2, -1 \pm 2\cdot 3) = (14,1,5)$ and $(-10,-7,-7)$. Ans.

(5): Required equation of line is $(x,y,z) = (3,-4,-2) + t(9,6,2)$. So, $(x-3, y+4, z+2) = t(9,6,2)$. So, $\frac{x-3}{9} = \frac{y+4}{6} = \frac{z+2}{2}$. Proved. Let $P(9t+3, 6t-4, 2t-2)$ be any point on the line. If $AP = 22$ then $(9t+3-3)^2 + (6t-4+4)^2 + (2t-2+2)^2 = (22)^2$. So, $t = \pm 2$.

So, $P(9(\pm 2)+3, 6(\pm 2)-4, 2(\pm 2)-2)$. So, required points are $(21,8,2)$ and $(-15,-16,-6)$. Ans.

(6):

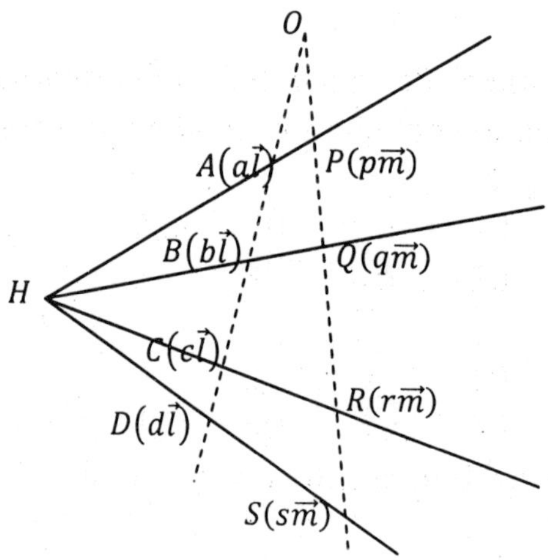

Let $\vec{l}$ be the unit vector along $\overrightarrow{OA}$, and $\vec{m}$ be the unit vector along $\overrightarrow{OP}$. To prove: $\frac{\frac{b-a}{d-b}}{\frac{c-a}{d-c}} = \frac{\frac{q-p}{s-q}}{\frac{r-p}{s-r}}$. We can suppose $\vec{h} = \lambda_1 a\vec{l} + \mu_1 p\vec{m} = \lambda_2 b\vec{l} + \mu_2 q\vec{m} = \lambda_3 c\vec{l} + \mu_3 r\vec{m} = \lambda_4 d\vec{l} + \mu_4 r\vec{m}$ where $1 = \lambda_1 + \mu_1 = \lambda_2 + \mu_2 = \lambda_3 + \mu_3 = \lambda_4 + \mu_4 \quad \cdots (*)$. So, $\lambda_1 a = \lambda_2 b$ and $(1-\lambda_1)p = (1-\lambda_2)q$, $\lambda_2 b = \lambda_4 d$ and $(1-\lambda_2)q = (1-\lambda_4)s$.

$LHS = \frac{b-a}{d-b}\frac{d-c}{c-a} = \frac{\frac{\lambda_1}{\lambda_2}a - a}{\frac{\lambda_2}{\lambda_4}b - b}\frac{\frac{\lambda_3}{\lambda_4}c - c}{\frac{\lambda_1}{\lambda_3}a - a} = \frac{\lambda_1-\lambda_2}{\lambda_2}\frac{\lambda_4}{\lambda_2-\lambda_4}\frac{a}{b}\frac{\lambda_3-\lambda_4}{\lambda_4}\frac{c}{a}\frac{\lambda_3}{\lambda_1-\lambda_3} = \frac{(\lambda_1-\lambda_2)(\lambda_3-\lambda_4)}{(\lambda_2-\lambda_4)(\lambda_1-\lambda_3)}$ and

$RHS = \frac{(\mu_1-\mu_2)(\mu_3-\mu_4)}{(\mu_2-\mu_4)(\mu_1-\mu_3)}$ (as above). Hence, on using $(*)$, $RHS = \frac{(\lambda_2-\lambda_1)(\lambda_4-\lambda_3)}{(\lambda_4-\lambda_2)(\lambda_3-\lambda_1)} = \frac{(\lambda_1-\lambda_2)(\lambda_3-\lambda_4)}{(\lambda_2-\lambda_4)(\lambda_1-\lambda_3)}$. Proved.

(7): Let $P\left(\frac{l\vec{a}+m\vec{b}}{l+m}\right), Q\left(\frac{m\vec{b}+n\vec{c}}{m+n}\right), R\left(\frac{n\vec{c}+t\vec{d}}{n+t}\right), S\left(\frac{t\vec{d}+l_1\vec{a}}{t+l_1}\right)$ be the points where the plane cuts the sides AB, BC, CD, DA respectively. Here P, Q, R, S are coplanar, so $\lambda_1\left(\frac{l\vec{a}+m\vec{b}}{l+m}\right) + \lambda_2\left(\frac{m\vec{b}+n\vec{c}}{m+n}\right) + \lambda_3\left(\frac{n\vec{c}+t\vec{d}}{n+t}\right) + \lambda_4\left(\frac{t\vec{d}+l_1\vec{a}}{t+l_1}\right) = \vec{0}$

where $\lambda_1 + \lambda_2 + \lambda_3 + \lambda_4 = 0$. Let us choose D as the origin. Then $\lambda_1\left(\frac{l\vec{a}+m\vec{b}}{l+m}\right) + \lambda_2\left(\frac{m\vec{b}+n\vec{c}}{m+n}\right) + \lambda_3\left(\frac{n\vec{c}}{n+t}\right) - (\lambda_1+\lambda_2+\lambda_3)\left(\frac{l_1\vec{a}}{t+l_1}\right) = \vec{0}$. So, $\left\{\frac{\lambda_1 l}{l+m} - \frac{(\lambda_1+\lambda_2+\lambda_3)l_1}{t+l_1}\right\}\vec{a} + \left\{\frac{\lambda_1 m}{l+m} + \frac{\lambda_2 m}{m+n}\right\}\vec{b} + \left\{\frac{\lambda_2 n}{m+n} + \frac{\lambda_3 n}{n+t}\right\}\vec{c} = \vec{0}$. Since $\vec{a}, \vec{b}, \vec{c}$ are non-coplanar, so, (coefficient of $\vec{a}$) $= 0 =$ (coefficient of $\vec{b}$) $=$ (coefficient of $\vec{c}$). So,

$l(t+l_1)\lambda_1 = (l+m)(\lambda_1+\lambda_2+\lambda_3)l_1$ $\cdots$ (1)

Also $(m+n)\lambda_1 + (l+m)\lambda_2 = 0$ $\cdots$ (2)

and $(n+t)\lambda_2 + (m+n)\lambda_3 = 0$ $\cdots$ (3)

To prove: $\frac{l}{m}\frac{m}{n}\frac{n}{t}\frac{t}{l_1} = 1$ that is, $l = l_1$. From (1), (2), (3), $l(t+l_1)\lambda_1 = l_1(l+m)\left\{\lambda_1 - \frac{(m+n)\lambda_1}{l+m} - \frac{n+t}{m+n}\frac{m+n}{\{-(l+m)\}}\lambda_1\right\}$ so,

$l(t+l_1) = l_1(l+m)\left\{\frac{l+m-(m+n)+n+t}{l+m}\right\}$ or $l(t+l_1) = l_1(l+t)$ or, $lt = l_1 t$ or, $l = l_1$. Proved.

(8): Required resultant force $= P_1\left(\frac{\overrightarrow{OA_1}}{OA_1}\right) + P_2\left(\frac{\overrightarrow{OA_2}}{OA_2}\right) + \cdots + P_n\left(\frac{\overrightarrow{OA_n}}{OA_n}\right) = \sum_i\left(\frac{P_i}{OA_i}\right)\overrightarrow{OA_i} = \left\{\sum_i \frac{P_i}{OA_i}\right\}\overrightarrow{OG}$ where G is the centroid of the points $A_1, A_2, \cdots, A_n$ with associated numbers $\frac{P_1}{OA_1}, \frac{P_2}{OA_2}, \cdots, \frac{P_n}{OA_n}$. Proved.

(9): Let $\vec{r} = \vec{a} + t\vec{d}$ $\cdots$ (1)

and $\vec{r} = \overrightarrow{a_1} + u\overrightarrow{d_1}$ $\cdots$ (2)

be two non-coplanar lines. So, $\overrightarrow{d_1} = \vec{d}$ and $\vec{a} - \overrightarrow{a_1} \neq t\vec{d} - u\overrightarrow{d_1}$ for any real values of t, u. Let $P(\vec{a} + t\vec{d})$ and $P_1(\vec{a} + t_1\vec{d})$ be any two points on the 1st line.

Let $Q(\vec{a_1}+u\vec{d_1})$ and $Q_1(\vec{a_1}+u_1\vec{d_1})$ be any two points on the 2nd line. To prove: PQ cannot be parallel to P_1Q_1. If not, otherwise, let $\overrightarrow{PQ} \parallel \overrightarrow{P_1Q_1}$. So, $(\vec{a_1}-\vec{a})+u\vec{d_1}-t\vec{d}=k\{(\vec{a_1}-\vec{a})+u_1\vec{d_1}-t_1\vec{d}\}$ for some scalar k. So, $(1-k)(\vec{a_1}-\vec{a})=(ku_1-u)\vec{d_1}-(kt_1-t)\vec{d}$ or, $\overrightarrow{AA_1}(=\vec{a_1}-\vec{a})=$ (a linear combination of $\vec{d}$ and $\vec{d_1}$), which is perpendicular to the line of shortest distance(S.D.) of the given two lines. Here A lies on (1) and A_1 lies on (2). Since A, A_1 are arbitrary points of lines (1) and (2) respectively, so every line joining two points of the given two lines is perpendicular to the line of S.D.; which is impossible. Proved.

(10): Here $\vec{m}=\frac{\vec{a}+\vec{b}}{2}, \vec{n}=\frac{\vec{c}+\vec{d}}{2}, \vec{b}-\vec{a}=\vec{c}-\vec{d}$. So, $2\vec{m}=\vec{a}+\vec{b}=\vec{a}+(\vec{a}+\vec{c}-\vec{d})$. So, $\frac{2\vec{m}+\vec{d}}{3}=\frac{2\vec{a}+\vec{c}}{3}\equiv\overrightarrow{OK}$. So, K divides DM in the ratio 2:1 and K divides AC in the ratio 1:2. So, DM cuts AC at its points of trisection K. Again $2\vec{n}=\vec{c}+(\vec{c}+\vec{a}-\vec{b})$. So, $\frac{2\vec{n}+\vec{b}}{3}=\frac{2\vec{c}+\vec{a}}{3}\equiv\overrightarrow{OL}$. So, L divides BN in the ratio 2:1 and L divides AC in the ratio 2:1. So, BN and AC intersect at the point of trisection L. Proved.

(11):

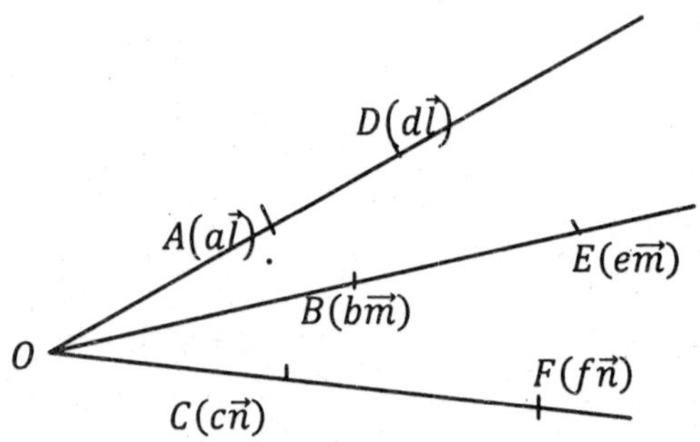

Here B, C, F, E are coplanar, so we can write $\lambda_1(b\vec{m})+\lambda_2(c\vec{n})+\lambda_3(f\vec{n})+\lambda_4(e\vec{m})=\vec{0}$ where $\lambda_1+\lambda_2+\lambda_3+\lambda_4=0$. So, $\frac{\lambda_1(b\vec{m})+\lambda_2(c\vec{n})}{\lambda_1+\lambda_2}=\frac{\lambda_3(f\vec{n})+\lambda_4(e\vec{m})}{\lambda_3+\lambda_4}\equiv\overrightarrow{OK_1}$, where K_1 is the point of intersection of BC and EF.

So, $K_1\left(\frac{\lambda_1(b\vec{m})+\lambda_2(c\vec{n})}{\lambda_1+\lambda_2}\right)$ or $K_1(t_1 b\vec{m}+(1-t_1)c\vec{n})$ where $t_1 \equiv \frac{\lambda_1}{\lambda_1+\lambda_2}$. Similarly $K_2\left(t_2(c\vec{n})+(1-t_2)(a\vec{l})\right)$ and $K_3\left(t_3(a\vec{l})+(1-t_3)(b\vec{m})\right)$. Let $\alpha\overrightarrow{k_1}+\beta\overrightarrow{k_2}+\gamma\overrightarrow{k_3}=\vec{0}$ or, $\alpha(t_1 b\vec{m}+(1-t_1)c\vec{n})+\beta\left(t_2(c\vec{n})+(1-t_2)(a\vec{l})\right)+\gamma\left(t_3(a\vec{l})+(1-t_3)(b\vec{m})\right)=\vec{0}$ or, $\{\lambda t_3 a+\beta(1-t_2)a\}\vec{l}+\{\alpha t_1 b+\gamma(1-t_3)b\}\vec{m}+\{\alpha(1-t_1)c+\beta t_3 c\}\vec{n}=\vec{0}$. If $\vec{l},\vec{m},\vec{n}$ are non-coplanar, so (coefficient of $\vec{l}$) = (coefficient of $\vec{m}$) = (coefficient of $\vec{n}$) = 0. So, $t_3\gamma+(1-t_2)\beta=0$ $\cdots(1)$

$t_1\alpha+(1-t_3)\gamma=0$ $\cdots(2)$

and $t_3\beta+(1-t_1)\alpha=0$ $\cdots(3)$

Now, on adding equations (1), (2) and (3), we get: $\alpha+\beta+\gamma=0$. So, K_1, K_2, K_3 are collinear points. Proved.

(12): Let the point of intersection of AB and DE be N, the point of intersection of BC and EF be L, the point of intersection of CA and FD be M. Given that L, M, N are collinear points. To prove: DA, EB, FC are concurrent. If $U(\vec{u})$ is the point of intersection of EB and FC, $V(\vec{v})$ is the point of intersection of FC and DA, $W(\vec{w})$ is the point of intersection of DA and EB then we have to prove that U, V, W are the same points, that is, $\vec{u}=\vec{v}=\vec{w}$. From symmetry, it is enough to prove that $\vec{v}=\vec{w}$. From the given conditions, we can write:

$$\begin{cases} \vec{l} \equiv r\vec{b}+(1-r)\vec{c}=r_1\vec{e}+(1-r_1)\vec{f} \\ \vec{m} \equiv s\vec{c}+(1-s)\vec{a}=s_1\vec{f}+(1-s_1)\vec{d} \\ \vec{n} \equiv t\vec{a}+(1-t)\vec{b}=t_1\vec{d}+(1-t_1)\vec{e} \quad \cdots(*) \end{cases}$$

for some scalars r, r_1, s, s_1, t, t_1. From question, $\exists\lambda,\mu,\nu$(not all zero) s.t. (i) $\lambda\vec{l}+\mu\vec{m}+\nu\vec{n}=\vec{0}$ and (ii) $\lambda+\mu+\nu=0$. Now, from $(*)$, $t\vec{a}-t_1\vec{d}=(t-1)\vec{b}+(1-t_1)\vec{e}$. Hence $\frac{t\vec{a}-t_1\vec{d}}{t-t_1}=\frac{(t-1)\vec{b}+(1-t_1)\vec{e}}{(t-1)+(1-t_1)}=\vec{w}$. Similarly $\frac{s\vec{c}-s_1\vec{f}}{s-s_1}=\frac{(s-1)\vec{a}+(1-s_1)\vec{d}}{(s-1)+(1-s_1)}=\vec{v}$. Now $\vec{v}=\vec{w}$ is equivalent to $\frac{t\vec{a}-t_1\vec{d}}{t-t_1}=\frac{(s-1)\vec{a}+(1-s_1)\vec{d}}{(s-1)+(1-s_1)}$ or $\{t(s-s_1)-(t-t_1)s+(t-t_1)\}\vec{a}+\{-t_1(s-s_1)-(t-t_1)+(t-t_1)s_1\}\vec{d}=\vec{0}$ or, $\{-ts_1+t_1s+t-t_1\}(\vec{a}-\vec{d})=$

$\vec{0}$. Since A and D are different points so, $t(1-s_1)-t_1(1-s)=0$ or, $\frac{t}{1-s}=\frac{t_1}{1-s_1}$. This is what we have to prove. Here, from (i), $\lambda(r\vec{b}+(1-r)\vec{c})+\mu(s\vec{c}+(1-s)\vec{a})+\nu(t\vec{a}+(1-t)\vec{b})=\vec{0}$ or, $(\mu(1-s)+\nu t)\vec{a}+(\nu(1-t)+\lambda r)\vec{b}+(\lambda(1-r)+\mu s)\vec{c}=\vec{0}$. So each of the coefficients are zero, and hence $(1-s)\mu+t\nu=0$. Similarly $(1-s_1)\mu+t_1\nu=0$. So, $\frac{1-s}{1-s_1}=\frac{t}{t_1}$ or, $\frac{t}{1-s}=\frac{t_1}{1-s_1}$. Proved.

(13): Let $A(\vec{a}), B(\vec{b}), C(\vec{c})$ be the vertices of ΔABC. Let us find scalars λ, μ, ν satisfying $\vec{a}+\lambda\left(\frac{\vec{b}-\vec{a}}{c}+\frac{\vec{c}-\vec{a}}{b}\right)=\vec{b}+\mu\left(\frac{\vec{b}-\vec{c}}{c}+\frac{\vec{a}-\vec{b}}{c}\right)=\vec{c}+\nu\left(\frac{\vec{c}-\vec{b}}{a}+\frac{\vec{a}-\vec{c}}{b}\right)$ that is;, $\left(1-\frac{\lambda}{c}-\frac{\lambda}{b}\right)\vec{a}+\frac{\lambda}{c}\vec{b}+\frac{\lambda}{b}\vec{c}=\frac{\mu}{c}\vec{a}+\left(1+\frac{\mu}{a}-\frac{\mu}{c}\right)\vec{b}-\frac{\mu}{a}\vec{c}=\frac{\nu}{b}\vec{a}-\frac{\nu}{a}\vec{b}+\left(1+\frac{\nu}{a}-\frac{\nu}{b}\right)\vec{c}$.

Let us put $\begin{cases}\left(1-\frac{\lambda}{c}-\frac{\lambda}{b}\right)=\frac{\mu}{c}\\ \frac{\lambda}{c}=\left(1+\frac{\mu}{a}-\frac{\mu}{c}\right)\\ \frac{\lambda}{b}=-\frac{\mu}{a}\end{cases}$ or, $\begin{cases}bc-b\lambda-c\lambda=b\mu\\ a\lambda=ca+c\mu-a\mu\\ a\lambda=-b\mu\end{cases}$. So, $bc-b\lambda-c\lambda=-a\lambda$ or, $\lambda=\frac{bc}{b+c-a}$, $\mu=\frac{-ca}{b+c-a}=\frac{ca}{a-b-c}$. Further, $\begin{cases}\frac{\mu}{c}=\frac{\nu}{b}\\ 1+\frac{\mu}{a}-\frac{\mu}{c}=-\frac{\nu}{a}\\ -\frac{\mu}{a}=1+\frac{\nu}{a}-\frac{\nu}{b}\end{cases}$ or,

$\begin{cases}b\mu=c\nu\\ ca+c\mu-a\mu=-c\nu\\ -b\mu=ab+b\nu-a\nu\end{cases}$. So, $-c\nu=ab+b\nu-a\nu$ or, $\nu=\frac{ab}{a-b-c}$ and $\mu=\frac{ca}{a-b-c}$.

So, required values of λ, μ, ν are $\frac{-bc}{a-b-c}, \frac{ca}{a-b-c}, \frac{ab}{a-b-c}$ respectively. So, the three angle bisectors intersect at a common point, say I_1. Here $\overrightarrow{OI_1}=\vec{a}+\frac{-bc}{a-b-c}\left(\frac{\vec{b}-\vec{a}}{c}+\frac{\vec{c}-\vec{a}}{b}\right)=\left(1+\frac{b}{a-b-c}+\frac{c}{a-b-c}\right)\vec{a}-\frac{b}{a-b-c}\vec{b}-\frac{c}{a-b-c}\vec{c}$ or, $\overrightarrow{OI_1}=\frac{a\vec{a}-b\vec{b}-c\vec{c}}{a-b-c}$. Similarly $\overrightarrow{OI_2}=\frac{b\vec{b}-c\vec{c}-a\vec{a}}{b-c-a}$, $\overrightarrow{OI_3}=\frac{c\vec{c}-a\vec{a}-b\vec{b}}{c-a-b}$. Ans.

(14): Here let us choose one particular diagonal whose two ends are $(0,0,0)$ and (a,a,a). Now we can suppose that the mid-points of the six edges of the cube, which do not meet this diagonal, are

$A\left(0,\frac{a}{2},a\right), B\left(\frac{a}{2}.0,a\right), C\left(0,a,\frac{a}{2}\right), D\left(a,0,\frac{a}{2}\right), E\left(\frac{a}{2},a,0\right), F\left(a,\frac{a}{2},0\right)$. So, equation of the plane ABC is $(x,y,z)-\left(0,\frac{a}{2},a\right)=\lambda\overrightarrow{BA}+\mu\overrightarrow{CA}$ or, $\left(x,y-\frac{a}{2},z-a\right)=\lambda\left(-\frac{a}{2},\frac{a}{2},0\right)+\mu\left(0,\frac{a}{2},-\frac{a}{2}\right)$. So, $\begin{cases} x=-\lambda\frac{a}{2} \\ y-\frac{a}{2}=\lambda\frac{a}{2}+\mu\frac{a}{2} \\ z-a=-\mu\frac{a}{2} \end{cases}$. On adding, $x+y+z-\frac{3a}{2}=0$ or, $x+y+z=\frac{3a}{2}$, which is the equation of the plane ABC. Clearly this equation is satisfied by points D,E,F. So, the six points A,B,C,D,E,F are coplanar. Proved

(15): Let $A(\vec{a}),B(\vec{b}),C(\vec{c}),D(\vec{d})$ be four vertices of a tetrahedron. The equation of the plane containing edge AB and bisecting edge CD is $\vec{r}=\vec{a}+\lambda(\vec{b}-\vec{a})+\mu\left(\frac{\vec{c}+\vec{d}}{2}-\vec{a}\right)$ or, $\vec{r}=(1-\lambda-\mu)\vec{a}+\lambda\vec{b}+\frac{\mu}{2}\vec{c}+\frac{\mu}{2}\vec{d}$ $\cdots(*)$. Let us put $1-\lambda-\mu=\lambda=\frac{\mu}{2}$. Then $2\lambda+\mu=1$ and $2\lambda-\mu=0$. So, $\lambda=\frac{1}{4}$ and $\mu=\frac{1}{2}$. So, $\frac{\vec{a}+\vec{b}+\vec{c}+\vec{d}}{4}$ is a point lying on $(*)$, that is, the plane $(*)$ passes through the centroid G of the tetrahedron. Similarly all the other 5 planes passes through G. So, all the 6 planes intersect at a common point G. Proved.

(16): Let two parallel planes be $(\vec{r}-\vec{a})\cdot\vec{d}=0$ $\cdots(1)$

and $(\vec{r}-\overrightarrow{a_1}).\overrightarrow{d_1}=0$ $\cdots(2)$

Let the 3rd plane be $(\vec{r}-\overrightarrow{a_2}).\overrightarrow{d_2}=0$ $\cdots(3)$

So $\vec{d}\times\overrightarrow{d_2}$ is a vector along the line of intersection of (1) and (3). Similarly $\overrightarrow{d_1}\times\overrightarrow{d_2}$ is a vector along the line of intersection of (2) and (3). Since (2) ∥ (1), so $\vec{d}\parallel\overrightarrow{d_1}$. So, $\vec{d}\times\overrightarrow{d_2}\parallel\overrightarrow{d_1}\times\overrightarrow{d_2}$. Proved.

(17): Let $OABC$ be any tetrahedron.

Let $O(0,0,0), A(a,0,0), B(b_1,b_2,0), C(c_1,c_2,c_3)$. Let us take OA and BC as the opposite edges. Here $\overrightarrow{OA}\times\overrightarrow{BC}=(a,0,0)\times(c_1-b_1,c_2-b_2,c_3)=(0,0-a(c_3),a(c_2-b_2))=(0,-ac_3,ac_2-ab_2)=a(0,-c_3,c_2-b_2)$. So, equation to a plane parallel to OA and BC is $0(x)-c_3y+(c_2-b_2)z=d$ or,

$-c_3y+(c_2-b_2)z=d \quad \cdots(*)$. Let $K_1(\lambda b_1, \lambda b_2, 0)$ be the point where $(*)$ intersect OB. So, $-c_3(\lambda b_2)=d$. Hence $\lambda=\frac{-d}{b_2c_3}$. So, $K_1\left(\frac{-d}{b_2c_3}, \frac{-db}{c_3}, 0\right)$.

Let $K_2(\mu c_1, \mu c_2, \mu c_3,)$ be the point where $(*)$ intersects OC. So, $-c_3\mu c_2+(c_2-b_2)(\mu c_3)=d$.

$\therefore \mu=\frac{d}{c_2c_3-b_2c_3-c_2c_3}=\frac{-d}{b_2c_3}$. So, $K_2\left(\frac{-dc_1}{b_2c_3}, \frac{-dc_2}{b_2c_3}, \frac{-d}{b_2}\right)$. Let $K_3(ta+(1-t)b_1, (1-t)b_2, 0)$ be the point where $(*)$ intersect AB. So, $-(1-t)b_2c_3+0=d$.

$\therefore t=1+\frac{d}{b_2c_3}$. So, $K_3\left(\left(1+\frac{d}{b_2c_3}\right)a+\frac{-d}{b_2c_3}b_1, \frac{-d}{c_3}, 0\right)$. Let $K_4(uc_1+(1-u)a, uc_2, uc_3)$ be the point where $(*)$ intersects AC. So, $-c_3(uc_2)+(c_2-b_2)uc_3=d$. $\therefore u=\frac{-d}{b_2c_3}$. So, $K_4\left(\frac{-dc_1}{b_2c_3}+\left(\frac{b_2c_3+d}{b_2c_3}\right)a, \frac{-dc_2}{b_2c_3}, \frac{-d}{b_2}\right)$.

So, $\overrightarrow{K_1K_2}=\left(\frac{-dc_1}{b_2c_3}, \frac{-dc_2}{b_2c_3}, \frac{-d}{b_2}\right)-\left(\frac{-db_1}{b_2c_3}, \frac{-d}{c_3}, 0\right)=\left(\frac{d(b_1-c_1)}{b_2c_3}, \frac{d(b_2-c_2)}{b_2c_3}, \frac{-d}{b_2}\right)$ and $\overrightarrow{K_4K_3}=\left(\frac{-dc_1}{b_2c_3}+\frac{(b_2c_3+d)a}{b_2c_3}-\left(1+\frac{d}{b_2c_3}\right)a+\frac{db_1}{b_2c_3}, \frac{-dc_2}{b_2c_3}+\frac{d}{c_3}, \frac{-d}{b_2}-0\right)=\left(\frac{d(b_1-c_1)}{b_2c_3}, \frac{d(b_2-c_2)}{b_2c_3}, \frac{-d}{b_2}\right)$. So, $K_1K_2K_3K_4$ is a parallelogram. Proved.

(18):

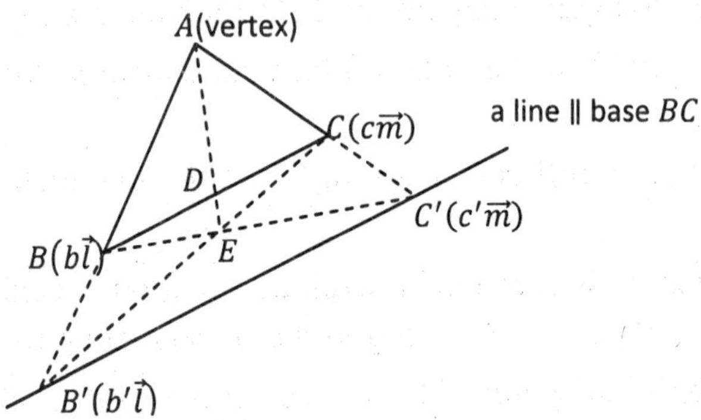

Let $A(\vec{0})$, $\overrightarrow{AB}\equiv b\vec{l}$, $\overrightarrow{AC}\equiv c\vec{m}$, $\overrightarrow{AB'}\equiv b'\vec{l}$, $\overrightarrow{AC'}\equiv c'\vec{m}$, where $\vec{l}, \vec{m}$ are unit vectors along $\overrightarrow{AB}$ and $\overrightarrow{AC}$ respectively. Here $\overrightarrow{BC}\parallel\overrightarrow{B'C'}$. So, $(c\vec{m}-b\vec{l})=\lambda(c'\vec{m}-b'\vec{l})$ for some λ. So, $\lambda=\frac{c}{c'}=\frac{b}{b'} \quad \cdots(*)$ or, $\frac{b\vec{l}+\lambda(c'\vec{m})}{1+\lambda}=\frac{c\vec{m}+\lambda(b'\vec{l})}{1+\lambda}=$

$\overrightarrow{OE} \equiv \vec{e}$. So, $\vec{e} = \frac{b\vec{l}+\frac{c}{c'}(c'\vec{m})}{1+\frac{c}{c'}}$ from $(*)$. So, $\frac{1}{2}\left(1+\frac{c}{c'}\right)\vec{e} = \frac{b\vec{l}+c\vec{m}}{2} = \overrightarrow{OD}$ where D is the mid-point of BC. So, $\overrightarrow{AE}$ divides BC at its mid-point D. Proved.

(19): Let $A(\vec{a}), B(\vec{b}), C(\vec{c}), D(\vec{d})$ be the vertices of a tetrahedron. Let us take BC, CA, AB as three coplanar edges. Here equations to the three straight lines are

$$\begin{cases} \vec{r} = \frac{1}{2}(\vec{b}+\vec{c}) + \lambda(\vec{a}+\vec{d}) & \cdots (1) \\ \vec{r} = \frac{1}{2}(\vec{c}+\vec{a}) + \lambda(\vec{b}+\vec{d}) & \cdots (2) \\ \vec{r} = \frac{1}{2}(\vec{a}+\vec{b}) + \lambda(\vec{c}+\vec{d}) & \cdots (3) \end{cases}$$

If we take $\frac{1}{2}$ for λ, $\frac{\vec{a}+\vec{b}+\vec{c}+\vec{d}}{2}$ is a point on line (1). Similarly $\frac{\vec{a}+\vec{b}+\vec{c}+\vec{d}}{2}$ is a point on line (2) and on line (3).

Hence (1), (2), (3) are concurrent and they meet at P where $\overrightarrow{OP} = \frac{\vec{a}+\vec{b}+\vec{c}+\vec{d}}{2} = 2\overrightarrow{OG}$, where G is the centroid of the tetrahedron. Proved.

3

PRODUCTS OF TWO VECTORS APPLICATIONS

Examples (page 48): **(1):** Here $(6,2,3)\cdot(2,-9,6) = 6(2) + 2(-9) + 3(6) = 12 - 18 + 18 = 12$. Ans. $\cos\theta = \frac{12}{\sqrt{6^2+2^2+3^2}\sqrt{2^2+(-9)^2+6^2}} = \frac{12}{77}$. Ans.

(2): For suitable coordinate system, we can suppose that OC and AD are two diagonals of the cube, where $O(0,0,0), A(a,0,0), C(a,a,a), D(0,a,a)$.

So, $\overrightarrow{OC} = (a,a,a)$, $\overrightarrow{AD} = (0,a,a) - (a,0,0) = (-a,a,a)$. So, $\cos\theta = \frac{a(-a)+a(a)+a(a)}{\sqrt{a^2+a^2+a^2}\sqrt{(-a)^2+a^2+a^2}} = \frac{1}{3}$. Proved.

(3): $\overrightarrow{AB} = (-3,0,-3) = (-3)(1,0,1)$, $\overrightarrow{AC} = (1,2,-7)$.

So, $\cos A = \frac{(-3,0,-3)\cdot(1,2,-7)}{|(-3,0,-3)||(1,2,-7)|} = \frac{1}{3}\sqrt{3}$. Also, $\overrightarrow{CB} = (-4,-2,4)$ and $\overrightarrow{CA} = (-1,-2,7)$ so, $\cos C = \frac{4+4+28}{\sqrt{36}\sqrt{54}} = \frac{1}{3}\sqrt{6}$. Ans. Further $\overrightarrow{BA}\cdot\overrightarrow{BC} = (3,0,3)\cdot(4,2,-4) = 12 - 12 = 0$. So, $\angle B = 90°$. Proved.

(4): Here (projection of PQ on RS) $= \overrightarrow{PQ}\cdot\frac{\overrightarrow{RS}}{|\overrightarrow{RS}|} = (-2,2,1)\cdot\frac{(1,-2,2)}{|(1,-2,2)|} = \frac{-4}{3}$.

Ans. Also, (projection of RS on PQ) $= \overrightarrow{RS}\cdot\frac{\overrightarrow{PQ}}{|\overrightarrow{PQ}|} = \frac{-4}{\sqrt{4+4+1}} = \frac{-4}{3}$. Hence required $\cos\theta = \frac{\overrightarrow{PQ}\cdot\overrightarrow{RS}}{|\overrightarrow{PQ}||\overrightarrow{RS}|} = \frac{-4}{(3)(3)} = \frac{-4}{9}$. Proved.

(5): This is solved in the text.

(6): Here $a^2+b^2+2\vec{a}\cdot\vec{b}=|\vec{a}+\vec{b}|^2=|\vec{a}-\vec{b}|^2$, from question. Hence $a^2+b^2+2\vec{a}\cdot\vec{b}=a^2+b^2-2\vec{a}\cdot\vec{b}$ or, $4\vec{a}\cdot\vec{b}=0$ or, $\vec{a}\cdot\vec{b}=0$. So, $\vec{a}$ is perpendicular to $\vec{b}$. <u>Proved</u>.

(7): This is solved in the text.

(8): This is solved in the text.

(9): This is solved in the text.

(10): This is solved in the text.

(11): This is solved in the text.

Examples (page 52): **(1):** This is solved in the text.

(2): This is solved in the text.

(3): This is solved in the text.

(4): Here $p=\vec{r}\cdot\vec{n}=(\vec{c}+t\vec{d})\cdot\vec{n}=\vec{c}\cdot\vec{n}+t(\vec{d}\cdot\vec{n})$. So, $t=\frac{p-\vec{c}\cdot\vec{n}}{\vec{d}\cdot\vec{n}}$.

So, required point is $\vec{c}+\left(\frac{p-\vec{c}\cdot\vec{d}}{\vec{d}\cdot\vec{n}}\right)\vec{d}$. <u>Proved</u>.

Examples (page 55): **(1):** Here equation of sphere is $F(\vec{r})=\vec{r}^2-2\vec{c}\cdot\vec{r}+k=0$ or, $|\vec{r}-\vec{c}|^2=\left(\sqrt{c^2-k}\right)^2$. Its centre is given by $\vec{c}$ and radius is $\sqrt{c^2-k}$. Let the line $\vec{r}=\vec{h}+t\vec{b}$ from H intersects the sphere at P and Q where $|\vec{b}|=1$. Let $S(\vec{s})$ be a point on this line such that P,Q are the harmonic mean of H and S. The locus of S is called the *polar plane of the point* H *with* the sphere. So, for some λ, let $P\left(\frac{\lambda\vec{h}+\vec{s}}{\lambda+1}\right)$, $Q\left(\frac{\lambda\vec{h}-\vec{s}}{\lambda-1}\right)$. Since P lies on the sphere, so $\left(\frac{\lambda\vec{h}+\vec{s}}{\lambda+1}\right)^2-2\vec{c}\cdot\left(\frac{\lambda\vec{h}+\vec{s}}{\lambda+1}\right)+k=0$ or, $\lambda^2\vec{h}^2+2\lambda\vec{h}\cdot\vec{s}+\vec{s}^2-2\lambda(\lambda+1)\vec{c}\cdot\vec{h}-2(\lambda+1)\vec{c}\cdot\vec{s}+k(\lambda^2+2\lambda+1)=0$ $\cdots(1)$

Similarly $\lambda^2\vec{h}^2-2\lambda\vec{h}\cdot\vec{s}+\vec{s}^2-2\lambda(\lambda-1)\vec{c}\cdot\vec{h}+2(\lambda-1)\vec{c}\cdot\vec{s}+k(\lambda^2-2\lambda+1)=0$ $\cdots(2)$

Now (1) − (2) gives $4\lambda\vec{h}\cdot\vec{s}-4\lambda\vec{c}\cdot\vec{h}-4\lambda\vec{c}\cdot\vec{s}+4k\lambda=0$ or, $\vec{h}\cdot\vec{s}-\vec{c}\cdot\vec{h}-\vec{c}\cdot\vec{s}+k=0$ or, $(\vec{h}-\vec{c})\cdot\vec{s}-\vec{c}\cdot\vec{h}+k=0$. So, required equation of locus of $S(\vec{s})$ is $(\vec{h}-\vec{c})\cdot\vec{r}-\vec{c}\cdot\vec{h}+k=0$ or, $\vec{h}\cdot\vec{r}-\vec{c}\cdot(\vec{h}+\vec{r})+k=0$ (*). This is an equation of polar plane of H with respect to the sphere. Proved. If $(*)$ passes through $G(\vec{g})$ then $\vec{h}\cdot\vec{g}-\vec{c}\cdot(\vec{h}+\vec{g})+k=0$ $\cdots(**)$. The polar plane of $G(\vec{g})$ with respect to sphere is $\vec{g}\cdot\vec{r}-\vec{c}\cdot(\vec{g}+\vec{r})+k=0$. This will pass through $H(\vec{h})$ if $\vec{g}\cdot\vec{h}-\vec{c}\cdot(\vec{g}+\vec{h})+k=0$. This condition is satisfied from $(**)$. Here the polar plane of G will pass through H. Proved.

(2): Let $S(\vec{s})$ be any point on the required locus. Now equation of a chord through $S(\vec{s})$ and parallel to unit vector $\vec{b}$ is $\vec{r}=\vec{s}+t\vec{b}$ $\cdots(1)$

Let us solve $F(\vec{r})\equiv\vec{r}^2-2\vec{c}\cdot\vec{r}+k=0$ and $\vec{r}=\vec{s}+t\vec{b}$ simultaneously. We get $t^2-2(\vec{s}-\vec{c})\cdot\vec{b}t+F(\vec{s})=0$. Let its two roots be t_1,t_2. Since S is the mid-point of the points where chord (1) intersects the sphere, so t_1,t_2 are equal in magnitude but opposite in signs. So, $2(\vec{s}-\vec{c})\cdot\vec{b}=t_1+t_2=0$ or, $(\vec{s}-\vec{c})\cdot\vec{b}=0$. So, the locus of $S(\vec{s})$ is $(\vec{r}-\vec{c})\cdot\vec{b}=0$. This is an equation of plane, called the *diametral plane bisecting chords parallel to* $\vec{b}$.

(3): Let $H(\vec{h})$ be any point from which length of tangents on spheres $F(\vec{r})=0$ and $F'(\vec{r})=0$ are equal. So, $\sqrt{\vec{h}^2-2\vec{c}\cdot\vec{h}+k}=\sqrt{\vec{h}^2-2\vec{c'}\cdot\vec{h}+k'}$ or, $2(\vec{c}-\vec{c'})\cdot\vec{h}=k-k'$. So, required equation of locus of $H(\vec{h})$ is $2\vec{r}\cdot(\vec{c}-\vec{c}')=k-k'$. Proved. This is called the *radical plane of the two spheres.* Here $F(\vec{r})-\lambda F'(\vec{r})=0$ $\cdots$ $(*)$ becomes $\vec{r}^2-2\vec{c}\cdot\vec{r}+k-\lambda\{\vec{r}^2-2\vec{c}'\cdot\vec{r}+k'\}=0$ or, $\vec{r}^2-2\left(\frac{\vec{c}-\lambda\vec{c}'}{1-\lambda}\right)\cdot\vec{r}+\frac{k-\lambda k'}{1-\lambda}=0$, which is a sphere whose centre is $\frac{\vec{c}-\lambda\vec{c}'}{1-\lambda}$ and radius $=\sqrt{\left(\frac{\vec{c}-\lambda\vec{c}'}{1-\lambda}\right)^2-\left(\frac{k-\lambda k'}{1-\lambda}\right)}$. So, $(*)$ represents a family of spheres, where λ is a parameter. Let $F(\vec{r})-\lambda_1F'(\vec{r})=0$ $\cdots(1)$

and $F(\vec{r})-\lambda_2F'(\vec{r})=0$ $\cdots(2)$

be any two members of the family of spheres.

Here (1) becomes

$\vec{r}^2-2\left(\frac{\vec{c}-\lambda_1\vec{c}'}{1-\lambda_1}\right)\cdot\vec{r}+\frac{k-\lambda k'}{1-\lambda}=0$. (2) becomes $\vec{r}^2-2\left(\frac{\vec{c}-\lambda_2\vec{c}'}{1-\lambda_2}\right)\cdot\vec{r}+\frac{k-\lambda_2k'}{1-\lambda_2}=0$. As above,equation of radical plane is $2\vec{r}\cdot\left(\frac{\vec{c}-\lambda_1\vec{c}'}{1-\lambda_1}-\frac{\vec{c}-\lambda_2\vec{c}'}{1-\lambda_2}\right)=\frac{k-\lambda_1k'}{1-\lambda_1}-\frac{k-\lambda_2k'}{1-\lambda_2}$

or, $2\vec{r}\cdot\left((1-\lambda_2)\vec{c}-\lambda_1(1-\lambda_2)\vec{c}'-(1-\lambda_1)\vec{c}+\lambda_2(1-\lambda_1)\vec{c}'\right)=(1-\lambda_2)k-\lambda_1(1-\lambda_2)k'-(1-\lambda_1)k+\lambda_2(1-\lambda_1)k'$ or, $2\vec{r}\cdot\left((\lambda_1-\lambda_2)\vec{c}+(\lambda_2-\lambda_1)\vec{c}'\right)=(\lambda_1-\lambda_2)k+(\lambda_2-\lambda_1)k'$ or, $2\vec{r}.(\vec{c}-\vec{c}')=k-k'$ which is free from λ_1,λ_2. So, any two spheres of the family $(*)$ have the same radical plane. Proved.

Examples (page 58): **(1):** Here $\vec{a}\times\vec{b}=\vec{0}$. This is equivalent to $(a_1,a_2,a_3)\times(b_1,b_2,b_3)=(0,0,0)$ or, $(a_2b_3-a_3b_2,a_3b_1-a_1b_3,a_1b_2-a_2b_1)=(0,0,0)$ or, $a_2b_3-a_3b_2=0, a_3b_1-a_1b_3=0$ and $a_1b_2-a_2b_1=0$ or, $\frac{a_2}{b_2}=\frac{a_3}{b_3}, \frac{a_3}{b_3}=\frac{a_1}{b_1}$ and $\frac{a_1}{b_1}=\frac{a_2}{b_2}$ or, $\frac{a_1}{b_1}=\frac{a_2}{b_2}=\frac{a_3}{b_3}$. Proved.

(2): One of the vector is $(4,-2,3)\times(5,1,-4)=\left((-2)(-4)-3(1),3(5)-4(-4),4(1)-(-2)(5)\right)=(5,31,14)$. Ans.

(3): $(6,2,-3)\times(4,-1,-2)=(-4-3,-12+12,-6-8)=(-7,0,-14)=-7(1,0,2)$ and $(2,5,-7)\times(-3,4,1)=(5+28,21-2,8+15)=(33,19,23)$. Ans.

(4): This is solved in the text.

(5): This is solved in the text.

(6): Required distance $=\left|\{(4,-2,3)-(3,1,-4)\}\times\frac{(6,-2,3)}{|(6,-2,3)|}\right|=\frac{1}{\sqrt{49}}|(1,-3,7)\times(6,-2,3)|=\frac{1}{\sqrt{49}}|(-9+14,42-3,-2+18)|=\frac{1}{7}\sqrt{25+1521+256}=\frac{1}{7}\sqrt{1802}$. Ans.

(7): This is solved in the text.

(8): Here required equation of plane is $\{(x,y,z)-(2,-4,5)\}\cdot\{(2,3,-4)\times(3,1,-2)\}=0$ or, $(x-2,y+4,z-5)\cdot(-2,-8,-7)=0$ or, $-2(x-2)-8(y+4)-7(z-5)=0$ or, $2x+8y+7z=7$. Ans.

(9): This is solved in the text.

(10): Required equation of plane is $\{(x,y,z)-(3,-5,-1)\}\cdot\left(\overrightarrow{AB}\times(9,-1,7)\right)=0$ or, $(x-3,y+5,z+1)\cdot\{(-4,10,8)\times(3,-1,7)\}=0$ or, $(x-3,y+5,z+1)\cdot\{(-2,5,4)\times(3,-1,7)\}=0$ or, $(35+4)(x-3)+(12+14)(y+5)+(2-15)(z+1)=0$ or, $3x+2y-z=0$. Ans.

(11): Here equation of straight line is $(\vec{r}-\vec{a})\times\frac{\vec{b}-\vec{a}}{|\vec{b}-\vec{a}|}=\vec{0}$. So, required perpendicular distance from $C(\vec{c})$ is equal to $\left|(\vec{c}-\vec{a})\times\frac{(\vec{b}-\vec{a})}{|\vec{b}-\vec{a}|}\right|=\frac{1}{|\vec{b}-\vec{a}|}|\vec{c}\times\vec{b}-\vec{c}\times\vec{a}-\vec{a}\times\vec{b}+\vec{0}|=\frac{|\vec{b}\times\vec{c}+\vec{c}\times\vec{a}+\vec{a}\times\vec{b}|}{|\vec{b}-\vec{a}|}$. Proved.

(12): This is solved in the text.

(13): Here $\vec{a}\times\vec{n}=\vec{b}\times\vec{n}$. So, $(\vec{a}\times\vec{n})\cdot\vec{b}=(\vec{b}\times\vec{n})\cdot\vec{b}=0$. So, $[\vec{a}\,\vec{n}\,\vec{b}]=0$. So, $\vec{a},\vec{n},\vec{b}$ are coplanar vectors, that is, incident ray, reflected ray and normal are coplanar. Further $\vec{a}\times\vec{n}=\vec{b}\times\vec{n}$ gives $(1)(1)\sin(\pi-i)=|\vec{a}\times\vec{n}|=|\vec{b}\times\vec{n}|=(1)(1)\sin r$ so, $\sin i=\sin r$ or, $\boxed{i=r}$.

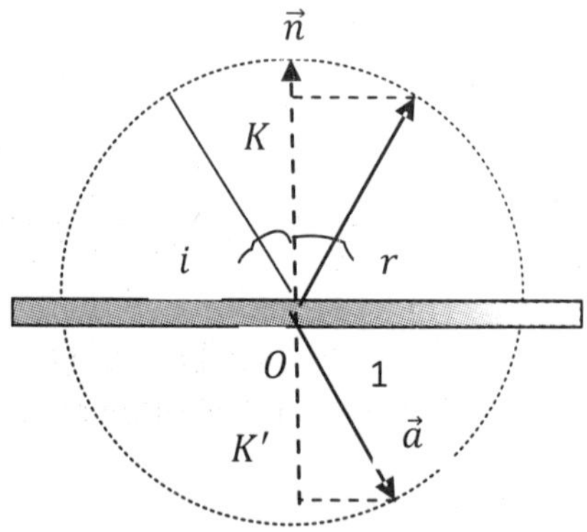

So, $\vec{a}\times\vec{n}=\vec{b}\times\vec{n}$ gives all laws of reflection. Conversely, let (i) $\vec{a},\vec{n},\vec{b}$ are coplanar (ii) $i=r$. To prove: $\vec{a}\times\vec{n}=\vec{b}\times\vec{n}$. Since $|\vec{a}\times\vec{n}|=(1)(1)\sin(\pi-i)=\sin i=\sin r$, from (ii), and $|\vec{b}\times\vec{n}|=(1)(1)\sin r$, so $\vec{a}\times\vec{n}$ and $\vec{b}\times\vec{n}$ are vectors, equal in magnitudes. Further, from (i), $\vec{a}\times\vec{n}$ has the direction normal to the plane Π containing $\vec{a},\vec{n},\vec{b}$. Similarly $\vec{b}\times\vec{n}$ is normal to the plane Π. So, $\vec{a}\times\vec{n}\parallel\vec{b}\times\vec{n}$. From figure, $\vec{a}\times\vec{n}$ and $\vec{b}\times\vec{n}$ have same sense of directions, so $\vec{a}\times\vec{n}=\vec{b}\times\vec{n}$. So $\vec{a}\times\vec{n}=\vec{b}\times\vec{n}$ is the law of reflection. Next let $\mu\vec{a}\times\vec{n}=\mu'\vec{c}\times\vec{n}\quad\cdots(*)$, so,

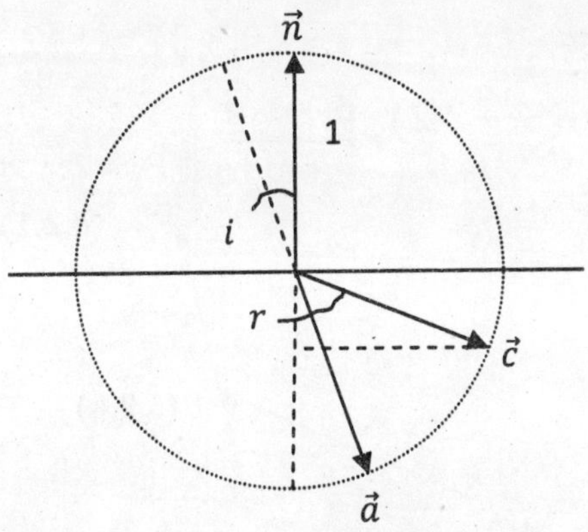

$\mu \sin i = \mu(1)(1)\sin(\pi - i) = \mu|\vec{a} \times \vec{n}| = \mu'|\vec{c} \times \vec{n}|$, from($*$), and $\mu'|\vec{c} \times \vec{n}| = \mu'(1)(1)\sin(\pi - r) = \mu' \sin r$. So, $\mu \sin i = \mu' \sin r$ (Snell's law). Since $\mu\vec{a} \times \vec{n} = \mu'\vec{c} \times \vec{n}$ so, $(\vec{a} \times \vec{n}) \cdot \vec{c} = \frac{\mu'}{\mu}(\vec{c} \times \vec{n}) \cdot \vec{c} = 0$. So, $\vec{a}, \vec{n}, \vec{c}$ are coplanar. Hence incident ray, refracted ray and normal are coplanar. Conversely, let (i) $\vec{a}, \vec{n}, \vec{c}$ are coplanar, (ii) $\mu \sin i = \mu' \sin r$. To prove: $\mu\vec{a} \times \vec{n} = \mu'\vec{c} \times \vec{n}$. Here $|\mu\vec{a} \times \vec{n}| = \mu(1)(1)\sin(\pi - i) = \mu \sin i = \mu' \sin r$ from (ii), and $|\mu'\vec{c} \times \vec{n}| = \mu'|\vec{c} \times \vec{n}| = \mu'(1)(1)\sin(\pi - r) = \mu' \sin r$. So, $\mu\vec{a} \times \vec{n}$ and $\mu'\vec{c} \times \vec{n}$ have equal magnitudes. Further let Π be the plane containing $\vec{a}, \vec{n}, \vec{c}$, from (i). Here $\mu\vec{a} \times \vec{n}$ has the same direction as that of $\vec{a} \times \vec{n}$. Also $\vec{a} \times \vec{n}$ is along normal to the plane Π. So, $\mu\vec{a} \times \vec{n}$ is normal to Π. So, $\mu\vec{a} \times \vec{n} \parallel \mu'\vec{c} \times \vec{n}$. ALso, from figure, $\mu\vec{a} \times \vec{n}$ and $\mu'\vec{c} \times \vec{n}$ have the same sense of directions, so $\mu\vec{a} \times \vec{n} = \mu'\vec{c} \times \vec{n}$. So, $\mu\vec{a} \times \vec{n} = \mu'\vec{c} \times \vec{n}$ is the law of refraction. Proved.

Here $\vec{b} = \vec{a} + 2(OK')\vec{n} = \vec{a} + 2(-\vec{a} \cdot \vec{n})\vec{n}$ (Reason: $\vec{a} \cdot \vec{n} = 1(1)\cos(\pi - r) = -\cos r = -OK'$). So, $\vec{b} = \vec{a} - 2(\vec{a} \cdot \vec{n})\vec{n}$. Proved.

Examples (page 65): **(1):** This is solved in the text.

(2): This is solved in the text.

(3): This is solved in the text.

(4): This is solved in the text.

(5): Here $\vec{\omega} = (5)\frac{(2,1,-2)}{\sqrt{4+1+4}} = \frac{5}{3}(2,1,-2)$. So, velocity of the particle at $(4,1,0)$ is $\frac{5}{3}(2,1,-2) \times \big((4,1,0) - (3,-1,-2)\big) = \frac{5}{3}(2,1,-2) \times (1,2,2) = \frac{5}{3}(6,-6,3) = 5(2,-2,1)$. Proved. Also velocity of particle at $(3,2,1)$ is $\frac{5}{3}(2,1,-2) \times$

$\big((3,2,1)-(3,-1,-2)\big)=\frac{5}{3}(2,1,-2)\times(0,3,3)=5(2,1,-2)\times(0,1,1)=5(1+2,0-2,2-0)=5(3,-2,2)$. Proved.

(6):

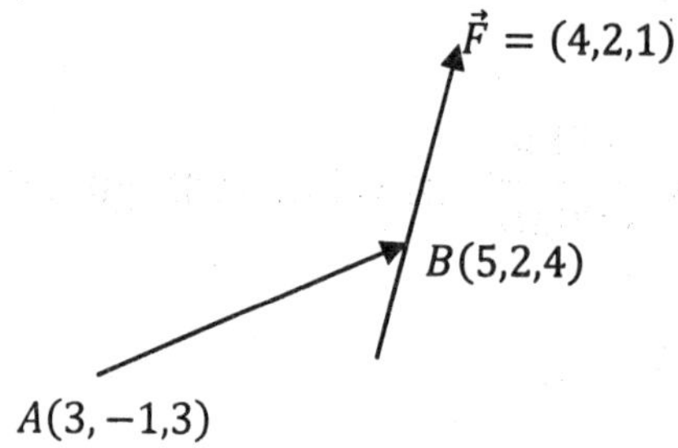

Here torque $=\overrightarrow{AB}\times\vec{F}=(2,3,1)\times(4,2,1)=(3-2,4-2,4-12)=(1,2,-8)$. Ans.

Exercises on Chapter 3

(1): This is solved in the text.

(2): This is solved in the text.

(3): Here vectorial area of triangle ABC is $\frac{1}{2}\overrightarrow{AB}\times\overrightarrow{AC}=\frac{1}{2}(\vec{b}-\vec{a})\times(\vec{c}-\vec{a})=\frac{1}{2}[\vec{b}\times\vec{c}-\vec{b}\times\vec{a}-\vec{a}\times\vec{c}+\vec{0}]=\frac{1}{2}(\vec{b}\times\vec{c}+\vec{c}\times\vec{a}+\vec{a}\times\vec{b})$. Proved. If A,B,C are collinear then (vectorial area of ΔABC) $=\vec{0}$. So, $\vec{b}\times\vec{c}+\vec{c}\times\vec{a}+\vec{a}\times\vec{b}=\vec{0}$. Proved.

(4):

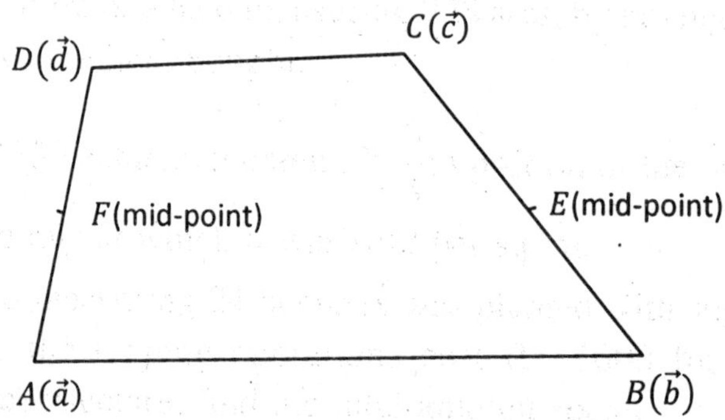

Here $E\left(\frac{\vec{b}+\vec{c}}{2}\right)$. So, area of ΔAED is $\frac{1}{2}\left(\vec{a}\times\frac{\vec{b}+\vec{c}}{2}+\frac{\vec{b}+\vec{c}}{2}\times\vec{d}+\vec{d}\times\vec{a}\right)=\frac{1}{4}\left(\vec{a}\times\vec{b}+\vec{a}\times\vec{c}+\vec{b}\times\vec{d}+\vec{c}\times\vec{d}+2\vec{d}\times\vec{a}\right)=\frac{1}{4}\left(\vec{a}\times\vec{b}+\vec{a}\times\vec{d}+\vec{b}\times\vec{c}+\vec{c}\times\right.$

$\vec{d} - 2\vec{a}\times\vec{d})$ (Reason: Here $AB \parallel CD$. So, $\vec{0} = \overrightarrow{AB}\times\overrightarrow{CD} = (\vec{b}-\vec{a})\times(\vec{d}-\vec{c}) = \vec{b}\times\vec{d} - \vec{b}\times\vec{c} - \vec{a}\times\vec{d} + \vec{a}\times\vec{c}$ or, $\vec{a}\times\vec{c} + \vec{b}\times\vec{d} = \vec{a}\times\vec{d} + \vec{b}\times\vec{c}$). So, (area of $\Delta AED) = \frac{1}{4}(\vec{a}\times\vec{b} + \vec{d}\times\vec{a} + \vec{b}\times\vec{c} + \vec{c}\times\vec{d})$. Also (area of trapezium $ABCD$) $= \frac{1}{2}\left((\vec{b}\times\vec{c} + \vec{c}\times\vec{a} + \vec{a}\times\vec{b}) + (\vec{a}\times\vec{c} + \vec{c}\times\vec{d} + \vec{d}\times\vec{a})\right) = \frac{1}{2}(\vec{a}\times\vec{b} + \vec{d}\times\vec{a} + \vec{b}\times\vec{c} + \vec{c}\times\vec{d}) = 2$(area of ΔAED). Proved.

(5): Required unit vector $= \pm\frac{(2,-1,1)\times(3,4,-1)}{|(2,-1,1)\times(3,4,-1)|} = \pm\frac{(1-4,3+2,8+3)}{\sqrt{9+25+121}} = \pm\frac{1}{\sqrt{155}}(-3,5,11)$. Ans. Also, $\sqrt{4+1+1}\sqrt{9+16+1}\sin\theta = \sqrt{9+25+121}$ so, $\sin\theta = \sqrt{\frac{155}{6\cdot 26}} = \sqrt{\frac{155}{156}}$. Ans.

(6): Here $\vec{\tau} = \left((2,-1,3)-(1,2,-1)\right)\times(3,0,1) = (1,-3,4)\times(3,0,1) = (-3,11,9)$. Ans.

(7): (Twice the vertical area of closed polygon $A_1A_2A_3\cdots A_nA_1) = 2\left\{\frac{1}{2}(\overrightarrow{a_1}\times\overrightarrow{a_2} + \overrightarrow{a_2}\times\overrightarrow{a_3} + \overrightarrow{a_3}\times\overrightarrow{a_1}) + \frac{1}{2}(\overrightarrow{a_1}\times\overrightarrow{a_3} + \overrightarrow{a_3}\times\overrightarrow{a_4} + \overrightarrow{a_4}\times\overrightarrow{a_1}) + \frac{1}{2}(\overrightarrow{a_1}\times\overrightarrow{a_4} + \overrightarrow{a_4}\times\overrightarrow{a_5} + \overrightarrow{a_5}\times\overrightarrow{a_1}) + \cdots + \frac{1}{2}(\overrightarrow{a_1}\times\overrightarrow{a_{n-1}} + \overrightarrow{a_{n-1}}\times\overrightarrow{a_n} + \overrightarrow{a_n}\times\overrightarrow{a_1})\right\} = \overrightarrow{a_1}\times\overrightarrow{a_2} + \overrightarrow{a_2}\times\overrightarrow{a_3} + \overrightarrow{a_3}\times\overrightarrow{a_4} + \cdots + \overrightarrow{a_{n-1}}\times\overrightarrow{a_n} + \overrightarrow{a_n}\times\overrightarrow{a_1}$ and (torque of $\overrightarrow{A_1A_2}$ about $B(\vec{b})$) + (torque of $\overrightarrow{A_2A_3}$ about $B(\vec{b})$) + (torque of $\overrightarrow{A_3A_4}$ about $B(\vec{b})$) + $\cdots$ + (torque of $\overrightarrow{A_{n-1}A_n}$ about $B(\vec{b})$) + (torque of $\overrightarrow{A_nA_1}$ about $B(\vec{b})$) $= (\overrightarrow{a_1}-\vec{b})\times(\overrightarrow{a_2}-\overrightarrow{a_1}) + (\overrightarrow{a_2}-\vec{b})\times(\overrightarrow{a_3}-\overrightarrow{a_2}) + (\overrightarrow{a_3}-\vec{b})\times(\overrightarrow{a_4}-\overrightarrow{a_3}) + \cdots + (\overrightarrow{a_{n-1}}-\vec{b})\times(\overrightarrow{a_n}-\overrightarrow{a_{n-1}}) + (\overrightarrow{a_n}-\vec{b})\times(\overrightarrow{a_1}-\overrightarrow{a_n}) = \overrightarrow{a_1}\times\overrightarrow{a_2} - \vec{b}\times(\overrightarrow{a_2}-\overrightarrow{a_1}) + \overrightarrow{a_2}\times\overrightarrow{a_3} - \vec{b}\times(\overrightarrow{a_3}-\overrightarrow{a_2}) + \overrightarrow{a_3}\times\overrightarrow{a_4} - \vec{b}\times(\overrightarrow{a_4}-\overrightarrow{a_3}) + \cdots + \overrightarrow{a_{n-1}}\times\overrightarrow{a_n} - \vec{b}\times(\overrightarrow{a_n}-\overrightarrow{a_{n-1}}) + \overrightarrow{a_n}\times\overrightarrow{a_1} - \vec{b}\times(\overrightarrow{a_1}-\overrightarrow{a_n}) = \overrightarrow{a_1}\times\overrightarrow{a_2} + \overrightarrow{a_2}\times\overrightarrow{a_3} + \overrightarrow{a_3}\times\overrightarrow{a_4} + \cdots + \overrightarrow{a_{n-1}}\times\overrightarrow{a_n} + \overrightarrow{a_n}\times\overrightarrow{a_1}$. This completes the proof.

(8): Let $\vec{d} \equiv \lambda(\vec{b}\times\vec{c}) + \mu(\vec{c}\times\vec{a}) + \nu(\vec{a}\times\vec{b})$ be the unit vector equally inclined to $\vec{a},\vec{b},\vec{c}$. So, $\left(\frac{1}{a}\vec{a}\right)\cdot\vec{d} = \left(\frac{1}{b}\vec{b}\right)\cdot\vec{d} = \left(\frac{1}{c}\vec{c}\right)\cdot\vec{d}$. So, $\frac{1}{a}\{\lambda[\vec{a}\,\vec{b}\,\vec{c}]\} = \frac{1}{b}\{\mu[\vec{b}\,\vec{c}\,\vec{a}]\} = \frac{1}{c}\{\nu[\vec{c}\,\vec{a}\,\vec{b}]\}$. So, $\frac{\lambda}{a} = \frac{\mu}{b} = \frac{\nu}{c} \equiv k$. So, $\vec{d} = k(a\vec{b}\times\vec{c} + b\vec{c}\times\vec{a} +$

$c\vec{a}\times\vec{b})$. So, required equation of straight line is $\vec{r}=\vec{t}+s\vec{d}$ or, $\vec{r}=\vec{t}+s'\{a(\vec{b}\times\vec{c})+b(\vec{c}\times\vec{a})+c(\vec{a}\times\vec{b})\}$. Ans.

(**Note:** We have seen that $a(\vec{b}\times\vec{c})+b(\vec{c}\times\vec{a})+c(\vec{a}\times\vec{b})$ is the vector equally inclined to three vectors $\vec{a},\vec{b},\vec{c}$.)

(9): Let 3 given coplanar straight lines be $\begin{cases}\vec{r}=\overrightarrow{a_1}+\lambda_1\overrightarrow{d_1} & \cdots(1)\\ \vec{r}=\overrightarrow{a_2}+\lambda_2\overrightarrow{d_2} & \cdots(2)\\ \vec{r}=\overrightarrow{a_3}+\lambda_3\overrightarrow{d_3} & \cdots(3)\end{cases}$

where $\overrightarrow{d_1},\overrightarrow{d_2},\overrightarrow{d_3}$ are unit vectors. Clearly $\frac{\overrightarrow{d_1}\times\overrightarrow{d_2}}{|\overrightarrow{d_1}\times\overrightarrow{d_2}|}=\frac{\overrightarrow{d_2}\times\overrightarrow{d_3}}{|\overrightarrow{d_2}\times\overrightarrow{d_3}|}=\frac{\overrightarrow{d_3}\times\overrightarrow{d_1}}{|\overrightarrow{d_3}\times\overrightarrow{d_1}|}\equiv\vec{n}$. So, $(\overrightarrow{d_1}\times\overrightarrow{d_2})\times\vec{n}=|\overrightarrow{d_1}\times\overrightarrow{d_2}|\vec{n}\times\vec{n}=\vec{0}$, etc. Here $\vec{n}$ is a unit vector perpendicular to the plane Π containing 3 lines (1), (2), (3). A line equally inclined to (1), (2), (3) has the direction of $d_1(\overrightarrow{d_2}\times\overrightarrow{d_3})+d_2(\overrightarrow{d_3}\times\overrightarrow{d_1})+d_3(\overrightarrow{d_1}\times\overrightarrow{d_2})$, see the note of previous Ex. 8. To prove: $\vec{n}\times\{d_1(\overrightarrow{d_2}\times\overrightarrow{d_3})+d_2(\overrightarrow{d_3}\times\overrightarrow{d_1})+d_3(\overrightarrow{d_1}\times\overrightarrow{d_2})\}=\vec{0}$.

Here $LHS=-d_1(\overrightarrow{d_2}\times\overrightarrow{d_3})\times\vec{n}-d_2(\overrightarrow{d_3}\times\overrightarrow{d_1})\times\vec{n}-d_3(\overrightarrow{d_1}\times\overrightarrow{d_2})\times\vec{n}=\vec{0}+\vec{0}+\vec{0}$, from (*). Proved.

(10): Let ABC be a right- angled triangle with angle $A=90°$. Let M be the mid-point of the hypotenuse BC. Let us take A as the origin. Let $B(\vec{b})$ and $C(\vec{c})$. So, $M\left(\frac{\vec{b}+\vec{c}}{2}\right)$. Let $P(\vec{x})$ be such that $\overline{AP}=\overline{BP}=\overline{CP}$. To prove: MP is perpendicular to the plane of ΔABC. Here $\vec{b}\cdot\vec{c}=0$ $\cdots(1)$.

Given: $|\vec{x}-\vec{0}|=|\vec{x}-\vec{b}|=|\vec{x}-\vec{c}|$ so, $x^2=x^2-2\vec{x}\cdot\vec{b}+b^2=x^2-2\vec{x}\cdot\vec{c}+c^2$. So, $0=b^2-2\vec{x}\cdot\vec{b}=c^2-2\vec{x}\cdot\vec{c}$. So, $\vec{b}\cdot\vec{b}-2\vec{x}\cdot\vec{b}=0$ or, $(\vec{b}-2\vec{x})\cdot\vec{b}=0$. Similarly $(\vec{c}-2\vec{x})\cdot\vec{c}=0$. Now we have to prove: $(\vec{b}\times\vec{c})\times\left\{\vec{x}-\frac{\vec{b}+\vec{c}}{2}\right\}=\vec{0}$. $LHS=(\vec{b}\cdot\vec{x})\vec{c}-(\vec{c}\cdot\vec{x})\vec{b}-\frac{1}{2}\{b^2\vec{c}-(\vec{b}\cdot\vec{c})\vec{b}+(\vec{b}\cdot\vec{c})\vec{c}-c^2\vec{b}\}=(\vec{b}\cdot\vec{x})\vec{c}-(\vec{c}.\vec{x})\vec{b}-\frac{1}{2}b^2\vec{c}+\frac{1}{2}c^2\vec{b}$, from (1). So, $LHS=\left(\vec{b}\cdot\vec{x}-\frac{1}{2}b^2\right)\vec{c}+\left(\frac{1}{2}c^2-\vec{c}\cdot\vec{x}\right)\vec{b}=(2\vec{b}\cdot\vec{x}-\vec{b}\cdot\vec{b})\left(\frac{1}{2}\vec{c}\right)+(\vec{c}\cdot\vec{c}-2\vec{c}\cdot\vec{x})\left(\frac{1}{2}\vec{b}\right)=0\left(\frac{1}{2}\vec{c}\right)+0\left(\frac{1}{2}\vec{b}\right)$, from (*). Proved.

(11):

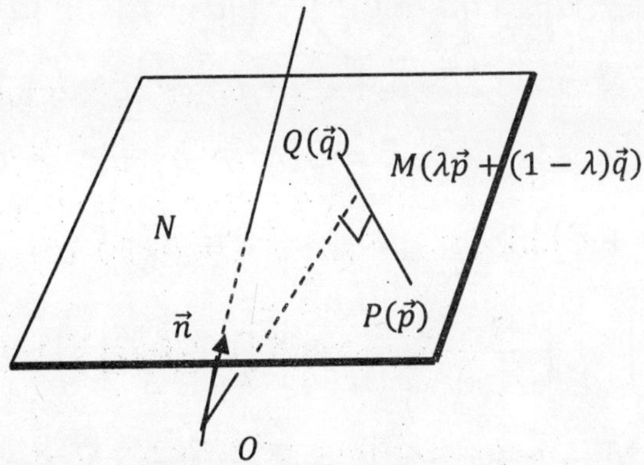

Let equation of the plane be $\vec{r}\cdot\vec{n}=p$ $\quad\cdots(1)$

Since $P(\vec{p})$ lies on the plane (1), so $\vec{p}\cdot\vec{n}=p$. Similarly $\vec{q}\cdot\vec{n}=p$. Let $M(\lambda\vec{p}+(1-\lambda)\vec{q})$. Given: $\overrightarrow{OM}\perp\overrightarrow{PQ}$. So, $\overrightarrow{OM}\cdot\overrightarrow{PQ}=0$ $\quad\cdots(2)$

$LHS=\left(\overrightarrow{ON}-\overrightarrow{OM}\right)\cdot\overrightarrow{PQ}=\overrightarrow{ON}\cdot\overrightarrow{PQ}-\overrightarrow{OM}\cdot\overrightarrow{PQ}=\overrightarrow{ON}\cdot\overrightarrow{PQ}$, from (2). So, $LHS=(p\vec{n})\cdot(\vec{q}-\vec{p})=p(\vec{q}\cdot\vec{n}-\vec{p}\cdot\vec{n})=p(p-p)=0=RHS$. Proved.

(12): Here $\vec{a}\cdot\vec{p}=0$ and $\vec{b}\cdot\vec{p}=0$. So, $\left(\vec{a}-\vec{b}\right)\cdot\vec{p}=\vec{a}\cdot\vec{p}-\vec{b}.\vec{p}=0-0=0$. So, $\vec{p}=\vec{0}$ or $\vec{p}$ is perpendicular to $\vec{a}-\vec{b}$. To prove: $\vec{p}\parallel\left(\vec{a}\times\vec{b}\right)$ or $\vec{p}\times\left(\vec{a}\times\vec{b}\right)=\vec{0}$. $LHS=\vec{p}\times\left(\vec{a}\times\vec{b}\right)=(\vec{p}\cdot\vec{b})\vec{a}-(\vec{p}\cdot\vec{a})\vec{b}=0\vec{a}-0\vec{b}=\vec{0}=RHS$. Proved.

(13): Let $A(\vec{a}),B\left(\vec{b}\right),C(\vec{c}),D\left(\vec{d}\right)$ be the vertices of a tetrahedron. Here P,Q,R,P_1,Q_1,R_1 are the mid-points of the edges shown in the figure. To prove: $(AB^2+BC^2+CA^2)+(AD^2+BD^2+CD^2)=4\left({PP_1}^2+{QQ_1}^2+{RR_1}^2\right)$.

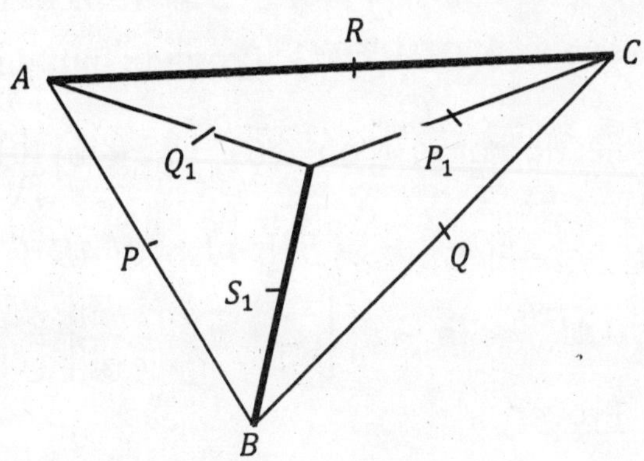

$LHS = \left(\left|\vec{b}-\vec{a}\right|^2 + \left|\vec{c}-\vec{b}\right|^2 + |\vec{a}-\vec{c}|^2\right) + \left(\left|\vec{d}-\vec{a}\right|^2 + \left|\vec{d}-\vec{b}\right|^2 + \left|\vec{d}-\vec{c}\right|^2\right) = a^2 + b^2 - 2\vec{a}\cdot\vec{b} + c^2 + b^2 - 2\vec{c}\cdot\vec{b} + c^2 + a^2 - 2\vec{c}\cdot\vec{a} + d^2 + a^2 - 2\vec{d}\cdot\vec{a} + d^2 + b^2 - 2\vec{d}\cdot\vec{b} + d^2 + c^2 - 2\vec{d}\cdot\vec{c} = 3(a^2+b^2+c^2+d^2) - 2\left(\vec{a}\cdot\vec{b} + \vec{b}\cdot\vec{c} + \vec{c}\cdot\vec{a} + \vec{d}\cdot(\vec{a}+\vec{b}+\vec{c})\right).$

$RHS = 4\left(\left|\frac{\vec{c}+\vec{d}}{2} - \frac{\vec{a}+\vec{b}}{2}\right|^2 + \left|\frac{\vec{b}+\vec{c}}{2} - \frac{\vec{d}+\vec{a}}{2}\right|^2 + \left|\frac{\vec{c}+\vec{a}}{2} - \frac{\vec{b}+\vec{d}}{2}\right|^2\right) = \left|-\vec{a}-\vec{b}+\vec{c}+\vec{d}\right|^2 + \left|-\vec{a}+\vec{b}+\vec{c}-\vec{d}\right|^2 + \left|\vec{a}-\vec{b}+\vec{c}-\vec{d}\right|^2 = 3\sum a^2 - 2\sum \vec{b}\cdot\vec{c}$ (after cancellation). Proved.

(14):

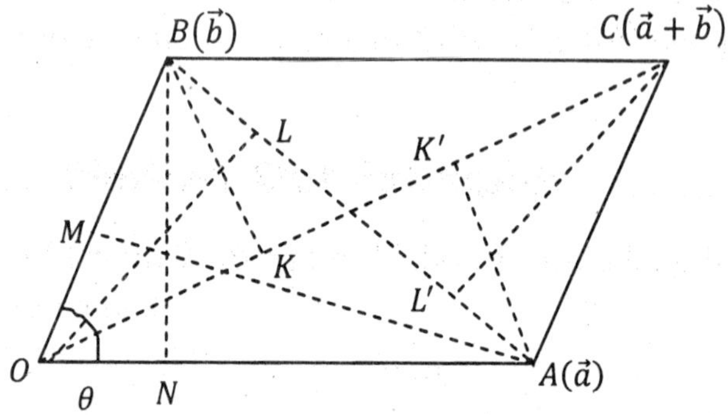

To prove: (i): $AB^2 + OC^2 = 2(OA^2 + OB^2)$ (ii): $OC^2 - AB^2 = 4(OA)(ON) = 4(OB)(OM)$ (iii): $|OA^2 - OB^2| = |(OC)(KK')| = |(AB)(LL')|$. Proof of (i): $LHS = \left|\vec{b}-\vec{a}\right|^2 + \left|\vec{a}+\vec{b}\right|^2 = b^2 + a^2 - 2\vec{a}\cdot\vec{b} + a^2 + b^2 - 2\vec{a}\cdot\vec{b} = 2(a^2+b^2) = 2(OA^2+OB^2)$. Proved. (ii): $LHS = \left|\vec{a}+\vec{b}\right|^2 - \left|\vec{b}-\vec{a}\right|^2 = a^2 + b^2 + 2\vec{a}\cdot\vec{b} - \left(b^2 + a^2 - 2\vec{a}\cdot\vec{b}\right) = 4\vec{a}\cdot\vec{b} = 4ab\cos\theta = 4a(b\cos\theta) = 4(OA)(ON) = 4b(a\cos\theta) = 4(OB)(OM)$. Proved. (iii): $LHS = |a^2 - b^2|$. $RHS = \left|\vec{a}+\vec{b}\right|$(projection of $\overrightarrow{AB}$ on $\overrightarrow{OC}$) $= \left|\vec{a}+\vec{b}\right|\left|\left((\vec{b}-\vec{a})\cdot\frac{\vec{a}+\vec{b}}{|\vec{a}+\vec{b}|}\right)\right| = \left|(\vec{b}-\vec{a})\cdot(\vec{b}+\vec{a})\right| = |b^2 - a^2| = |a^2 - b^2|$. Also, $|(AB)(LL')| = \left|\vec{b}-\vec{a}\right|$(projection of $\overrightarrow{OC}$ on $\overrightarrow{AB}$) $= \left|\vec{b}-\vec{a}\right|\left|(\vec{a}+\vec{b})\cdot\frac{\vec{b}-\vec{a}}{|\vec{b}-\vec{a}|}\right| = \left|(\vec{b}+\vec{a})\cdot(\vec{b}-\vec{a})\right| = |b^2 - a^2| = |a^2 - b^2|$. Proved.

(15): Let $OABC$ be a regular tetrahedron, so $|\vec{a}| = |\overrightarrow{OA}| = |\overrightarrow{OB}| = |\vec{b}| = |\overrightarrow{OC}| = |\vec{c}| = |\vec{b} - \vec{c}| = |\vec{c} - \vec{a}| = |\vec{a} - \vec{b}| = k(\text{let})$ $\cdots(1)$

Let us take vertex O. Its opposite face is ABC. Here centroid G_1 of face ABC is given by $\vec{g_1} = \frac{\vec{a}+\vec{b}+\vec{c}}{3}$. To prove: $\vec{g_1} \parallel (\vec{b} \times \vec{c} + \vec{c} \times \vec{a} + \vec{a} \times \vec{b})$.

$LHS = \frac{1}{3}(\vec{a} + \vec{b} + \vec{c}) \times (\vec{b} \times \vec{c} + \vec{c} \times \vec{a} + \vec{a} \times \vec{b}) = \frac{1}{3}[\{(\vec{c}\cdot\vec{a})\vec{b} - (\vec{a}\cdot\vec{b})\vec{c} + a^2\vec{c} - (\vec{c}\cdot\vec{a})\vec{a} + (\vec{a}\cdot\vec{b})\vec{a} - a^2\vec{b}\} + \{(\vec{a}\cdot\vec{b})\vec{c} - (\vec{b}\cdot\vec{c})\vec{a} + b^2\vec{a} - (\vec{a}\cdot\vec{b})\vec{b} + (\vec{b}\cdot\vec{c})\vec{b} - b^2\vec{c}\} + \{(\vec{b}\cdot\vec{c})\vec{a} - (\vec{c}\cdot\vec{a})\vec{b} + c^2\vec{b} - (\vec{b}\cdot\vec{c})\vec{c} + (\vec{c}\cdot\vec{a})\vec{c} - c^2\vec{a}\}]$.

Here $|\vec{b}| = |\vec{a}| = |\vec{a} - \vec{b}|$ implies $b^2 = a^2 = a^2 + b^2 - 2\vec{a}\cdot\vec{b}$.

So, $2\vec{a}\cdot\vec{b} = b^2 = a^2$ or, $2ab\cos\theta = b^2 = a^2$ $\therefore \cos\theta = \frac{1}{2}$ $\therefore \theta = 60°$. So, in a regular tetrahedron, angle between two intersecting edges is $60°$. So, $\vec{a}\cdot\vec{b} = \vec{b}\cdot\vec{c} = \vec{c}\cdot\vec{a} = \frac{a^2}{2}$. So, $LHS = 0 = RHS$. Let α be the angle between faces OBC and OCA. So,

$$\cos\alpha = \frac{(\vec{b}\times\vec{c})\cdot(\vec{c}\times\vec{a})}{|\vec{b}\times\vec{c}||\vec{c}\times\vec{a}|} = \frac{(\vec{b}\cdot\vec{c})(\vec{c}\cdot\vec{a}) - c^2(\vec{b}\cdot\vec{a})}{(bc\sin 60°)(ca\sin 60°)} = \left|\frac{(bc\cos 60°)(ca\cos 60°) - c^2(ab\cos 60°)}{abc^2(\sin 60°)^2}\right| =$$

$$\left|\frac{\frac{1}{4}-\frac{1}{2}}{\frac{3}{4}}\right| = \frac{1}{3}. \text{ So, } \alpha = \cos^{-1}\frac{1}{3}. \text{ \underline{Ans}.}$$

Let us take a face OAB and an edge OC. Let the required angle be β.

$$\text{So, } \cos(90° - \beta) = \frac{(\vec{a}+\vec{b})\cdot\vec{c}}{|\vec{a}+\vec{b}||\vec{c}|} = \frac{\vec{a}\cdot\vec{c}+\vec{b}\cdot\vec{c}}{\left(\sqrt{a^2+b^2+2\vec{a}\cdot\vec{b}}\right)c}\frac{ac\cos 60° + bc\cos 60°}{\left(\sqrt{a^2+b^2+2ab\cos 60°}\right)c}$$

$$\therefore \sin\beta = \frac{2\cos 60°}{\sqrt{2(1+\cos 60°)}} = \frac{1}{\sqrt{3}}. \text{ So, } \beta = \sin^{-1}\frac{1}{\sqrt{3}}. \text{ \underline{Ans}.}$$

Let $G\left(\frac{\vec{o}+\vec{a}+\vec{b}+\vec{c}}{4}\right)$ be the centroid of the regular tetrahedron. Let φ be the angle between $\overrightarrow{GO}$ and $\overrightarrow{GA}$. So,

$$\cos\varphi = \frac{-\frac{1}{4}(\vec{a}+\vec{b}+\vec{c})\cdot\left\{\vec{a} - \frac{\vec{a}+\vec{b}+\vec{c}}{4}\right\}}{\left|\frac{\vec{a}+\vec{b}+\vec{c}}{4}\right|\left|\vec{a} - \frac{\vec{a}+\vec{b}+\vec{c}}{4}\right|} = -\frac{(\vec{a}+\vec{b}+\vec{c})\cdot(3\vec{a}-\vec{b}-\vec{c})}{|\vec{a}+\vec{b}+\vec{c}||3\vec{a}-\vec{b}-\vec{c}|} =$$

$$-\frac{3k^2-k^2-k^2+(2\vec{a}\cdot\vec{b}-2\vec{b}\cdot\vec{c}+2\vec{c}\cdot\vec{a})}{\sqrt{k^2+k^2+k^2+2k^2\left(\frac{1}{2}\right)+2k^2\left(\frac{1}{2}\right)+2k^2\left(\frac{1}{2}\right)}\sqrt{9k^2+k^2+k^2-6k^2\left(\frac{1}{2}\right)+2k^2\left(\frac{1}{2}\right)-6k^2\left(\frac{1}{2}\right)}} = -\frac{k^2+2k^2\left(\frac{1}{2}\right)}{\sqrt{6}k\sqrt{6}k} =$$

$\frac{-2}{6} = \frac{-1}{3}$. So, $\varphi = \cos^{-1}\left(\frac{-1}{3}\right)$. Thus angle between face and interecting edge is $\cos^{-1}\left(\frac{-1}{3}\right)$. Ans.

(16): Let $OABC$ be any orthrocentric tetrahedron. So, $\begin{cases} \vec{a}\cdot(\vec{b}-\vec{c}) = 0 & \cdots(1) \\ \vec{b}\cdot(\vec{c}-\vec{a}) = 0 & \cdots(2) \\ \vec{c}\cdot(\vec{a}-\vec{b}) = 0 & \cdots(3) \end{cases}$

Here (3) follows from (1) and (2). Equation of the straight line through $O(\vec{0})$ and perpendicular to the plane ABC is $\vec{r} = \vec{0} + t(\vec{b}\times\vec{c} + \vec{c}\times\vec{a} + \vec{a}\times\vec{b})$ $\cdots(4)$

Also, equation of plane ABC is $\vec{r}\cdot\frac{\vec{b}\times\vec{c}+\vec{c}\times\vec{a}+\vec{a}\times\vec{b}}{|\vec{b}\times\vec{c}+\vec{c}\times\vec{a}+\vec{a}\times\vec{b}|} = p$ $\cdots(5)$

Let us solve (4) and (5): $t(\vec{b}\times\vec{c} + \vec{c}\times\vec{a} + \vec{a}\times\vec{b})\cdot\left\{\frac{\vec{b}\times\vec{c}+\vec{c}\times\vec{a}+\vec{a}\times\vec{b}}{|\vec{b}\times\vec{c}+\vec{c}\times\vec{a}+\vec{a}\times\vec{b}|}\right\} = p$ $\cdots(5)$

so, $t = \frac{p}{|\vec{b}\times\vec{c}+\vec{c}\times\vec{a}+\vec{a}\times\vec{b}|}$. So, the foot of perpendicular drawn from O on face ABC is $K\left(\frac{p}{|\vec{b}\times\vec{c}+\vec{c}\times\vec{a}+\vec{a}\times\vec{b}|}(\vec{b}\times\vec{c} + \vec{c}\times\vec{a} + \vec{a}\times\vec{b})\right)$. Here it remains to prove: AK is perpendicular to BC, etc. Here $\overrightarrow{AK}\cdot\overrightarrow{BC} = \left\{\frac{p(\vec{b}\times\vec{c}+\vec{c}\times\vec{a}+\vec{a}\times\vec{b})}{|\vec{b}\times\vec{c}+\vec{c}\times\vec{a}+\vec{a}\times\vec{b}|} - \vec{a}\right\}.(\vec{c} - \vec{b}) = \frac{p}{|\vec{b}\times\vec{c}+\vec{c}\times\vec{a}+\vec{a}\times\vec{b}|}\{[\vec{a}\,\vec{b}\,\vec{c}] - [\vec{c}\,\vec{a}\,\vec{b}]\} - \vec{a}\cdot\vec{c} + \vec{a}\cdot\vec{b} = 0 + \vec{a}\cdot(\vec{b}-\vec{c})$. So, $\overrightarrow{AK}\cdot\overrightarrow{BC} = 0$, from (1). Proved.

(17): Let the required equation of plane be $\vec{r}\cdot(\vec{a}-\lambda\vec{b}) = \vec{r}\cdot\vec{a} - \lambda(\vec{r}\cdot\vec{b}) = p - \lambda q$ or, $\vec{r}\cdot(\vec{a}-\lambda\vec{b}) = p - \lambda q$. Since it passes through $O(0,0,0)$, so, $0 = p - \lambda q$ or, $\lambda = \frac{p}{q}$. So the required equation of plane is $\vec{r}\cdot\left(\vec{a} - \frac{p}{q}\vec{b}\right) = 0$ or $\vec{r}\cdot(q\vec{a} - p\vec{b}) = 0$. Ans.

(18): Here equation of the plane is $\vec{r}\cdot(5,2,-7) = -9$ or, $\vec{r}\cdot\vec{n} = p$ where $0 < p = \frac{9}{\sqrt{25+4+49}} = \frac{9}{\sqrt{78}}$, $\vec{n} = \frac{1}{\sqrt{78}}(-5,-2,7)$. Here $|\vec{n}| = 1$. Let the required distance of $(1,-1,3)$ from the above plane be d_1. Then $\{(1,-1,3)+d_1\vec{n}\}\cdot\vec{n}=p$ or, $d_1=p-(1,-1,3)\cdot\left\{\frac{1}{\sqrt{78}}(-5,-2,7)\right\}=\frac{9}{\sqrt{78}}-\frac{-5+2+21}{\sqrt{78}}=\frac{-9}{\sqrt{78}}$. So, required distance $=\frac{9}{\sqrt{78}}$. Also, let d_2 be the distance of $(3,3,3)$ from the plane. So, $\left((3,3,3)+d_2\vec{n}\right)\cdot\vec{n}=p$.

Hence $d_2=p-(3,3,3)\cdot\vec{n}=\frac{9}{\sqrt{78}}-(3,3,3)\cdot\left\{\frac{1}{\sqrt{78}}(-5,-2,7)\right\}=\frac{9}{\sqrt{78}}$. Ans.

Here d_1 and d_2 have the opposite signs so, the points are on opposite sides of the plane. Here $(3,3,3)$ and origin lie on the same side of the plane. Also the points are equidistance from the plane. Proved.

(19): Let us put $z=0$. Then $\begin{cases}3x-y-12=0\\ x+4y+5=0\end{cases}$ so, $x=\frac{-5+48}{12+1}$, $y=\frac{-12-15}{12+1}$. So, $x=\frac{43}{12}$, $y=\frac{-27}{13}$. So, $\left(\frac{43}{13},\frac{-27}{13},0\right)$ is a point on the given straight line. Direction of the line is along $(3,-1,1)\times(1,4,-2)=(2-4,1+6,12+1)=(-2,7,13)$. So, required equation of line is $\vec{r}=\left(\frac{43}{13},\frac{-27}{13},0\right)+\lambda(-2,7,13)$. Required equation of plane is $\{\vec{r}-(1,2,-1)\}\cdot(-2,7,13)=0$ or, $(-2)(x-1)+7(y-2)+13(z+1)=0$ or, $-2x+7y+13z+1=0$. Ans.

(20): This is solved in the text.

(21): Here equation of diagonal is $\vec{r}=\vec{0}+t(1,1,1)$ so, required perpendicular distance $=\frac{|(1,0,0)\times(1,1,1)|}{|(1,1,1)|}=\frac{|(0,0-1,1-0)|}{\sqrt{1+1+1}}=\sqrt{\frac{2}{3}}=\frac{1}{3}\sqrt{6}$. Ans.

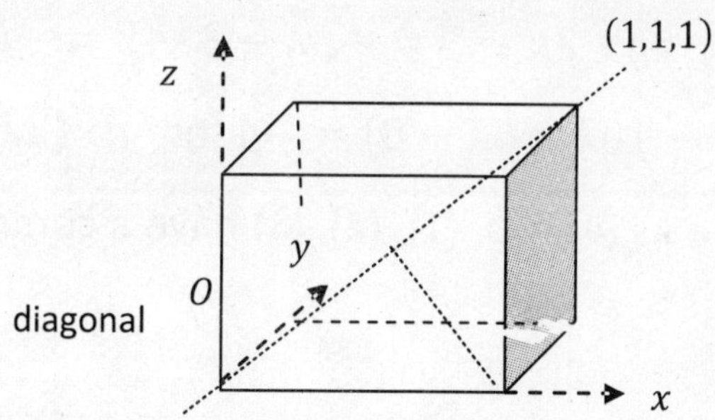

(22): Let $\frac{x}{\lambda}+\frac{y}{\mu}+\frac{z}{\nu}=1$ be the equation of a plane with fixed origin. Here x-intercept $=\lambda$, y-intercept $=\mu$, $z-$ intercept $=\nu$. So, $D\equiv$ (distance of the plane from origin) $=\frac{1}{\sqrt{\left(\frac{1}{\lambda}\right)^2+\left(\frac{1}{\mu}\right)^2+\left(\frac{1}{\nu}\right)^2-1}}$. So, $\frac{1}{\lambda^2}+\frac{1}{\mu^2}+\frac{1}{\nu^2}=\frac{1}{D^2}+1$, which is a constant. Proved.

(23): Let the required centre be (ξ, η, ζ). Given that perpendicular distances of (ξ, η, ζ) from each of the planes are equal. So, $\zeta = \eta = \xi = \frac{a-\xi-\eta-\zeta}{\sqrt{1^2+1^2+1^2}}$ or, $\sqrt{3}\zeta = a - 3\zeta$ $\therefore \zeta = \frac{a}{3+\sqrt{3}}$. So, $\xi = \zeta = \frac{a}{3+\sqrt{3}} = \frac{a}{6}(3-\sqrt{3})$. So, required centre is $\left(\frac{a}{6}(3-\sqrt{3}), \frac{a}{6}(3-\sqrt{3}), \frac{a}{6}(3-\sqrt{3})\right) = \frac{a}{6}(3-\sqrt{3})(\vec{\imath}+\vec{\jmath}+\vec{k})$. Ans.

(24): The plane (1) can be written as $\vec{r}\cdot(2,5,3) = 0$, plane (2) can be written as $\vec{r}\cdot(1,-1,4) = 2$, and plane (3) can be written as $\vec{r}\cdot(0,\ 7,-5) = -4$. Equation of the plane containing the line of intersection of (1) and (2) is $\lambda\vec{r}\cdot(2,5,3) + \vec{r}\cdot(1,-1,4) = \lambda(0) + 2$ or $\vec{r}\cdot(2\lambda+1, 5\lambda-1, 3\lambda+4) = 2$ ⋯ (4)

Similarly, equation of the plane containing line (3), (1) is $\vec{r}\cdot(2\mu, 5\mu+7, 3\mu-5) = -4$ ⋯ (5)

Here, if (4) and (5) are the same plane then $\frac{2\lambda+1}{2\mu} = \frac{5\lambda-1}{5\mu+7} = \frac{3\lambda+4}{3\mu-5}$. From the first equality, $10\lambda\mu + 14\lambda + 5\mu + 7 = 10\lambda\mu - 2\mu$ or, $14\lambda + 7\mu + 7 = 0$ or, $2\lambda + \mu + 1 = 0$. From the second equality, $6\lambda\mu - 10\lambda + 3\mu - 5 = 6\lambda\mu + 8\mu$ or, $10\lambda + 5\mu + 1 = 0$ or, $2\lambda + \mu + 1 = 0$. So, $2\mu = -(4\lambda+2)$, $5\mu + 7 = -10\lambda - 5 + 7 = 2 - 10\lambda$ and $3\mu - 5 = -6\lambda - 3 - 5 = -6\lambda - 8$. So, (5) becomes $\vec{r}\cdot(-4\lambda-2, 2-10\lambda, -6\lambda-8) = -4$ or, $\vec{r}\cdot(2\lambda+1, 5\lambda-1, 3\lambda+4) = 2$, which is the same as (4). So, (1), (2), (3) have a common line of intersection. Proved.

(25): Here direction of the line of intersection of planes $\begin{cases} x+2y+3z=0 \\ 3x+2y+z=0 \end{cases}$ is along $(1,2,3)\times(3,2,1) = (2-6, 9-1, 2-6) = (-4,8,-4)$ or $(1,-2,1)$. So, angle of this line with the x-axis is $\cos^{-1}\frac{1}{\sqrt{1+4+1}} = \cos^{-1}\frac{1}{\sqrt{6}}$, angle of this line with the y and z-axes are $\cos^{-1}\frac{2}{\sqrt{6}}$ and $\cos^{-1}\frac{1}{\sqrt{6}}$ respectively. So, this line is equally inclined with the z and x-axes. Proved. If $\cos^{-1}\left|\frac{-2}{\sqrt{6}}\right| = \theta$ then

$\cos\theta = \frac{2}{\sqrt{6}} = \sqrt{\frac{2}{3}}$. So, $\cos 2\theta = 2\cos^2\theta - 1 = \frac{4}{3} - 1 = \frac{1}{3}$. So, $\sec 2\theta = 3$. So,

$\theta = \frac{1}{2}\sec^{-1} 3$. So, the acute angle which the line makes with the y-axis is

$\cos^{-1}\left|\frac{-2}{\sqrt{6}}\right| = \theta = \frac{1}{2}\sec^{-1} 3$. Proved.

(26): Let $A(\vec{a})$ and $B(\vec{b})$ be any two fixed points. Let $P(\vec{r})$ be any point on the required locus. So, $\{\vec{r}\cdot 2(\vec{b}-\vec{a})\} + a^2 - b^2 = |\vec{r}-\vec{a}|^2 - |\vec{r}-\vec{b}|^2 =$ constant, k. So, $\vec{r}\cdot(\vec{b}-\vec{a}) = \frac{k-a^2+b^2}{2}$, which is a plane for which line AB is normal to the plane. Ans.

(27): Case 1: *when forces are coplanar*: Let two coplanar forces be $\vec{F}$ and $t\vec{F}$.

Let $\vec{F}$ be acting on the line $\vec{r} = \vec{a} + \lambda\vec{F}$ and $t\vec{F}$ be acting on the line $\vec{r} = \vec{b} + \mu\vec{F}$. Let $P(\vec{s})$ be a point on the required locus. So, $(\vec{a}-\vec{s})\times\vec{F} = -\{(\vec{b}-\vec{s})\times t\vec{F}\}$. So, $(1+t)\vec{s}\times\vec{F} = (\vec{a}+t\vec{b})\times\vec{F}$ or, $\vec{s}\times\vec{F} = \frac{1\vec{a}+t\vec{b}}{1+t}\vec{F}$

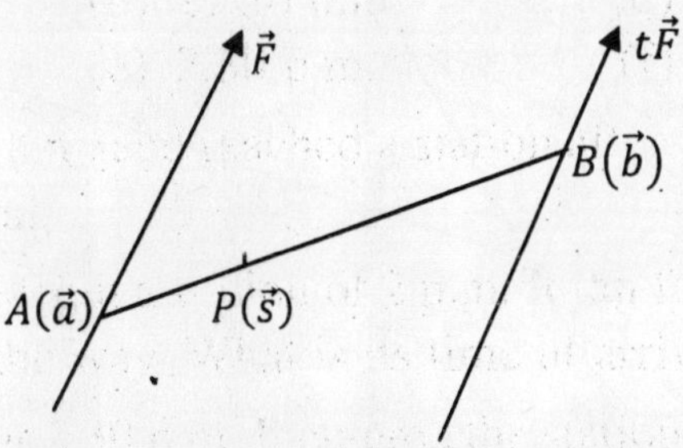

or, $\left(\vec{s} - \frac{1\vec{a}+t\vec{b}}{1+t}\right)\times\vec{F} = \vec{0}$ so, $\vec{s} - \frac{\vec{a}+t\vec{b}}{1+t} = \nu\vec{F}$, where ν is any scalar. So, required equation of locus of $P(\vec{s})$ is $\vec{r} = \frac{\vec{a}+t\vec{b}}{1+t} + \nu\vec{F}$, which is a straight line parallel to $\vec{F}$ and passing through a point on AB which divides AB in the ratio $t:1$.

Case 2: *when forces are intersecting:* Let force $\overrightarrow{F_1}$ be acting on the line $\vec{r} = \vec{a} + \lambda\overrightarrow{F_1}$ and let force $\overrightarrow{F_2}$ be acting on the line $\vec{r} = \vec{a} + \mu\overrightarrow{F_2}$.

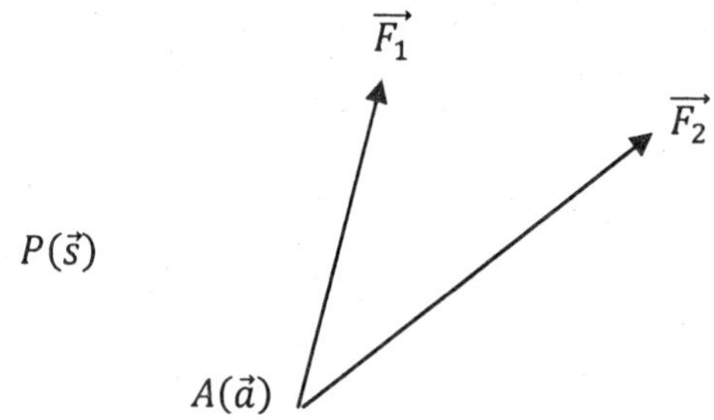

Let $P(\vec{s})$ be any point on the required locus. So, $(\vec{a}-\vec{s})\times\vec{F_1}=(\vec{a}-\vec{s})\times\vec{F_2}$ or, $(\vec{a}-\vec{s})\times(\vec{F_1}-\vec{F_2})=\vec{0}$. So, let $\vec{a}-\vec{s}=-t(\vec{F_1}-\vec{F_2})$ where t is a scalar. Hence $\vec{s}=\vec{a}+t(\vec{F_1}-\vec{F_2})$. So, required equation of locus of $P(\vec{s})$ is $\vec{s}=\vec{a}+\ t(\vec{F_1}-\vec{F_2})$, which is a straight line passing through the point $A(\vec{a})$ and parallel to $\vec{F_1}-\vec{F_2}$. Ans.

(28):

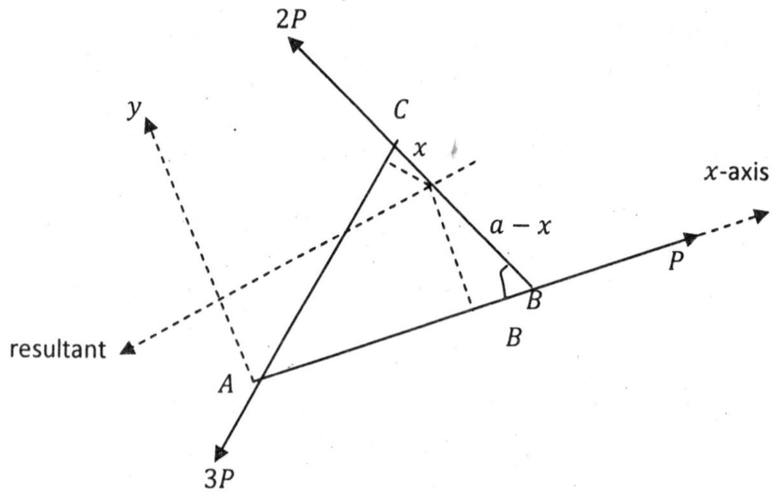

Let us choose A as the origin and $\overrightarrow{AB}$ as the positive direction of the x-axis. So, required resultant $=(P)(1,0)+(2P)(-\cos B,\ \sin B)+(3P)(-\cos A,-\sin A)=(P)(1-2\cos B-3\cos A,\ 2\sin B-3\sin A)$, so, the magnitude of resultant $=P\sqrt{(1-2\cos B-3\cos A)^2+(2\sin B-3\sin A)^2}$ where $A=B=C=60°$ (from question). So, magnitude of resultant $=P\sqrt{\left(1-\frac{5}{2}\right)^2+\left(-\frac{\sqrt{3}}{2}\right)^2}=\sqrt{3}P$.

Here resultant force $= P\left(1-\frac{5}{2},-\frac{\sqrt{3}}{2}\right) = P\left(-\frac{3}{2},\frac{-\sqrt{3}}{2}\right) = -P\sqrt{3}\left(\frac{\sqrt{3}}{2},\frac{1}{2}\right) =$ $-P\sqrt{3}(\cos 30°, \sin 30°)$ which acts parallel to $\overrightarrow{D'A}$, where D' is the mid-point of BC. On taking torque about D, $3P(x\sin 60°) + P((a-x)\sin 60°) = 0$ or, $3x + (a-x) = 0$ or, $x = -\frac{a}{2}$. So, the required resultant divides CB at D, where $BD:BC = 3:2$. Ans .

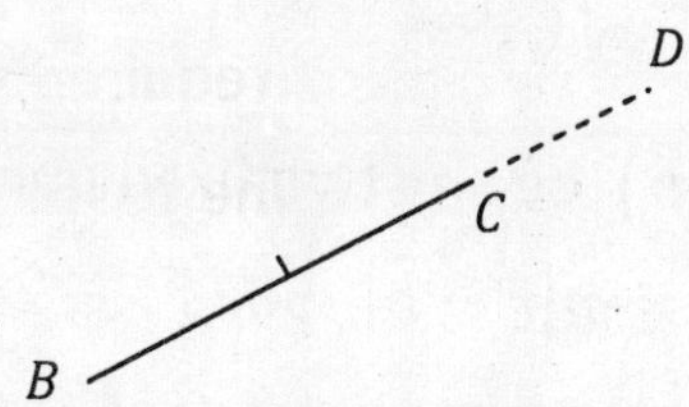

(29): Required work done $= \{(4,1,-3)+(3,1,-1)\}\cdot\{(5,4,1)-(1,2,3)\} =$ $(7,2,-4)\cdot(4,2,-2) = 7(4)+2(2)+(-4)(-2) = 40$ unit. Ans.

(30): Let $|\vec{r}-\vec{a}| = k$ $\quad\cdots(1)$

be any sphere, whose centre is $\vec{a}$ and radius is $k(>0)$. Let $R'\left(\vec{a}-\frac{1}{2}\vec{d}\right)$ and $R\left(\vec{a}+\frac{1}{2}\vec{d}\right)$ be two extremities of a diameter,where $|\vec{d}| = 2k$ $\quad\cdots(2)$

Let $P(\vec{r})$ be any point on the sphere. Now to prove: $R'P \perp RP$, i.e., $\left\{\vec{r}-\left(\vec{a}-\frac{\vec{d}}{2}\right)\right\}\cdot\left\{\vec{r}-\left(\vec{a}+\frac{\vec{d}}{2}\right)\right\} = 0$. $LHS = |\vec{r}-\vec{a}|^2 - \left|\frac{1}{2}\vec{d}\right|^2 = k^2 - \frac{1}{4}(4k^2)$, from (1), (2). Proved.

(31): Here the centre of the sphere is $\frac{1}{2}(\vec{g}+\vec{h})$ and radius $= \frac{1}{2}|\vec{g}-\vec{h}|$.
So, required equation of sphere is $\left|\vec{r}-\frac{1}{2}(\vec{g}+\vec{h})\right| = \frac{1}{2}|\vec{g}-\vec{h}|$
or, $\vec{r}\cdot\vec{r}-(\vec{g}+\vec{h})\cdot\vec{r}+\frac{1}{4}(\vec{g}\cdot\vec{g}+\vec{h}\cdot\vec{h}+2\vec{g}\cdot\vec{h}) = \frac{1}{4}(\vec{g}\cdot\vec{g}+\vec{h}\cdot\vec{h}-2\vec{g}\cdot\vec{h})$.
So, $\vec{r}\cdot\vec{r}-\vec{g}\cdot\vec{r}-\vec{h}\cdot\vec{r}+\vec{g}\cdot\vec{h} = 0$ or, $(\vec{r}-\vec{g})\cdot(\vec{r}-\vec{h}) = 0$. Proved.

(32): Let n points be $A_1(\overrightarrow{a_1}), A_2(\overrightarrow{a_2}), \cdots, A_n(\overrightarrow{a_n})$. Let $P(\vec{r})$ be any point on the locus. So, $(\vec{r}-\overrightarrow{a_1})\cdot(\vec{r}-\overrightarrow{a_1})+(\vec{r}-\overrightarrow{a_2})\cdot(\vec{r}-\overrightarrow{a_2})+\cdots+(\vec{r}-\overrightarrow{a_n})\cdot$

$(\vec{r}-\vec{a_n}) = $ constant $\equiv k^2$. So, $n\vec{r}\cdot\vec{r} - 2(\vec{a_1}+\vec{a_2}+\cdots+\vec{a_n})\cdot\vec{r} + (a_1{}^2 + a_2{}^2 + \cdots + a_n{}^2 - k^2) = 0$

or, $\vec{r}\cdot\vec{r} - 2\left(\frac{\vec{a_1}+\vec{a_2}+\cdots+\vec{a_n}}{n}\right)\cdot\vec{r} + \frac{1}{n}(a_1{}^2 + a_2{}^2 + \cdots + a_n{}^2 - k^2) = 0$, which represents a sphere, whose centre is $\left(\frac{\vec{a_1}+\vec{a_2}+\cdots+\vec{a_n}}{n}\right)$ (that is, the mean centre of $\vec{a_1}, \vec{a_2}, \cdots, \vec{a_n}$) and radius $= \sqrt{\left\{\frac{1}{n}\left(\sum_{i=1}^{n}\vec{a_i}\right)\right\}^2 - \frac{1}{n}\left(\sum_{i=1}^{n}|\vec{a_i}|^2 - k^2\right)}$. Proved.

(33): Let $A(\vec{a})$ and $B(\vec{b})$ be any two fixed points. Let $P(\vec{r})$ be any point on the locus. Given $1|\vec{r}-\vec{a}| = n|\vec{r}-\vec{b}|$. So, $\vec{r}^2 - 2\vec{a}\cdot\vec{r} + \vec{a}^2 = n^2(\vec{r}^2 - 2\vec{b}\cdot\vec{r} + \vec{b}^2)$ or, $(n^2-1)\vec{r}^2 - 2(\vec{a} - n^2\vec{b})\cdot\vec{r} + (n^2b^2 - a^2) = 0$, which represents a sphere, whose centre is $\frac{\vec{a}-n^2\vec{b}}{n^2-1}$ and radius $= \sqrt{\frac{1}{(n^2-1)^2}\left|\vec{a}-n^2\vec{b}\right|^2 + a^2 - n^2b^2}$

Let $\vec{r}^2 - 2\left(\frac{\vec{a}-n_1{}^2\vec{b}}{n_1{}^2-1}\right)\cdot\vec{r} + \frac{n_1{}^2b^2-a^2}{n_1{}^2-1} = 0$ $\cdots(1)$

and $\vec{r}^2 - 2\left(\frac{\vec{a}-n_2{}^2\vec{b}}{n_2{}^2-1}\right)\cdot\vec{r} + \frac{n_2{}^2b^2-a^2}{n_2{}^2-1} = 0$ $\cdots(2)$

be two such spheres. Let $P(\vec{s})$ be any point on the radical plane.

So, $\sqrt{\vec{s}^2 - 2\left(\frac{\vec{a}-n_1{}^2\vec{b}}{n_1{}^2-1}\right)\cdot\vec{s} + \frac{n_1{}^2b^2-a^2}{n_1{}^2-1}}$ = (length of tangent from $P(\vec{s})$ to the sphere (1))
= (length of tangent from $P(\vec{s})$ to the sphere (2)) =
$\sqrt{\vec{s}^2 - 2\left(\frac{\vec{a}-n_2{}^2\vec{b}}{n_2{}^2-1}\right)\cdot\vec{s} + \frac{n_2{}^2b^2-a^2}{n_2{}^2-1}}$. On squaring both sides and cancelling, $2(n_2{}^2-1)(n_1{}^2\vec{b}-\vec{a})\cdot\vec{s} + (n_2{}^2-1)(n_1{}^2b^2-a^2) - 2(n_1{}^2-1)(n_2{}^2\vec{b}-\vec{a})\cdot\vec{s} - (n_1{}^2-1)(n_2{}^2b^2-a^2) = 0$. So, $2[n_2{}^2(n_1{}^2\vec{b}-\vec{a}) - (n_1{}^2\vec{b}-\vec{a}) - n_1{}^2(n_2{}^2\vec{b}-\vec{a}) + n_2{}^2\vec{b}-\vec{a}]\cdot\vec{s} + n_2{}^2(n_1{}^2b^2-a^2) - (n_1{}^2b^2-a^2) - n_1{}^2(n_2{}^2b^2-a^2) + n_2{}^2b^2 - a^2 = 0$. Hence $2(n_1{}^2-n_2{}^2)\vec{a}\cdot\vec{s} + (n_1{}^2-n_2{}^2)a^2 - (n_1{}^2-n_2{}^2)b^2 = 0$ or, $2\vec{a}\cdot\vec{s} = b^2 - a^2$ or, $\vec{s}\cdot\vec{a} = \frac{b^2-a^2}{2}$, which is a plane free from n_1, n_2. Proved.

[**Note**: *Here normal to the radical plane is along $\overrightarrow{OA}$.]*

(34): Let $|\vec{r}-\vec{a}|=k$ be any sphere. Let $P(\vec{p})$. So, $|\vec{p}-\vec{a}|=k$ $\cdots(1)$

Let $Q\left(\frac{1}{\lambda+1}\vec{p}\right)$, where $OP:OQ=\frac{1+\lambda}{1}$, which is a constant.

From (1), $\left|\frac{1}{1+\lambda}\vec{p}-\frac{\vec{a}}{\lambda+1}\right|=\frac{k}{\lambda+1}$, so the required locus of Q is $\left|\vec{r}-\frac{\vec{a}}{\lambda+1}\right|=\frac{k}{\lambda+1}$.
This is a sphere whose centre is $\frac{1}{1+\lambda}\vec{a}$ and radius is $\frac{k}{\lambda+1}$. Ans

(35): Let S: $F(\vec{r})\equiv\vec{r}^2-2\vec{c}\cdot\vec{r}+k=0$ be any fixed sphere and let $D(\vec{d})$ be a fixed point outside the sphere. Let a line through D be in the direction of unit vector $\hat{b}$ intersect the sphere at a point P. If $DP\equiv t_0$ then $P(\vec{d}+t_0\hat{b})$, whic lies on the sphere S. So, $(\vec{d}+t_0\hat{b})^2-2\vec{c}\cdot(\vec{d}+t_0\hat{b})+k=0$ $\cdots$

Here the locus of $Q\left(\vec{d}+\frac{1}{t_0}\hat{b}\right)$, where $\hat{b}$ is a parameter, is what we mean by *inverse of sphere with respect to point* D. Let $Q(\vec{s})$. So, $\vec{s}=\vec{d}+\frac{1}{t_0}\hat{b}$ or,

$t_0^2(\vec{s}-\vec{d})^2=1$ $\cdots(2)$

From (1), $(\vec{d}^2-2\vec{c}.\vec{d}+k)-2\vec{c}\cdot t_0\hat{b}+2\vec{d}\cdot t_0\hat{b}+t_0^2=0$ or, $F(\vec{d})-2(\vec{c}-\vec{d})\cdot t_0\hat{b}+t_0^2=0$ or, $F(\vec{d})-2(\vec{c}-\vec{d})\cdot t_0\{(\vec{s}-\vec{d})t_0\}+t_0^2=0$, from (2). Hence $F(\vec{d})+[1-2(\vec{c}-\vec{d})\cdot(\vec{s}-\vec{d})]t_0^2=0$ or, $F(\vec{d})+[1-2(\vec{c}-\vec{d})\cdot(\vec{s}-\vec{d})]\frac{1}{(\vec{s}-\vec{d})^2}=0$, from (2). Thus the equation of locus of $Q(\vec{s})$ is $\left(F(\vec{d})\right)(\vec{s}-\vec{d})^2+[1-2(\vec{c}-\vec{d})\cdot(\vec{s}-\vec{d})]=0$ or, $(\vec{s}-\vec{d})^2-2\frac{(\vec{c}-\vec{d})}{F(\vec{d})}\cdot(\vec{s}-\vec{d})+\frac{1}{F(\vec{d})}=0$,

which is a sphere whose centre is $\vec{d}+\frac{(\vec{c}-\vec{d})}{F(\vec{d})}$ and radius is $\sqrt{\left\{\frac{(\vec{c}-\vec{d})}{F(\vec{d})}\right\}^2-\frac{1}{F(\vec{d})}}$,

provided $F(\vec{d})\neq 0$, that is, point D is outside the sphere. Ans. If point D is on the sphere, that is, $F(\vec{d})=0$ then, $1-2(\vec{c}-\vec{d})\cdot(\vec{s}-\vec{d})=0$, which shows that the locus of $Q(\vec{s})$ is a plane. Ans.

(36): Let $|\vec{r}-\vec{a}|=k$ $\cdots(1)$

be any sphere. Let $M(\vec{m})$ and $N(\vec{n})$ be any two points. Equation of polar plane of $M(\vec{m})$ is $\vec{m}\cdot\vec{r}-\vec{a}\cdot(\vec{m}+\vec{r})+a^2-k=0$ or, $\vec{r}\cdot(\vec{m}-\vec{a})=\vec{a}\cdot\vec{m}+k-a^2$ $\cdots(2)$

So, the length of perpendicular from $N(\vec{n})$ on (2) is $\frac{\vec{a}\cdot\vec{m}+k-a^2-\vec{n}\cdot(\vec{m}-\vec{a})}{|\vec{m}-\vec{a}|}=\frac{\vec{a}\cdot(\vec{m}+\vec{n})-\vec{m}\cdot\vec{n}+k-a^2}{|\vec{m}-\vec{a}|}$, etc. So, (length of perpendicular from N on polar plane of M)(distance of N from centre) = (length of perpendicular from M on polar plane of N) (distance of M from centre). So, (length of perpendicular from N on polar plane of M): (length of perpendicular from M on polar plane of N) = istance of M from centre): (distance of N from centre). Proved.

: Let $F(\vec{r})\equiv\vec{r}^2-2\vec{c}\cdot\vec{r}+k$, $F'(\vec{r})\equiv\vec{r}^2-2\vec{c'}\cdot\vec{r}+k'$ and $F''(\vec{r})\equiv\vec{r}^2-$ $\cdot\vec{r}+k''$. Let $F''(\vec{r})=0$ cuts $F(\vec{r})=0$ and $F'(\vec{r})=0$ orthogonally. So,

$c\cdot\vec{c''}=k+k''$ $\cdots(1)$

and $2\vec{c'}\cdot\vec{c''}=k'+k''$ $\cdots(2)$

Here another sphere is $F(\vec{r})-\lambda F'(\vec{r})=0$ or, $(1-\lambda)\vec{r}^2-2(\vec{c}-\lambda\vec{c'})\cdot\vec{r}+(k-\lambda k')=0$ or, $\vec{r}^2-\left(\frac{\vec{c}-\lambda\vec{c'}}{1-\lambda}\right)\cdot\vec{r}+\frac{k-\lambda k'}{1-\lambda}=0$. This sphere will cut the sphere $F''(\vec{r})=0$ orthogonally, if $2\left(\frac{\vec{c}-\lambda\vec{c'}}{1-\lambda}\right)\cdot\vec{c''}=\frac{k-\lambda k'}{1-\lambda}+k''$. $LHS=\frac{2(\vec{c}\cdot\vec{c''}-\lambda\vec{c'}\cdot\vec{c''})}{1-\lambda}=\frac{k+k''-\lambda(k'+k'')}{1-\lambda}=\frac{(k-\lambda k')+k''-\lambda k''}{1-\lambda}=\frac{k-\lambda k'}{1-\lambda}+k''=RHS$. Proved.

(38): Let $\vec{r}^2-2\vec{c}\cdot\vec{r}+k_1=0$ $\cdots(1)$

and $\vec{r}^2-2\vec{c}\cdot\vec{r}+k_2=0$ $\cdots(2)$

be any two concentric spheres, having the same centre $\vec{c}$. Let $L\left(\vec{c}-\frac{\vec{d}}{2}\right)$ and $M\left(\vec{c}+\frac{\vec{d}}{2}\right)$ be two extremities of a diameter of (2), where

$|\vec{d}| = 2\sqrt{c^2 - k_2}$ $\cdots(*)$. Let $S(\vec{s})$ be any point on (1).

So, $\vec{s}^2 - 2\vec{c}\cdot\vec{s} + k_1 = 0$ $\cdots(3)$

Hence $LS^2 + MS^2 = \left(\vec{s} - \vec{c} + \frac{\vec{d}}{2}\right)\cdot\left(\vec{s} - \vec{c} + \frac{\vec{d}}{2}\right) + \left(\vec{s} - \vec{c} - \frac{\vec{d}}{2}\right)\cdot\left(\vec{s} - \vec{c} - \frac{\vec{d}}{2}\right) =$ $2\left[(\vec{s}-\vec{c})^2 + \left(\frac{d}{2}\right)^2\right] = 2\left[(-k_1 + \vec{c}\cdot\vec{c}) + \frac{1}{4}d^2\right]$, from (3). So, from $(*)$,

$LS^2 + MS^2 = 2[2c^2 - k_1 - k_2]$, which is free from $\vec{s}$. Proved.

(39): $RHS = \vec{r} + \left[\frac{\gamma-1}{v^2}\vec{v}\cdot\vec{r} - \gamma t\right]\vec{v} + \left[\frac{\gamma-1}{v^2}\vec{v}\cdot\left[\vec{r} + \frac{\gamma-1}{v^2}\vec{v}\cdot\vec{r} - \gamma t\right]\vec{v} + \gamma\gamma\left(t - \frac{\vec{v}\cdot\vec{r}}{c^2}\right)\right]\vec{v} = \vec{r} + \left[\frac{\gamma-1}{v^2}\vec{v}\cdot\vec{r} - \gamma t + \frac{\gamma-1}{v^2}\vec{v}\cdot\vec{r} + \frac{(\gamma-1)^2}{v^4}(\vec{v}\cdot\vec{r})v^2 - \frac{\gamma t(\gamma-1)}{v^2}v^2 + \gamma^2 t - \frac{\gamma^2}{c^2}\vec{v}\cdot\vec{r}\right]\vec{v} = \vec{r} + \left[(\vec{v}\cdot\vec{r})\left\{2\frac{\gamma-1}{v^2} + \frac{(\gamma-1)^2}{v^2} - \frac{\gamma^2}{c^2}\right\} - \gamma t - \gamma^2 t + \gamma t + \gamma^2 t\right]\vec{v} =$

$\vec{r} + \left[(\vec{v}\cdot\vec{r})\left\{\frac{\gamma^2-1}{v^2} - \frac{\gamma^2}{c^2}\right\}\right]\vec{v} = \vec{r} + \vec{v}\cdot\vec{r}\left\{\frac{1}{v^2}\left(\frac{c^2}{c^2-v^2} - 1\right) - \frac{1}{c^2-v^2}\right\}\vec{v} =$

$\vec{r} + \vec{v}\cdot\vec{r}\left\{\frac{1}{c^2-v^2} - \frac{1}{c^2-v^2}\right\}\vec{v} = \vec{r} = LHS$. Proved. Next $RHS =$ $\gamma\left[\gamma\left(t - \frac{\vec{v}\cdot\vec{r}}{c^2}\right) + \frac{\vec{v}}{c^2}\cdot\left\{\vec{r} + \left(\frac{\gamma-1}{v^2}\vec{v}\cdot\vec{r} - \gamma t\right)\vec{v}\right\}\right] = \gamma\left[\gamma t - \frac{\gamma}{c^2}\vec{v}\cdot\vec{r} + \frac{1}{c^2}\vec{v}\cdot\vec{r} + \frac{\gamma-1}{v^2c^2}(\vec{v}\cdot\vec{r})v^2 - \frac{\gamma t}{c^2}v^2\right] = \gamma\left[\gamma t - \vec{v}\cdot\vec{r}\left(\frac{\gamma}{c^2} + \frac{1-\gamma}{c^2} - \frac{1}{c^2}\right) - \gamma t\frac{v^2}{c^2}\right] = \gamma\left[\gamma t - \frac{\gamma t v^2}{c^2}\right] =$ $\gamma^2 t\left(1 - \frac{v^2}{c^2}\right) = \gamma^2 t\left(\frac{1}{\gamma}\right)^2 = t = LHS$. Proved.

(40): Let $\overrightarrow{a_1}, \overrightarrow{a_2}, \overrightarrow{a_3}, \cdots$ be the position vectors of n points $A_1, A_2, A_3, \cdots$. Let $A_1, A_2, A_3, \cdots$ be coplanar, and let $\lambda_1\overrightarrow{a_1} + \lambda_2\overrightarrow{a_2} + \lambda_3\overrightarrow{a_3} + \cdots + \lambda_{n-1}\overrightarrow{a_{n-1}} = \overrightarrow{a_n}$ or,

$\lambda_1\overrightarrow{a_1} + \lambda_2\overrightarrow{a_2} + \lambda_3\overrightarrow{a_3} + \cdots + \lambda_{n-1}\overrightarrow{a_{n-1}} + \lambda_n\overrightarrow{a_n} = \vec{0}$ where $\lambda_n = -1$ $\cdots(1)$

Let $\vec{r}\cdot\vec{n} = p$ be the equation of the plane in normal form. So, $\overrightarrow{a_1}\cdot\vec{n} = p$, $\overrightarrow{a_2}\cdot\vec{n} = p, \cdots$. So, $0 = \vec{0}\cdot\vec{n} = (\lambda_1\overrightarrow{a_1} + \lambda_2\overrightarrow{a_2} + \cdots)\cdot\vec{n} = \lambda_1(\overrightarrow{a_1}\cdot\vec{n}) + \lambda_2(\overrightarrow{a_2}\cdot\vec{n}) + \cdots = \lambda_1 p + \lambda_2 p + \cdots + \lambda_{n-1}p + (-1)p = (\lambda_1 + \lambda_2 + \cdots + \lambda_{n-1} + (-1))p$. So $\lambda_1 + \lambda_2 + \cdots + \lambda_{n-1} = 1$, since $p \neq 0$ (or, equivalently, $\vec{0}$ does not lie on the plane). Proved.

(41): Our equation is $\vec{x}^2 + \frac{\vec{b}}{a} \cdot \vec{x} + \frac{c}{a} = 0$ or, $\left(\vec{x} + \frac{1}{2a}\vec{b}\right)^2 = \frac{-c}{a} + \frac{b^2}{4a^2} = \frac{b^2-4ac}{4a^2}$ or, $\left|\vec{x} - \frac{-\vec{b}}{2a}\right| = \frac{\sqrt{b^2-4ac}}{2a}$ which represents a sphere, whose centre is $-\frac{1}{2a}\vec{b}$ and radius $= \frac{1}{2}\sqrt{b^2 - 4ac}$ (> 0). If $a = 0$ then the given equation represents a plane whose unit normal vector is $\frac{\vec{b}}{b}$ and length of perpendicular drawn from origin on the plane is $\frac{c}{b}$. If $a \neq 0$ and $b^2 < 4ac$, then we get no solution. If $a \neq 0$ and $b^2 = 4ac$ then the only solution is $\frac{-b}{2a}$. If $a \neq 0$ and $4bc < b^2$ then $\vec{x} = \frac{-\vec{b}}{2a} + \frac{1}{2a}\sqrt{b^2 - 4ac}\,\vec{e}$, where $\vec{e}$ denotes arbitrary unit vector. Ans.

4

PRODUCTS OF THREE OR FOUR VECTORS APPLICATIONS

Example (page 72): Let $\vec{r} = \lambda\vec{a} + \mu\vec{b} + \nu\vec{c}$. So, $[\vec{r}\ \vec{b}\ \vec{c}] = \vec{r}\cdot(\vec{b}\times\vec{c}) = (\lambda\vec{a} + \mu\vec{b} + \nu\vec{c})\cdot(\vec{b}\times\vec{c}) = \lambda\vec{a}\cdot(\vec{b}\times\vec{c}) + \mu\vec{b}\cdot(\vec{b}\times\vec{c}) + \nu\vec{c}\cdot(\vec{b}\times\vec{c}) = \lambda[\vec{a}\ \vec{b}\ \vec{c}] + \mu(0) + \nu(0) = \lambda[\vec{a}\ \vec{b}\ \vec{c}]$. Since $\vec{a},\vec{b},\vec{c}$ are non-coplanar, so, $[\vec{a}\ \vec{b}\ \vec{c}] \neq 0$. So, $\lambda = \frac{[\vec{r}\ \vec{b}\ \vec{c}]}{[\vec{a}\ \vec{b}\ \vec{c}]}$. Similarly, $\mu = \frac{[\vec{a}\ \vec{r}\ \vec{c}]}{[\vec{a}\ \vec{b}\ \vec{c}]}$, $\nu = \frac{[\vec{a}\ \vec{b}\ \vec{r}]}{[\vec{a}\ \vec{b}\ \vec{c}]}$. So, $[\vec{a}\ \vec{b}\ \vec{c}]\vec{r} = [\vec{r}\ \vec{b}\ \vec{c}]\vec{a} + [\vec{a}\ \vec{r}\ \vec{c}]\vec{b} + [\vec{a}\ \vec{b}\ \vec{r}]\vec{c} = [\vec{r}\ \vec{b}\ \vec{c}]\vec{a} + [\vec{r}\ \vec{c}\ \vec{a}]\vec{b} + [\vec{r}\ \vec{a}\ \vec{b}]\vec{c}$. Proved.

Example (page 73): (1): Let $P(x,y,z)$ be any point on the plane. Since $\overrightarrow{AP}, \overrightarrow{AB}, \overrightarrow{AC}$ are 3 coplanar vectors, so $[\overrightarrow{AP}\ \overrightarrow{AB}\ \overrightarrow{AC}] = 0$.

Hence $\begin{vmatrix} x-2 & y-3 & z+1 \\ 2 & 2 & 3 \\ 1 & 3 & 6 \end{vmatrix} = 0$ or, $(12-9)(x-2) + (3-12)(y-3) + (6-2)(z+1) = 0$. Ans.

(2): This is solved in the text.

(3): This is solved in the text.

(4): Here A,B,C,D are 4 coplanar points, that is, $\overrightarrow{AB}$, $\overrightarrow{AC}$, $\overrightarrow{AD}$ are coplanar, that is, $\begin{vmatrix} -4 & -6 & -2 \\ -1 & 4 & 3 \\ -8 & -1 & 3 \end{vmatrix} = 0$. Let us verify it: $LHS = -4(12+3) - 6(-24+3) - 2(1+32) = -60 + 126 - 66 = 0 = RHS$. Proved.

Example (page 75): (1): Equation of line AB is $\vec{r} = (3,2,1) + \lambda(1,3,4)$ and equation of line CD is $\vec{r} = (4,2,-2) + \mu(1,3,4)$. Let us calculate $[\overrightarrow{AC}\ \overrightarrow{AB}\ \overrightarrow{CD}] =$

$\begin{vmatrix} 1 & 0 & -3 \\ 1 & 3 & 4 \\ 2 & 3 & 1 \end{vmatrix} = 1(3-12) - 3(3-6) = -9 + 9 = 0$. So, AB intersects CD.

Let $P(3+\lambda, 2+3\lambda, 1+4\lambda) = (4+2\mu, 2+3\mu, -2+\mu)$. So,

$$\begin{cases} \lambda - 2\mu - 1 = 0 & \cdots(1) \\ 3\lambda - 2\mu = 0 & \cdots(2) \\ 4\lambda - \mu + 3 = 0 & \cdots(3) \end{cases}$$

From (2) and (3), $\frac{\lambda}{-9} = \frac{\mu}{-9} = \frac{1}{-3+12}$. $\therefore \lambda = -1,\ \mu = -1$. Now $[LHS \text{ of } (1)]_{(\lambda,\mu)=(-1,-1)} = (-1) - 2(-1) - 1 = 0 = RHS$ of (1). So, $\lambda = -1$, $\mu = -1$ satisfy all the 3 equations. So, AB, CD intersect. Also $P(3-1, 2+3(-1), 1+4(-1)) = P(2,-1,-3)$. Ans. Here equation of line AD and BC are $\vec{r} = (3,2,1) + \lambda(3,3,-2)$ and $\vec{r} = (4,5,5) + \mu(0,-3,-7)$. Let us calculate λ and μ such that $(3\lambda+3, 3\lambda+2, -2\lambda+1) = (4, -3\mu+5, -7\mu+5)$. So,

$$\begin{cases} 3\lambda - 1 = 0 & \cdots(4) \\ 3\lambda + 3\mu - 3 = 0 & \cdots(5) \\ -2\lambda + 7\mu - 4 = 0 & \cdots(6) \end{cases}$$

From (4), (5): $\lambda = \frac{1}{3}$, $\mu = 1 - \frac{1}{3} = \frac{2}{3}$. So $[LHS \text{ of } (6)]_{(\lambda,\mu)=\left(\frac{1}{3},\frac{2}{3}\right)} = \frac{-2}{3} + \frac{14}{3} - 4 = 0 = RHS$ of (6). So, $\lambda = \frac{1}{3}$, $\mu = \frac{2}{3}$ satisfy all the three equations (4), (5), (6). So, AD and BC intersect. Here required point of intersection is $Q(3\lambda+3, 3\lambda+2, -2\lambda+1)_{\lambda=\frac{1}{3}} = Q\left(4, 3, \frac{1}{3}\right)$. Ans.

(2): This is solved in the text.

(3): This is solved in the text.

(4): Required shortest distance $= \left|\frac{\{(-16,6,4)-(-1,2,-3)\}\times\{(4,9,7)-(1,-1,3)\}}{|\{(-16,6,4)-(-1,2,-3)\}\times\{(4,9,7)-(1,-1,3)\}|} \cdot \overrightarrow{AC}\right| = \left|\frac{(-15,4,7)\times(3,10,4)}{|(-15,4,7)\times(3,10,4)|} \cdot (2,-3,6)\right| = \left|\frac{(-54,81,-162)\cdot(2,-3,6)}{|(-54,81,-162)|}\right| = \left|\frac{(-2,3,-6)\cdot(2,-3,6)}{\sqrt{4+9+36}}\right| = |-7| = 7$ unit. Ans.

Example (page 77): (1): Here $\vec{d} = \frac{\overrightarrow{AB}}{AB} = \frac{\vec{b}-\vec{a}}{|\vec{b}-\vec{a}|}$. Ans. Also, $\vec{m} = \vec{a} \times \vec{d} = \frac{\vec{a}\times\vec{b} - \vec{a}\times\vec{a}}{|\vec{b}-\vec{a}|} = \frac{\vec{a}\times\vec{b}}{|\vec{b}-\vec{a}|}$. Ans.

(2): (i): $LHS = \vec{p} \times \vec{d} = (\vec{r} + \overrightarrow{PN}) \times \vec{d}$ where $P(\vec{r})$ is any point on the line. So, $LHS = \vec{r} \times \vec{d} + \overrightarrow{PN} \times \vec{d} = \vec{m} + \vec{0}$ (since $\overrightarrow{PN} \parallel \vec{d}$). Proved.

(ii): $\vec{d}\times\vec{m}=\vec{d}\times(\vec{p}\times\vec{d})$, from (i). So, $\vec{d}\times\vec{m}=(\vec{d}\cdot\vec{d})\vec{p}-(\vec{d}\cdot\vec{p})\vec{d}=1\vec{d}-0\vec{d}=\vec{p}$, since $\vec{d}\perp\vec{p}$. Proved.

(iii): $\vec{m}\times\vec{p}=(\vec{p}\times\vec{d})\times\vec{p}$, from (1). So, $\vec{m}\times\vec{p}=(\vec{p}\cdot\vec{p})\vec{d}-(\vec{d}\cdot\vec{p})\vec{p}=p^2\vec{d}-\vec{0}=p^2\vec{d}$. Proved.

(iv): From (i), $|\vec{m}|=|\vec{p}||\vec{d}|\sin 90^\circ=p(1)(1)=p=|\vec{p}|$. Proved.

(3): Equation of the line $\vec{d},\vec{m}$ is $\vec{r}=\vec{d}\times\vec{m}+\lambda\vec{d}$. So, length of perpendicular $=|\overrightarrow{NB}\times\vec{d}|$ where $\overline{ON}=\vec{d}\times\vec{m}$. Hence length of perpendicular $|\{\vec{b}-(\vec{d}\times\vec{m})\}\times\vec{d}|=|\vec{b}\times\vec{d}-(\vec{d}\times\vec{m})\times\vec{d}|=|\vec{b}\times\vec{d}-\{(\vec{d}\cdot\vec{d})\vec{m}-(\vec{d}\cdot\vec{m})\vec{d}\}|=\left|\vec{b}\times\vec{d}-\vec{m}+\left(\vec{d}\cdot(\vec{r}\times\vec{d})\right)\vec{d}\right|$ where $P(\vec{r})$ is any point on the line. So, length of perpendicular $=|\vec{b}\times\vec{d}-\vec{m}|$. Proved.

(4): This is solved in the text.

(5): This is solved in the text.

(6): This is solved in the text.

(7): Required equation of plane is $[\vec{r}-\vec{d}\times\vec{m}\;\;\vec{d}\times\vec{m}-\vec{d}\times\overrightarrow{m'}\;\;\vec{d}]=0$ or, $[\vec{r}\;\;\vec{d}\times(\vec{m}-\overrightarrow{m'})\;\;\vec{d}]+[\vec{d}\times\vec{m}\;\;\vec{d}\times\overrightarrow{m'}\;\;\vec{d}]$ or, $[\vec{r}\;\;\vec{d}\times(\vec{m}-\overrightarrow{m'})\;\;\vec{d}]+(\vec{d}\times\vec{m})\cdot(\overrightarrow{m'}-(\overrightarrow{m'}\cdot\vec{d})\vec{d})=0$ or, $[\vec{r}\;\;\vec{d}\times(\vec{m}-\overrightarrow{m'})\;\;\vec{d}]+[\vec{d}\;\vec{m}\;\overrightarrow{m'}]=0$. Here normal to the plane is along $\vec{d}\times(\vec{d}\times\vec{m}-\vec{d}\times\overrightarrow{m'})=\vec{d}\times\{\vec{d}\times(\vec{m}-\overrightarrow{m'})\}=(\vec{d}\times\vec{m}-\vec{d}\times\overrightarrow{m'})\vec{d}-1(\vec{m}-\overrightarrow{m'})=(0-0)\vec{d}+\overrightarrow{m'}-\vec{m}=\overrightarrow{m'}-\vec{m}$ Proved.

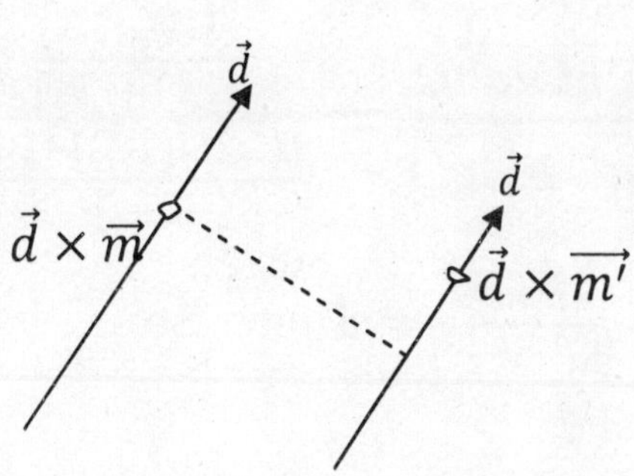

Required length of perpendicular $=|(\vec{d}\times\vec{m})\times\vec{d}-\overrightarrow{m'}|$ (see Example (3), page 77) $=|1\vec{m}-(\vec{m}\cdot\vec{d})\vec{d}-\overrightarrow{m'}|=|\vec{m}-\overrightarrow{m'}|$. Proved.

Example (page 80): (1): $LHS = \vec{a} \times (\vec{r} \times \vec{a}) = (\vec{a} \cdot \vec{a})\vec{r} - (\vec{a} \cdot \vec{r})\vec{a}$. $RHS = (\vec{a} \times \vec{r}) \times \vec{a} = (\vec{a} \cdot \vec{a})\vec{r} - (\vec{a} \cdot \vec{r})\vec{a}$. Proved. Now $LHS = (\vec{c} \cdot \vec{a})\vec{b} - (\vec{a} \cdot \vec{b})\vec{c} + (\vec{a} \cdot \vec{b})\vec{c} - (\vec{b} \cdot \vec{c})\vec{a} + (\vec{b} \cdot \vec{c})\vec{a} - (\vec{c} \cdot \vec{a})\vec{b} = \vec{0} = RHS$. Proved.

(2): This is solved in the text.

(3): This is solved in the text.

(4): This is solved in the text.

(5): This is solved in the text.

Example (page 80): (1): This is solved in the text.

(2): This is solved in the text.

(3): This is solved in the text.

(4): Multiplying scalarly by $\vec{a}$: $(\vec{x} \cdot \vec{b})(\vec{c} \cdot \vec{a}) = \vec{d} \cdot \vec{a}$. So, our equation is $\vec{x} \times \vec{a} = \vec{d} - \frac{\vec{d} \cdot \vec{a}}{\vec{c} \cdot \vec{a}}\vec{c}$ if $\vec{c} \cdot \vec{a} \neq \vec{0}$. So, $\vec{x} = \frac{1}{a^2}\left(\vec{a} \times \left(\vec{d} - \frac{\vec{d} \cdot \vec{a}}{\vec{c} \cdot \vec{a}}\vec{c}\right)\right) + \lambda\vec{a}$. Ans. Also, $\vec{x} = \frac{1}{a^2}\vec{a} \times \left(\frac{(\vec{c} \cdot \vec{a})\vec{d} - (\vec{d} \cdot \vec{a})\vec{c}}{\vec{c} \cdot \vec{a}}\right) + \lambda\vec{a}$ or, $\vec{x} = \frac{\vec{a} \times \{(\vec{c} \times \vec{d}) \times \vec{a}\}}{a^2(\vec{c} \cdot \vec{a})} + \lambda\vec{a}$. Ans.

(5): This is solved in the text.

Example (page 83): Here $(\vec{b} \times \vec{c}) \times (\vec{c} \times \vec{a}) = \{(\vec{b} \times \vec{c}) \cdot \vec{a}\}\vec{c} - \{(\vec{b} \times \vec{c}) \cdot \vec{c}\}\vec{a} = [\vec{a}\ \vec{b}\ \vec{c}]\vec{c}$. Proved. $LHS = [\vec{b} \times \vec{c}\ \ \vec{c} \times \vec{a}\ \ \vec{a} \times \vec{b}] = \{(\vec{b} \times \vec{c}) \times (\vec{c} \times \vec{a})\} \cdot (\vec{a} \times \vec{b}) = [\vec{a}\ \vec{b}\ \vec{c}]\{\vec{c} \cdot (\vec{a} \times \vec{b})\} = [\vec{a}\ \vec{b}\ \vec{c}]^2$. Proved.

Example (page 84): (i): Here $\frac{\overrightarrow{b'} \times \overrightarrow{c'}}{[\overrightarrow{a'}\ \overrightarrow{b'}\ \overrightarrow{c'}]} = \frac{\frac{(\vec{c} \times \vec{a}) \times (\vec{a} \times \vec{b})}{[\vec{a}\ \vec{b}\ \vec{c}][\vec{a}\ \vec{b}\ \vec{c}]}}{\frac{\{(\vec{c} \times \vec{a}) \times (\vec{a} \times \vec{b})\} \cdot (\vec{b} \times \vec{c})}{[\vec{a}\ \vec{b}\ \vec{c}]^3}} = \frac{\frac{[\vec{a}\ \vec{b}\ \vec{c}]\vec{a}}{[\vec{a}\ \vec{b}\ \vec{c}]^2}}{\frac{[\vec{a}\ \vec{b}\ \vec{c}]^2}{[\vec{a}\ \vec{b}\ \vec{c}]^3}}$, see the Example of page 83. So, $\frac{\overrightarrow{b'} \times \overrightarrow{c'}}{[\overrightarrow{a'}\ \overrightarrow{b'}\ \overrightarrow{c'}]} = \vec{a}$, etc. So, $\vec{a}$, $\vec{b}$, $\vec{c}$ are the reciprocal system of $\overrightarrow{a'}$, $\overrightarrow{b'}$, $\overrightarrow{c'}$. Proved.

(ii): Here $(\vec{u} \times \vec{v}) \cdot \vec{n} = |\vec{u} \times \vec{v}|(1)\sin\theta = |\vec{u}||\vec{v}|\sin\theta = \sin\theta$. So, required reciprocal system of vectors are $\frac{\vec{v} \times \vec{n}}{\sin\theta}$, $\frac{\vec{n} \times \vec{u}}{\sin\theta}$, $\frac{\vec{u} \times \vec{v}}{\sin\theta}\ (= \vec{n})$. Proved.

Exercises on Chapter 4

(1): This is solved in the text.

(2): *Lami's Theorem:* Let $F_1\overrightarrow{a_1},\ F_2\overrightarrow{a_2},\ F_3\overrightarrow{a_3}$ be three concurrent forces and let they are in equilibrium. So, $F_1\overrightarrow{a_1} + F_2\overrightarrow{a_2} + F_3\overrightarrow{a_3} = \vec{0}$. Multiplying vectorially by $\overrightarrow{a_1}$: $F_2\overrightarrow{a_1}\times\overrightarrow{a_2} + F_3\overrightarrow{a_1}\times\overrightarrow{a_3} = \vec{0}$. So, $F_2\sin\theta_3 = F_2|\overrightarrow{a_1}\times\overrightarrow{a_2}| = F_3|\overrightarrow{a_3}\times\overrightarrow{a_1}| = F_3\sin\theta_2$, where θ_3 is the angle between $\overrightarrow{a_1}$ and $\overrightarrow{a_2}$, θ_2 is the angle between $\overrightarrow{a_3}$ and $\overrightarrow{a_1}$. So, $\frac{F_1}{\sin\theta_1} = \frac{F_2}{\sin\theta_2}$. Similarly, $\frac{F_2}{\sin\theta_2} = \frac{F_3}{\sin\theta_3}$. Hence $\frac{F_1}{\sin\theta_1} = \frac{F_2}{\sin\theta_2} = \frac{F_3}{\sin\theta_3}$. Proved.

(3): Here $\overrightarrow{AB}, \overrightarrow{AC}, \overrightarrow{AD}$ are coplanar vectors, so $\overrightarrow{AB}\cdot\left(\overrightarrow{AC}\times\overrightarrow{AD}\right) = 0$ or, $(1, k-2, 4)\cdot\{(1,0,-3)\times(3,3,-2)\} = 0$ or, $(1, k-2, 4)\cdot(9,-7,3) = 0$ or, $9 - 7(k-2) + 4(3) = 0$ or, $k = \frac{35}{7} = 5$. Ans.

(4): Required volume $= \frac{1}{3}$(area of base ABC)(h), where h is the height of point D above the base ABC.

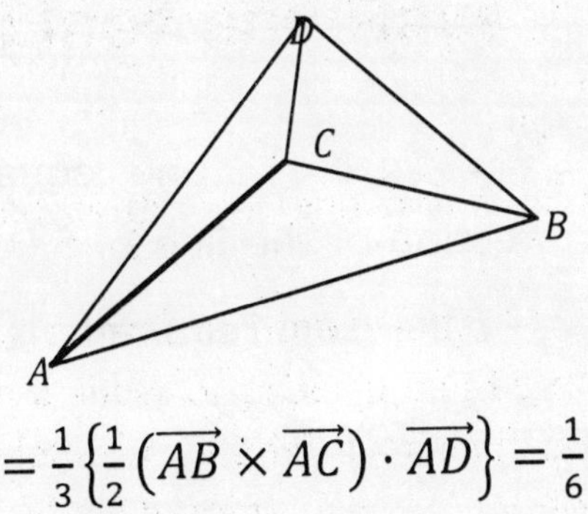

So, required volume $= \frac{1}{3}\left\{\frac{1}{2}\left(\overrightarrow{AB}\times\overrightarrow{AC}\right)\cdot\overrightarrow{AD}\right\} = \frac{1}{6}\left[\vec{b}-\vec{a}\ \ \vec{c}-\vec{a}\ \ \vec{d}-\vec{a}\right] = \frac{1}{6}\begin{vmatrix} 2 & 2 & 6 \\ 1 & 3 & 2 \\ -1 & 5 & 5 \end{vmatrix} = \frac{1}{3}\begin{vmatrix} 1 & 1 & 3 \\ 1 & 3 & 2 \\ 1 & 5 & 5 \end{vmatrix} = \frac{1}{3}[1(15-10) - 1(5+2) + 3(5+3)] = 7\frac{1}{3}$ square unit. Ans.

(5): Required equation of plane is $\left[\vec{r}-\vec{0}\ \ \vec{b}\times\vec{c}\ \ \vec{a}\right] = 0$ or, $\vec{r}\cdot\left(\left(\vec{b}\times\vec{c}\right)\times\vec{a}\right) = 0$. Ans.

(6): Required equation of plane is $\left[\vec{r}-\vec{a}\ \ \vec{a}-\vec{a'}\ \ \vec{b}\right] = 0$. Ans.

(7): Required equation of plane is $[\vec{r}-\vec{a}\ \vec{b}\ \vec{c}]=0$. Ans.

(8): Required equation of plane is $[\vec{r}-\vec{a}\ \vec{a}\ \vec{a'}\ ]=0$ or, $[\vec{r}\ \vec{a}\ \vec{a'}\ ]=$ $[\vec{r}-\vec{a}\ \vec{a}\ \vec{a'}\ ]=0$ or, $\vec{r}\perp(\vec{a}\times\vec{a'})$. Proved.

Interpretation: A line perpendicular to two intersecting lines is perpendicular to any line in the plane containing the two given lines.

(9):

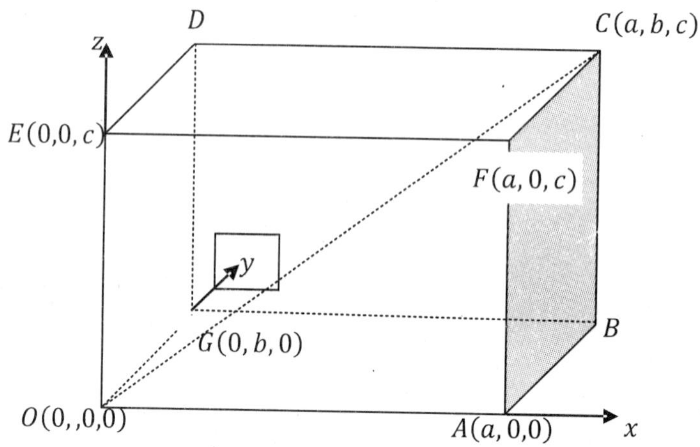

Equation of diagonal OC is $\vec{r}=t(a,b,c)$, and equation of edge EF is $\vec{r}=(0,0,c)+s(a,0,0)$. Here direction of the line of S.D. is given by $(a,b,c)\times(a,0,0)=(0,ca,-ab)=a(0,c,-b)$. So, required length of S.D. $=\left|\overrightarrow{FC}\cdot\frac{(0,c,-b)}{|(0,c,-b)|}\right|=\left|\frac{(0,-b,0)\cdot(0,c,-b)}{\sqrt{c^2+b^2}}\right|=\frac{bc}{\sqrt{b^2+c^2}}$. Proved.

(10): Required length of S.D. $=\left|(\vec{a}-\vec{0})\cdot\frac{\vec{k}\times\vec{b}}{|\vec{k}\times\vec{b}|}=\frac{[\vec{a}\ \vec{k}\ \vec{b}\,]}{|\vec{k}\times\vec{b}|}\right|$. Ans. Equation of the plane containing the line $\vec{r}=\vec{a}+s\vec{b}$ and the line of S.D. is

$$[\vec{r}-\vec{a}\ \vec{b}\ \vec{k}\times\vec{b}\,]=0 \qquad \cdots(1)$$

Similarly equation of the plane containing the line $\vec{r}=t\vec{k}$ and the line of S.D. is

$$[\vec{r}\ \vec{k}\ \vec{k}\times\vec{b}\,]=0 \qquad \cdots(2)$$

Here (1) and (2) together represents the required equation of the line of S.D.

Next let $2a$ be the length of each side of the cube. Here $ABCD$ forms a regular tetrahedron, because

$AB = AC = AD = BC = BD = CD = \sqrt{4a^2 + 4a^2 + 0} = 2\sqrt{2}a$. Here equation of edge AD is $\vec{r} = (a, a, -a) + \lambda(-2a, -2a, 0)$ or, $\vec{r} = (a, a, -a) + \lambda'(1,1,0)$. Again equation of edge BC is $\vec{r} = (a, -a, a) + \mu(2a, -2a, 0)$ or, $\vec{r} = (a, -a, a) + \mu'(1, -1,0)$. So, unit vector

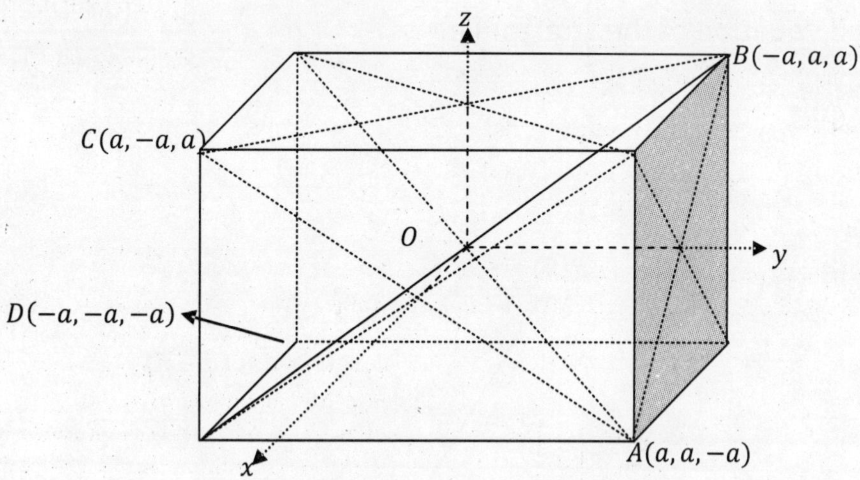

along one of S.D. is $\frac{(1,1,0)\times(1,-1,0)}{|(1,1,0)\times(1,-1,0)|} = \frac{(0,0,2)}{2} = (0,0,-1)$. So, required length of S.D.

$= |\{(a, -a, a) - (a, a, -a)\} \cdot (0,0,-1)| = |0(0) + (-2a)(0) + (2a)(-1)| = 2a$. Also, $\frac{1}{2}$(diagonal of the square described on the edge $(= 2\sqrt{2}a)$) $= \frac{1}{2}\left(\sqrt{2}(2\sqrt{2}a)\right) = 2a$. Proved.

(11): Here line $\vec{r} = t\vec{k}$ passes through origin $O(\vec{0})$ and is along $\vec{k}$. Another line $\vec{r} = \vec{a} + s\vec{b}$ passes through point $A(\vec{a})$ and is along $\vec{b}$. So, length of S.D. $= (\vec{a} - \vec{0}) \cdot \frac{\vec{k}\times\vec{b}}{|\vec{k}\times\vec{b}|} = \frac{[\vec{a}\,\vec{k}\,\vec{b}]}{|\vec{k}\times\vec{b}|}$. Ans. Equation of the plane containing the line of S.D. and the first line, is $(\vec{r} - \vec{0}) \cdot \left(\vec{k} \times (\vec{k} \times \vec{b})\right) = 0$ or, $\vec{r} \cdot (b_3\vec{k} - \vec{b}) = 0$ or,

$\vec{r} \cdot (-b_1\vec{i} - b_2\vec{j}) = 0$ or, $b_1x + b_2y = 0$ $\cdots(1)$

Equation of the plane containing the line of S.D. and the second line, is $(\vec{r}-\vec{a})\cdot\left(\vec{b}\times(\vec{k}\times\vec{b})\right)=0$ or, $(\vec{r}-\vec{a})\cdot(b^2\vec{k}-b_3\vec{b})=0$ or, $(x-a_1, y-a_2, z-a_3)\cdot(-b_3b_1, -b_2b_3, {b_1}^2+{b_2}^2)=0$ $\cdots(2)$

Let us put $x=0$. Then, from (1), $y=0$ and from (2),

$$z=\frac{a_1b_3b_1+a_2b_2b_3-a_3({b_1}^2+{b_2}^2)}{-({b_1}^2+{b_2}^2)}=a_3-\frac{(a_1b_1+a_2b_2)b_3}{{b_1}^2+{b_2}^2}=\vec{a}\cdot\vec{k}-\frac{\{\vec{a}\cdot\vec{b}-(\vec{a}\cdot\vec{k})(\vec{b}\cdot\vec{k})\}(\vec{b}\cdot\vec{k})}{\vec{b}^2-(\vec{b}\cdot\vec{k})^2}.$$

So required equation of the straight line is $\vec{r}=\left(\vec{a}\cdot\vec{k}-\frac{\{\vec{a}\cdot\vec{b}-(\vec{a}\cdot\vec{k})(\vec{b}\cdot\vec{k})\}(\vec{b}\cdot\vec{k})}{\vec{b}^2-(\vec{b}\cdot\vec{k})^2}\right)+\lambda(\vec{k}\times\vec{b})$. Ans.

(12):

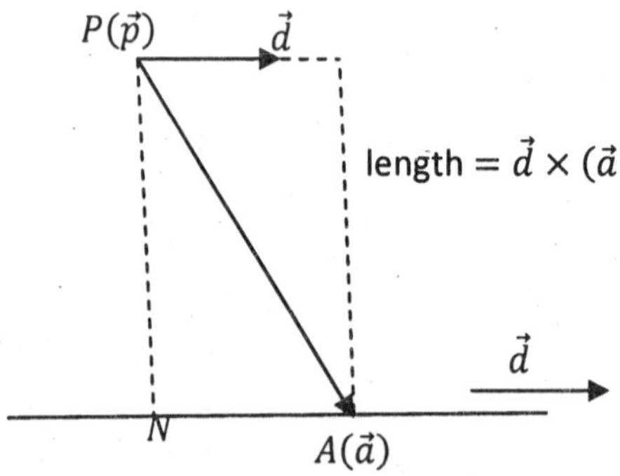

We see that $\{\vec{d}\times(\vec{a}-\vec{p})\}\times\vec{d}$ and $\overrightarrow{PN}$ have the same directions. Also $\left|\{\vec{d}\times(\vec{a}-\vec{p})\}\times\vec{d}\,\right|=\left|\vec{d}\times(\vec{a}-\vec{p})(1)\sin 90°\right|=\left|\overrightarrow{PN}\right|$, so, $\overrightarrow{PN}=\{\vec{d}\times(\vec{a}-\vec{p})\}\times\vec{d}$. Proved.

(13): Normal to the plane containing the required line and the line $\vec{r}-\vec{a}=s\vec{b}$ is along $\left((\vec{a}+s\vec{b})-\vec{c}\right)\times\vec{b}=(\vec{a}-\vec{c})\times\vec{b}$. Similarly, normal to the plane containing the required line and the line $\vec{r}-\vec{a'}=t\vec{b'}$ is along $(\vec{a'}-\vec{c})\times\vec{b'}$. Since the line of intersection (that is, required line) has the direction of $\{(\vec{a}-\vec{c})\times\vec{b}\}\times\{(\vec{a'}-\vec{c})\times\vec{b'}\}$. So, required equation of the line is $\vec{r}=\vec{c}+\lambda(\{(\vec{a}-\vec{c})\times\vec{b}\}\times\{(\vec{a'}-\vec{c})\times\vec{b'}\})$. Ans.

(14): Vector $(\vec{c}-\overrightarrow{a'})\times\vec{b}$ is perpendicular to the required line and $\vec{a}$ is perpendicular to the required line, so $\vec{a}\times\{(\vec{c}-\overrightarrow{a'})\times\vec{b}\}$ is a vector along the required line. So, equation of the required line is $\vec{r}=\vec{c}+\lambda\vec{a}\times\{(\vec{c}-\overrightarrow{a'})\times\vec{b}\}$. Ans.

(15): Let $\vec{n}$ be the unit vector normal to the given plane. Let $S(\vec{s})$ be the point which divides the line segment $P(\vec{a}+t\vec{b})$ and $Q(\overrightarrow{a'}+t'\overrightarrow{b'})$ in the ratio $m:n$.

So $\vec{s}=\frac{m(\overrightarrow{a'}+t'\overrightarrow{b'})+n(\vec{a}+t\vec{b})}{m+n}$ where $\{\overrightarrow{a'}+t'\overrightarrow{b'}-(\vec{a}+t\vec{b})\}\cdot\vec{n}=0$. So,

$t'=\frac{(\vec{a}-\overrightarrow{a'})\cdot\vec{n}+t(\vec{b}\cdot\vec{n})}{\overrightarrow{b'}\cdot\vec{n}}$. So, $\vec{s}=\frac{m\overrightarrow{a'}+n\vec{a}}{m+n}+\frac{m}{m+n}\left\{\frac{(\vec{a}-\overrightarrow{a'})\cdot\vec{n}}{\overrightarrow{b'}\cdot\vec{n}}+t\frac{\vec{b}\cdot\vec{n}}{\overrightarrow{b'}\cdot\vec{n}}\right\}\overrightarrow{b'}+\frac{nt}{m+n}\vec{b}$. So,

$\vec{s}=\left\{\frac{m\overrightarrow{a'}+n\vec{a}}{m+n}+\frac{m}{m+n}\left(\frac{(\vec{a}-\overrightarrow{a'})\cdot\vec{n}}{\overrightarrow{b'}\cdot\vec{n}}\right)\overrightarrow{b'}\right\}+t\left\{\frac{m}{m+n}\left(\frac{\vec{b}\cdot\vec{n}}{\overrightarrow{b'}\cdot\vec{n}}\right)\overrightarrow{b'}+\frac{n}{m+n}\vec{b}\right\}$, which represents a straight line passing through a fixed point

$\frac{m\overrightarrow{a'}+n\vec{a}}{m+n}+\frac{m}{m+n}\left(\frac{(\vec{a}-\overrightarrow{a'})\cdot\vec{n}}{\overrightarrow{b'}\cdot\vec{n}}\right)\overrightarrow{b'}$ and is parallel to vector $\frac{m}{m+n}\left(\frac{\vec{b}\cdot\vec{n}}{\overrightarrow{b'}\cdot\vec{n}}\right)\overrightarrow{b'}+\frac{n}{m+n}\vec{b}$.

Proved.

(16): Let $\vec{r}=\vec{a}+t\vec{b}$ $\cdots(1)$

and $\vec{r}=\overrightarrow{a'}+t'\overrightarrow{b'}$ $\cdots(2)$

be two straight lines. Let $S(\vec{s})$ be the middle-point of the line segment PQ where $P(\vec{a}+t\vec{b})$ and $Q(\overrightarrow{a'}+t'\overrightarrow{b'})$. So, $\vec{s}=\frac{\vec{a}+t\vec{b}+\overrightarrow{a'}+t'\overrightarrow{b'}}{2}$ or, $\vec{s}=\frac{\vec{a}+\overrightarrow{a'}}{2}+\left\{t\left(\frac{\vec{b}}{2}\right)+t'\left(\frac{\overrightarrow{b'}}{2}\right)\right\}$ $\cdots(*)$ which represents a plane passing through a fixed point $\frac{\vec{a}+\overrightarrow{a'}}{2}$ and normal to the plane is parallel to $\frac{\vec{b}}{2}\times\frac{\overrightarrow{b'}}{2}$ or $\vec{b}\times\overrightarrow{b'}$. Since $\vec{b}\times\overrightarrow{b'}$ is the direction of the line of shortest distance, so the required plane intersects the line of S.D. perpendicularly. Since the plane $(*)$ bisects every line terminated by (1) and (2) and the line segment of S.D. is also terminated by (1) and (2), so the plane $(*)$ will bisect the segment of S.D. at its mid-point. Proved.

(17): Let $\vec{a},\vec{b},\vec{c}$ be three given non-coplanar vectors. So, $[\vec{a}\,\vec{b}\,\vec{c}]\neq 0$. Let $\vec{d}$ be any fourth vector such that $\vec{d}\cdot\vec{a},\vec{d}\cdot\vec{b},\vec{d}\cdot\vec{c}$ are known. We have to write $\vec{d}$ in terms of the known quantities. Let $\vec{d} = \lambda\vec{a} + \mu\vec{b} + \nu\vec{c}$. So, $\vec{d}\cdot\vec{a} = \lambda\vec{a}\cdot\vec{a} + \mu\vec{b}\cdot\vec{a} + \nu\vec{c}\cdot\vec{a}$. Similarly $\vec{d}\cdot\vec{b} = \lambda\vec{a}\cdot\vec{b} + \mu\vec{b}\cdot\vec{b} + \nu\vec{c}\cdot\vec{b}$ and $\vec{d}\cdot\vec{c} = \lambda\vec{a}\cdot\vec{c} + \mu\vec{b}\cdot\vec{c} + \nu\vec{c}\cdot\vec{c}$. So, $\vec{d} = \frac{\Delta_1}{\Delta}\vec{a} + \frac{\Delta_2}{\Delta}\vec{b} + \frac{\Delta_3}{\Delta}\vec{c}$ where $\Delta \equiv \begin{vmatrix} \vec{a}\cdot\vec{a} & \vec{b}\cdot\vec{a} & \vec{c}\cdot\vec{a} \\ \vec{a}\cdot\vec{b} & \vec{b}\cdot\vec{b} & \vec{c}\cdot\vec{b} \\ \vec{a}\cdot\vec{c} & \vec{b}\cdot\vec{c} & \vec{c}\cdot\vec{c} \end{vmatrix}$, $\Delta_1 \equiv \begin{vmatrix} \vec{d}\cdot\vec{a} & \vec{b}\cdot\vec{a} & \vec{c}\cdot\vec{a} \\ \vec{d}\cdot\vec{b} & \vec{b}\cdot\vec{b} & \vec{c}\cdot\vec{b} \\ \vec{d}\cdot\vec{c} & \vec{b}\cdot\vec{c} & \vec{c}\cdot\vec{c} \end{vmatrix}$, $\Delta_2 \equiv \begin{vmatrix} \vec{a}\cdot\vec{a} & \vec{d}\cdot\vec{a} & \vec{c}\cdot\vec{a} \\ \vec{a}\cdot\vec{b} & \vec{d}\cdot\vec{b} & \vec{c}\cdot\vec{b} \\ \vec{a}\cdot\vec{c} & d\cdot\vec{c} & \vec{c}\cdot\vec{c} \end{vmatrix}$, $\Delta_3 \equiv \begin{vmatrix} \vec{a}\cdot\vec{a} & \vec{b}\cdot\vec{a} & \vec{d}\cdot\vec{a} \\ \vec{a}\cdot\vec{b} & \vec{b}\cdot\vec{b} & \vec{d}\cdot\vec{b} \\ \vec{a}\cdot\vec{c} & \vec{b}\cdot\vec{c} & \vec{d}\cdot\vec{c} \end{vmatrix}$. Ans.

(18): Let $ABCD$ be a tetrahedron, where $A(\vec{a}), B(\vec{b}), C(\vec{c})$ and $D(\vec{d})$. Let us take two edges AD and its opposite edge BC. Equation off a passing through A, D and $\frac{\vec{b}+\vec{c}}{2}$ is $\left[\vec{r}-\vec{a}\;\vec{d}-\vec{a}\;\frac{\vec{b}+\vec{c}}{2}-\vec{d}\right] = 0$ or $\{(\vec{r}-\vec{a}) \times (\vec{d}-\vec{a})\}\cdot(\vec{b}+\vec{c}-2\vec{d}) = 0 \quad \cdots (*)$. Let us put $\frac{\vec{a}+\vec{b}+\vec{c}+\vec{d}}{4}$ for $\vec{r}$ in the LHS of $(*)$. We get $\left\{\frac{1}{4}(\vec{b}+\vec{c}+\vec{d}-3\vec{a}) \times (\vec{d}-\vec{a})\right\}\cdot(\vec{b}+\vec{c}-2\vec{d}) = \frac{1}{4}\{\vec{b}\times\vec{d} + \vec{c}\times\vec{d} - 3\vec{a}\times\vec{d} + \vec{a}\times\vec{b} - \vec{c}\times\vec{a} - \vec{d}\times\vec{a}\}\cdot(\vec{b}+\vec{c}-2\vec{d}) = \frac{1}{4}\{0 + [\vec{b}\;\vec{c}\;\vec{d}] + 3[\vec{a}\;\vec{b}\;\vec{d}] - [[\vec{a}\;\vec{b}\;\vec{c}] -] [\vec{a}\;\vec{b}\;\vec{d}] - [\vec{b}\;\vec{c}\;\vec{d}] + 3[\vec{a}\;\vec{c}\;\vec{d}] + [\vec{a}\;\vec{b}\;\vec{c}] - [\vec{a}\;\vec{c}\;\vec{d}] - 2[\vec{a}\;\vec{b}\;\vec{d}] - 2[\vec{a}\;\vec{c}\;\vec{d}]\} = \frac{1}{4}(0) = RHS$ of $(*)$. So, from symmetry, centroid $G\left(\frac{\vec{a}+\vec{b}+\vec{c}+\vec{d}}{4}\right)$ lies on each of the six planes. Proved.

(19): Let $OABC$ be any tetrahedron where $O(\vec{0}), A(\vec{a}), B(\vec{b}), C(\vec{c})$. Here equation of plane OAB is $\vec{r}\cdot(\vec{a}\times\vec{b}) = 0$, equation of plane OBC is $\vec{r}\cdot(\vec{b}\times\vec{c}) = 0$, equation of plane OCA is $\vec{r}\cdot(\vec{c}\times\vec{a}) = 0$ and equation of plane ABC is $\vec{r}\cdot(\vec{b}\times\vec{c} + \vec{c}\times\vec{a} + \vec{a}\times\vec{b}) = [\vec{a}\;\vec{b}\;\vec{c}]$. Also equation of the angle bisector plane of edge OA is $\frac{\vec{r}\cdot(\vec{c}\times\vec{a})}{|\vec{c}\times\vec{a}|} = \frac{\vec{r}\cdot(\vec{a}\times\vec{b})}{|\vec{a}\times\vec{b}|}$ or, $\vec{r}\cdot\left(\frac{\vec{a}\times\vec{b}}{|\vec{a}\times\vec{b}|} - \frac{\vec{c}\times\vec{a}}{|\vec{c}\times\vec{a}|}\right) = 0$.

Similarly equation of the angle bisector plane of edge OB is $\vec{r}\cdot\left(\frac{\vec{b}\times\vec{c}}{|\vec{b}\times\vec{c}|} - \frac{\vec{a}\times\vec{b}}{|\vec{a}\times\vec{b}|}\right) =$

$0 \quad \cdots (*)$ and equation of the angle bisector plane of edge OC is $\vec{r} \cdot \left(\frac{\vec{c}\times\vec{a}}{|\vec{c}\times\vec{a}|} - \frac{\vec{b}\times\vec{c}}{|\vec{b}\times\vec{c}|}\right) = 0$. Also, equation of the angle bisector plane of edge AB is $\frac{\vec{r}\cdot(\vec{b}\times\vec{c}+\vec{c}\times\vec{a}+\vec{a}\times\vec{b})-[\vec{a}\ \vec{b}\ \vec{c}\,]}{|\vec{b}\times\vec{c}+\vec{c}\times\vec{a}+\vec{a}\times\vec{b}|} = \frac{\vec{r}\cdot(\vec{a}\times\vec{b})}{|\vec{a}\times\vec{b}|}$ or, $\vec{r}\cdot\left(|\vec{a}\times\vec{b}|(\vec{b}\times\vec{c}+\vec{c}\times\vec{a}+\vec{a}\times\vec{b}) - |\vec{b}\times\vec{c}+\vec{c}\times\vec{a}+\vec{a}\times\vec{b}|(\vec{a}\times\vec{b})\right) = |\vec{a}\times\vec{b}|[\vec{a}\ \vec{b}\ \vec{c}\,]$. Multiplying by $|\vec{b}\times\vec{c}||\vec{c}\times\vec{a}|$ we get, $\vec{r}\cdot\left(\,|\vec{b}\times\vec{c}||\vec{c}\times\vec{a}||\vec{a}\times\vec{b}|(\vec{b}\times\vec{c}+\vec{c}\times\vec{a}+\vec{a}\times\vec{b}) - |\vec{b}\times\vec{c}||\vec{c}\times\vec{a}||\vec{b}\times\vec{c}+\vec{c}\times\vec{a}+\vec{a}\times\vec{b}|(\vec{a}\times\vec{b})\right) = |\vec{b}\times\vec{c}||\vec{c}\times\vec{a}||\vec{a}\times\vec{b}|[\vec{a}\ \vec{b}\ \vec{c}\,] \quad \cdots (1)$.
Similarly equation of angle bisector plane of edge BC is , $\vec{r}\cdot\left(\,|\vec{b}\times\vec{c}||\vec{c}\times\vec{a}||\vec{a}\times\vec{b}|(\vec{b}\times\vec{c}+\vec{c}\times\vec{a}+\vec{a}\times\vec{b}) - |\vec{c}\times\vec{a}||\vec{a}\times\vec{b}||\vec{b}\times\vec{c}+\vec{c}\times\vec{a}+\vec{a}\times\vec{b}|(\vec{b}\times\vec{c})\right) = |\vec{b}\times\vec{c}||\vec{c}\times\vec{a}||\vec{a}\times\vec{b}|[\vec{a}\ \vec{b}\ \vec{c}\,] \quad \cdots (2)$. Now (1) – (2) gives $\vec{r}\cdot(|\vec{c}\times\vec{a}||\vec{b}\times\vec{c}+\vec{c}\times\vec{a}+\vec{a}\times\vec{b}|)\left(|\vec{a}\times\vec{b}|(\vec{b}\times\vec{c}) - |\vec{b}\times\vec{c}|(\vec{a}\times\vec{b})\right)$ or, $\vec{r}\cdot\left(\frac{\vec{b}\times\vec{c}}{|\vec{b}\times\vec{c}|} - \frac{\vec{a}\times\vec{b}}{|\vec{a}\times\vec{b}|}\right) = 0 \quad \cdots (3)$, which is same as $(*)$. Thus this equation is redundant out of six linear equations corresponding to the six planes. Similarly two other equations are also redundant. Now there are three linear equations in three variables. Its solution gives the required point of intersection of six planes. Proved.

(20): $LHS = [\vec{a}\times\vec{b}\ \ \vec{c}\times\vec{d}\ \ \vec{e}\times\vec{f}\,] = \{(\vec{a}\times\vec{b})\times(\vec{c}\times\vec{d})\}\cdot(\vec{e}\times\vec{f}) = \left\{\left((\vec{a}\times\vec{b})\cdot\vec{d}\right)\vec{c} - \left((\vec{a}\times\vec{b})\cdot\vec{c}\right)\vec{d}\right\}\cdot(\vec{e}\times\vec{f}) = [\vec{a}\ \vec{b}\ \vec{d}\,][\vec{c}\ \vec{e}\ \vec{f}\,] - [\vec{a}\ \vec{b}\ \vec{c}\,][\vec{d}\ \vec{e}\ \vec{f}\,]$. Again $LHS = [\vec{a}\times\vec{b}\ \ \vec{c}\times\vec{d}\ \ \vec{e}\times\vec{f}\,] = (\vec{a}\times\vec{b})\cdot\{(\vec{c}\times\vec{d})\times(\vec{e}\times\vec{f})\} = (\vec{a}\times\vec{b})\cdot\{[\vec{c}\ \vec{d}\ \vec{f}\,]\vec{e} - [\vec{c}\ \vec{d}\ \vec{e}\,]\vec{f}\} = [\vec{a}\ \vec{b}\ \vec{e}\,][\vec{c}\ \vec{d}\ \vec{f}\,] - [\vec{a}\ \vec{b}\ \vec{f}\,][\vec{c}\ \vec{d}\ \vec{e}\,]$. Proved. Again, $LHS = [\vec{a}\times\vec{b}\ \ \vec{c}\times\vec{d}\ \ \vec{e}\times\vec{f}\,] = (\vec{c}\times\vec{d})\cdot\{(\vec{e}\times\vec{f})\times(\vec{a}\times\vec{b})\} = (\vec{c}\times\vec{d})\cdot\{[\vec{e}\ \vec{f}\ \vec{b}\,]\vec{a} - [\vec{e}\ \vec{f}\ \vec{a}\,]\vec{b}\} = [\vec{c}\ \vec{d}\ \vec{a}\,][\vec{e}\ \vec{f}\ \vec{b}\,] - [\vec{c}\ \vec{d}\ \vec{b}\,][\vec{e}\ \vec{f}\ \vec{a}\,] = [\vec{c}\ \vec{d}\ \vec{a}\,][\vec{b}\ \vec{e}\ \vec{f}\,] - [\vec{c}\ \vec{d}\ \vec{b}\,][\vec{a}\ \vec{e}\ \vec{f}\,]$. Proved.

(21): Here four planes are $lx + my + nz = p \quad \cdots (1)$, $\begin{cases} lx + my = 0 & \cdots (2) \\ my + nz = 0 & \cdots (3) \\ nz + lx = 0 & \cdots (4) \end{cases}$

Let us solve (1), (2), (3) simultaneously. We get $z = \frac{p}{n}$, $x = \frac{p}{l}$, $y = \left(-\frac{l}{m}\right)\left(\frac{p}{l}\right) = -\frac{p}{m}$. So, one of the vertex is $\left(\frac{p}{l}, -\frac{p}{m}, \frac{p}{n}\right) (\equiv \vec{a})$. Other vertices are $\left(-\frac{p}{l}, \frac{p}{m}, \frac{p}{n}\right) (\equiv \vec{b})$, $\left(\frac{p}{l}, \frac{p}{m}, -\frac{p}{n}\right) (\equiv \vec{c})$, $(0,0,0)$. So, required volume $=$

$$\frac{1}{6}\left|[\vec{a}\ \vec{b}\ \vec{c}]\right| = \left|\frac{1}{6}\begin{vmatrix} \frac{p}{l} & -\frac{p}{m} & \frac{p}{n} \\ -\frac{p}{l} & \frac{p}{m} & \frac{p}{n} \\ \frac{p}{l} & \frac{p}{m} & -\frac{p}{n} \end{vmatrix}\right| = \left|\frac{1}{6}p^3\begin{vmatrix} \frac{1}{l} & -\frac{1}{m} & \frac{1}{n} \\ -\frac{1}{l} & \frac{1}{m} & \frac{1}{n} \\ \frac{1}{l} & \frac{1}{m} & -\frac{1}{n} \end{vmatrix}\right| =$$

$$\left|\frac{1}{6}p^3\begin{vmatrix} 0 & 0 & \frac{2}{n} \\ 0 & \frac{2}{m} & 0 \\ \frac{1}{l} & \frac{1}{m} & -\frac{1}{n} \end{vmatrix}\right|,\ R_1 \to R_1 + R_2,\ R_2 \to R_2 + R_3.$$ So, required volume

$$= \left|\frac{1}{6}p^3\left\{\frac{2}{n}\left(0 - \frac{2}{lm}\right)\right\}\right| = \left|\frac{2}{3}\left(\frac{p^3}{lmn}\right)\right|.$$ Proved.

(22): Let $x\vec{a} + y\vec{b} + z\vec{c} = \vec{d}$. On multiplying scalarly by $\vec{b} \times \vec{c}$: $x[\vec{a}\ \vec{b}\ \vec{c}] = [\vec{d}\ \vec{b}\ \vec{c}]$. So, $x = \frac{[\vec{d}\ \vec{b}\ \vec{c}]}{[\vec{a}\ \vec{b}\ \vec{c}]}$. Similarly $y = \frac{[\vec{a}\ \vec{d}\ \vec{c}]}{[\vec{a}\ \vec{b}\ \vec{c}]}$, $z = \frac{[\vec{a}\ \vec{b}\ \vec{d}]}{[\vec{a}\ \vec{b}\ \vec{c}]}$. Let the three equations be $\begin{cases} a_1x + b_1x + c_1z = d_1 \\ a_2x + b_2x + c_2z = d_2 \\ a_3x + b_3x + c_3z = d_3 \end{cases}$ or, $(a_1x + b_1x + c_1z,\ a_2x + b_2x + c_2z, a_3x + b_3x + c_3z) = (d_1, d_2, d_3)$. So, $x\vec{a} + y\vec{b} + z\vec{c} = \vec{d}$ where $x = \begin{vmatrix} d_1 & b_1 & c_1 \\ d_2 & b_2 & c_2 \\ d_3 & b_3 & c_3 \end{vmatrix} \Big/ \begin{vmatrix} a_1 & b_1 & c_1 \\ a_2 & b_2 & c_2 \\ a_3 & b_3 & c_3 \end{vmatrix}$, $y = \begin{vmatrix} a_1 & d_1 & c_1 \\ a_2 & d_2 & c_2 \\ a_3 & d_3 & c_3 \end{vmatrix} \Big/ \begin{vmatrix} a_1 & b_1 & c_1 \\ a_2 & b_2 & c_2 \\ a_3 & b_3 & c_3 \end{vmatrix}$, $z = \begin{vmatrix} a_1 & b_1 & d_1 \\ a_2 & b_2 & d_2 \\ a_3 & b_3 & d_3 \end{vmatrix} \Big/ \begin{vmatrix} a_1 & b_1 & c_1 \\ a_2 & b_2 & c_2 \\ a_3 & b_3 & c_3 \end{vmatrix}$. Ans.

(23): Let the three planes forming trihedral angle are

$$\begin{cases} \vec{r} \cdot \overrightarrow{n_1} = 0 & \cdots (1) \\ \vec{r} \cdot \overrightarrow{n_2} = 0 & \cdots (2) \\ \vec{r} \cdot \overrightarrow{n_3} = 0 & \cdots (3) \end{cases}$$

They form trihedral angle at the origin. Given: Let one of the lines be $\vec{r} = t\{\lambda(\vec{n_3} \times \vec{n_1}) + \mu(\vec{n_1} \times \vec{n_2})\}$ where $\{\lambda(\vec{n_3} \times \vec{n_1}) + \mu(\vec{n_1} \times \vec{n_2})\} \cdot (\vec{n_2} \times \vec{n_3}) = 0$. So, $\lambda\big((\vec{n_3} \cdot \vec{n_2})(\vec{n_1} \cdot \vec{n_3}) - (\vec{n_1} \cdot \vec{n_2})(\vec{n_3} \cdot \vec{n_3})\big) + \mu\big((\vec{n_1} \cdot \vec{n_2})(\vec{n_2} \cdot \vec{n_3}) - (\vec{n_1} \cdot \vec{n_3})(\vec{n_2} \cdot \vec{n_2})\big) = 0$. So, $\vec{r} = t\left\{\lambda(\vec{n_3} \times \vec{n_1}) - \frac{\lambda\big((\vec{n_3}\cdot\vec{n_2})(\vec{n_1}\cdot\vec{n_3}) - (\vec{n_1}\cdot\vec{n_2})(\vec{n_3}\cdot\vec{n_3})\big)}{(\vec{n_1}\cdot\vec{n_2})(\vec{n_2}\cdot\vec{n_3}) - (\vec{n_1}\cdot\vec{n_3})(\vec{n_2}\cdot\vec{n_2})}(\vec{n_1} \times \vec{n_2})\right\}$ or, $\vec{r} = t'\{\{(\vec{n_1} \cdot \vec{n_2})(\vec{n_2} \cdot \vec{n_3}) - (\vec{n_1} \cdot \vec{n_3})(\vec{n_2} \cdot \vec{n_2})\}(\vec{n_3} \times \vec{n_1}) + \{(\vec{n_1} \cdot \vec{n_2})(\vec{n_3} \cdot \vec{n_3}) - (\vec{n_2} \cdot \vec{n_3})(\vec{n_3} \cdot \vec{n_1})\}(\vec{n_1} \times \vec{n_2})\}$ where $t' \equiv t\lambda$. So, one line is of the form $\vec{r} = t'\vec{d'}$. Similarly other two lines are $\vec{r} = t''\{\{(\vec{n_2} \cdot \vec{n_3})(\vec{n_3} \cdot \vec{n_1}) - (\vec{n_1} \cdot \vec{n_2})(\vec{n_3} \cdot \vec{n_3})\}(\vec{n_1} \times \vec{n_2}) + \{(\vec{n_2} \cdot \vec{n_3})(\vec{n_1} \cdot \vec{n_1}) - (\vec{n_3} \cdot \vec{n_1})(\vec{n_1} \cdot \vec{n_2})\}(\vec{n_2} \times \vec{n_3})\}$ (or $\vec{r} = t''\vec{d''}$) and $\vec{r} = t'''\{\{(\vec{n_3} \cdot \vec{n_1})(\vec{n_1} \cdot \vec{n_2}) - (\vec{n_2} \cdot \vec{n_3})(\vec{n_1} \cdot \vec{n_1})\}(\vec{n_2} \times \vec{n_3}) + \{(\vec{n_3} \cdot \vec{n_1})(\vec{n_2} \cdot \vec{n_2}) - (\vec{n_1} \cdot \vec{n_2})(\vec{n_2} \cdot \vec{n_3})\}(\vec{n_3} \times \vec{n_1})\}$ (or $\vec{r} = t'''\vec{d'''}$).

Here $\vec{d'} + \vec{d''} + \vec{d'''} = \vec{0}$. So, the three lines are coplanar. Proved.

(24): $LHS = (\vec{a} \cdot \vec{b})(\vec{c} \cdot \vec{d}) - (\vec{b} \cdot \vec{d})(\vec{c} \cdot \vec{a}) + (\vec{b} \cdot \vec{c})(\vec{a} \cdot \vec{d}) - (\vec{c} \cdot \vec{d})(\vec{a} \cdot \vec{b}) + (\vec{c} \cdot \vec{a})(\vec{b} \cdot \vec{d}) - (\vec{a} \cdot \vec{d})(\vec{b} \cdot \vec{c}) = 0 = RHS$. Proved.

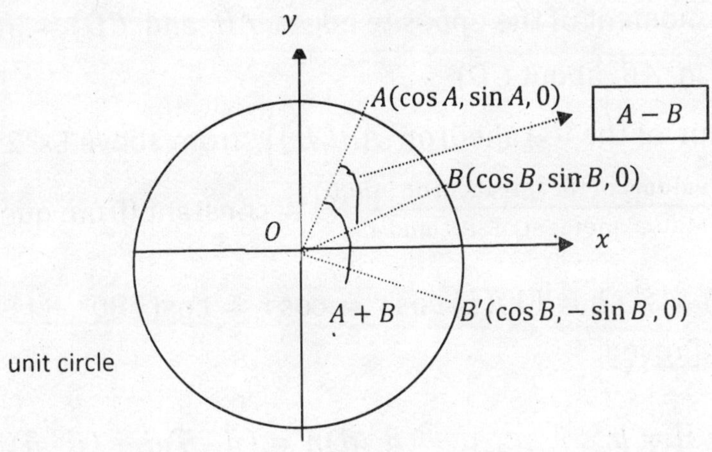

$LHS = \sin(A+B)\sin(A-B) = \left(\overrightarrow{OB'} \times \overrightarrow{OA}\right) \cdot \left(\overrightarrow{OB} \times \overrightarrow{OA}\right) = \left(\overrightarrow{OB'} \cdot \overrightarrow{OB}\right)\left(\overrightarrow{OA} \cdot \overrightarrow{OA}\right) - \left(\overrightarrow{OB'} \cdot \overrightarrow{OA}\right)\left(\overrightarrow{OA} \cdot \overrightarrow{OB}\right) = \cos^2 B - \sin^2 B - (\cos A \cos B - \sin A \sin B)(\cos A \cos B + \sin A \sin B) = \cos^2 B \sin^2 A - \sin^2 B \cos^2 A = \sin^2 A - \sin^2 B$. Proved.

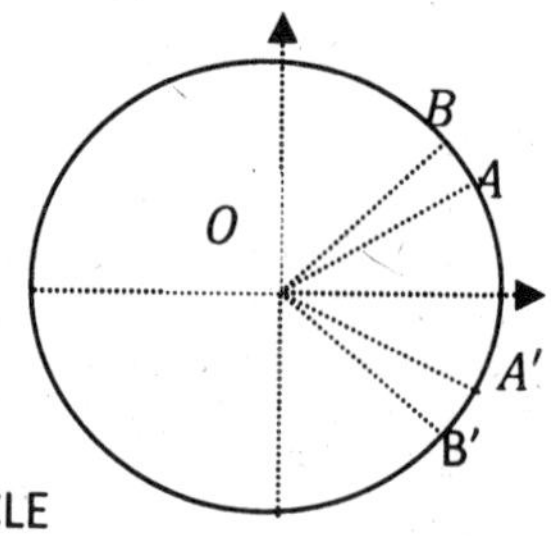

Here
$\frac{1}{2}(\cos 2B - \cos 2A) = \frac{1}{2}\left(\overrightarrow{OB}\cdot\overrightarrow{OB'} - \overrightarrow{OA}\cdot\overrightarrow{OA'}\right) =$
$\frac{1}{2}\big((\cos B, \sin B, 0)(\cos B, -\sin B, 0) - (\cos A, \sin A, 0)(\cos A, -\sin A, 0)\big) =$
$\frac{1}{2}\big(\cos^2 B - \sin^2 B - (\cos^2 A - \sin^2 A)\big) = \sin^2 A - \sin^2 B$. Proved.

(25): Required moment
$= \frac{\overrightarrow{CD}}{CD}\left(\overrightarrow{CA}\times\overrightarrow{AB}\right) = \frac{1}{l}\left(\overrightarrow{AB}\times\overrightarrow{AC}\right)\cdot\overrightarrow{CD} = \frac{\overrightarrow{CD}}{l}\{2(\text{area of } \Delta ABC)\} = \frac{1}{l}\overrightarrow{CD}\cdot$
$\left(\overrightarrow{CA}\times\overrightarrow{CB}\right) = \frac{1}{l}(6V)$, where $V \equiv$ volume of the tetrahedron $ABCD$. Proved.

(26): (Mutual moment of the opposite edges AB and CD) = (moment of unit vector localized in $\overrightarrow{AB}$ about $\overrightarrow{CD}$)
$= \frac{1}{AB}\left\{\frac{6}{CD}(\text{ volume of the tetrahedron } ABCD)\right\}$, from above Ex. 25. So,
$(AB)(CD) = \frac{6(\text{ volume of the tetrahedron } ABCD)}{\text{Mutual moment of } AB \text{ and } CD}$ = constant (from question). Proved.

(27): (i) $LHS = \vec{n}\cdot\vec{b} = (1)(1)\cos r = \cos r = \cos(180° - i) = -\cos i =$
$-\vec{n}\cdot\vec{a} = RHS$. Proved.

(ii) Since $\vec{a}\times\vec{n} = \vec{b}\times\vec{n}$ so, $\vec{a} - (\vec{n}\cdot\vec{a})\vec{n} = (\vec{n}\cdot\vec{n})\vec{a} - (\vec{n}\cdot\vec{a})\vec{n} = \vec{n}\times$
$(\vec{a}\times\vec{n}) = \vec{n}\times\left(\vec{b}\times\vec{n}\right) = (\vec{n}\cdot\vec{n})\vec{b} - \left(\vec{n}\cdot\vec{b}\right)\vec{n} = \vec{b} + (\vec{n}\cdot\vec{a})\vec{n}$, from (i). So,

$\vec{b} = \vec{a} - 2(\vec{n}\cdot\vec{a})\vec{n}$. Proved. Given $\vec{a} = \frac{(1,-3,1)}{\sqrt{1+9+1}} = \frac{1}{\sqrt{11}}(1,-3,1)$, $\vec{n} = \frac{(1,1,1)}{\sqrt{1+1+1}} =$
$\frac{1}{\sqrt{3}}(1,1,1)$. So, $\vec{b} = \frac{1}{\sqrt{11}}(1,-3,1) - 2\left(\frac{1}{\sqrt{3}}(1,1,1)\cdot\frac{1}{\sqrt{11}}(1,-3,1)\right)\frac{1}{\sqrt{3}}(1,1,1) =$

$\frac{1}{\sqrt{11}}(1,-3,1)-2\left(\frac{1}{\sqrt{33}}\right)(1-3+1)\frac{1}{\sqrt{3}}(1,1,1)=\frac{1}{\sqrt{11}}(1,-3,1)+\frac{2}{3\sqrt{11}}(1,1,1)=$ $\frac{1}{3\sqrt{11}}(5,-7,5)=\frac{1}{3\sqrt{11}}(5\vec{\imath}-7\vec{\jmath}+5\vec{k})$. So, $\vec{b}$ has the direction along (5, −7, 5). Ans.

(28): Let $z=0$ be any reflecting plane. Let $\vec{d}=\lambda\vec{\imath}+\mu\vec{\jmath}$ be any vector in the reflecting plane, where $|\vec{d}|=1$. Let $\vec{a}$ be the unit vector along incident ray and $\vec{b}$ be the unit vector along reflected ray. Here $\hat{n}=(0,0,1)$. So, $\vec{b}=\vec{a}-2(\vec{a}\cdot\vec{n})\,\vec{n}$. So, $\vec{b}\cdot\vec{d}=\{(a_1,a_2,a_3)-2((a_1,a_2,a_3)\cdot(0,0,1))\vec{n}\}\cdot(\lambda,\mu,0)=(a_1,a_2,-a_3)\cdot(\lambda,\mu,0)=\lambda a_1+\mu a_2$, and $\vec{a}\cdot\vec{d}=(a_1,a_2,a_3)\cdot(\lambda,\mu,0)=\lambda a_1+\mu a_2$. So, $\cos\theta=(1)(1)\cos\theta=\vec{b}\cdot\vec{d}=\vec{a}\cdot\vec{d}=(1)(1)\cos\alpha=\cos\alpha$. So, $\theta=\alpha$. Thus the (angle between incident ray and $\vec{d}$) = (the angle between reflected ray and $\vec{d}$). Proved. Here equation of incident ray is $\frac{x}{a_1}=\frac{y}{a_2}=\frac{z}{a_3}$ or, $\vec{r}=t\vec{a}$ $\quad\cdots(1)$

Here equation of normal is $\vec{r}=t_1\vec{k}$ $\quad\cdots(2)$

Equation of the reflected ray is $\vec{r}=t_2\{\vec{a}-2(\vec{a}\cdot\vec{n})\,\vec{n}\}=t_2\{(a_1,a_2,a_3)-2a_3(0,0,1)\}$ or, $\vec{r}=t_2(a_1,a_2,-a_3)$ $\quad\cdots(3)$

Let $\frac{x-b_1}{l_1}=\frac{y-b_2}{l_2}=\frac{z-b_3}{0}$ or, $\vec{r}=(b_1,b_2,0)+t_3(l_1,l_2,0)$ $\quad\cdots(4)$

be any straight line in the reflecting plane. So, (length of the S. D. between (1) and (4)) $=\left|(b_1,b_2,b_3)\cdot\frac{(a_1,a_2,a_3)\times(l_1,l_2,0)}{|(a_1,a_2,a_3)\times(l_1,l_2,0)|}\right|=\left|\frac{b_1(-a_3l_2)+b_2(a_3l_1)+0}{\sqrt{(a_3l_2)^2+(a_3l_1)^2+(a_1l_2-a_2l_1)^2}}\right|=$ $\frac{|a_3(b_2l_1-b_1l_2)|}{\sqrt{(a_3l_2)^2+(a_3l_1)^2+(a_1l_2-a_2l_1)^2}}$. Now, (the length of S.D. between (3) and (4)) $=\left|(b_1,b_2,b_3)\cdot\frac{(a_1,a_2,-a_3)\times(l_1,l_2,0)}{|(a_1,a_2,-a_3)\times(l_1,l_2,0)|}\right|=\frac{1}{\sqrt{(-a_3l_2)^2+(-a_3l_1)^2+(a_1l_2-a_2l_1)^2}}|b_1(a_3l_2)+b_2(-a_3l_1)|$. Proved.

(29): Let $ABCD$ and $A'B'C'D'$ be two tetrahedra. Let the four lines AA',BB',CC',DD' be concurrent, say, at O, the origin. Now, from question, we can suppose that $A(\vec{a}),B(\vec{b}),C(\vec{c}),D(\vec{d}),\ A'(\lambda_1\vec{a}),B'(\lambda_2\vec{b}),C'(\lambda_3\vec{c}),D'(\lambda_4\vec{d})$ and $\vec{d}=r\vec{a}+s\vec{b}+t\vec{c}$ for some scalars $r,s,t,\lambda_1,\lambda_2,\lambda_3,\lambda_4$. Here equation of the plane face ABC is $\vec{r}\cdot(\vec{b}\times\vec{c}+\vec{c}\times\vec{a}+\vec{a}\times\vec{b})=[\vec{a}\ \vec{b}\ \vec{c}]$ and equation of the plane face $A'B'C'$ is $\vec{r}\cdot(\lambda_2\vec{b}\times\lambda_3\vec{c}+\lambda_3\vec{c}\times\lambda_1\vec{a}+\lambda_1\vec{a}\times\lambda_2\vec{b})=$

$\left[\lambda_1\vec{a}\ \lambda_2\vec{b}\ \lambda_2\vec{c}\right]$ or, $\vec{r}\cdot\left(\mu_1\vec{b}\times\vec{c}+\mu_2\vec{c}\times\vec{a}+\mu_3\vec{a}\times\vec{b}\right)=\left[\vec{a}\ \vec{b}\ \vec{c}\right]$, where $\mu_i\equiv\frac{1}{\lambda_i}$, $i=1,2,3,4$. So, the direction of the line of intersection of these two planes is along $\left(\vec{b}\times\vec{c}+\vec{c}\times\vec{a}+\vec{a}\times\vec{b}\right)\times\left(\mu_1\vec{b}\times\vec{c}+\mu_2\vec{c}\times\vec{a}+\mu_3\vec{a}\times\vec{b}\right)$, which is parallel to $(\mu_2-\mu_3)\vec{a}+(\mu_3-\mu_1)\vec{b}+(\mu_1-\mu_2)\vec{c}$ (obtained after calculation). Similarly two other such lines are along the vectors $(\mu_3-\mu_4)\vec{b}+(\mu_4-\mu_2)\vec{c}+(\mu_2-\mu_3)\vec{d}$ and $(\mu_4-\mu_1)\vec{c}+(\mu_1-\mu_3)\vec{d}+(\mu_3-\mu_4)\vec{a}$. If we put the value of $\vec{d}$, in terms of $\vec{a}$, $\vec{b}$, $\vec{c}$, our second vector becomes $(\mu_3-\mu_4)\vec{b}+(\mu_4-\mu_2)\vec{c}+(\mu_2-\mu_3)\left(r\vec{a}+s\vec{b}+t\vec{c}\right)=(\mu_2-\mu_3)r\vec{a}+\left((\mu_3-\mu_4)+(\mu_2-\mu_3)s\right)\vec{b}+\left((\mu_4-\mu_2)+t(\mu_2-\mu_3)\right)\vec{c}$. Similarly the third vector becomes $\left((\mu_1-\mu_3)r+(\mu_3-\mu_4)\right)\vec{a}+(\mu_1-\mu_3)s\vec{b}+\left((\mu_1-\mu_3)t+(\mu_4-\mu_1)\right)\vec{c}$. Now the scalar triple product of these three vectors is the product of $\left[\vec{a}\ \vec{b}\ \vec{c}\right]$ and

$$\begin{vmatrix} \mu_2-\mu_3 & \mu_3-\mu_1 & \mu_1-\mu_2 \\ (\mu_2-\mu_3)r & (\mu_3-\mu_4)+(\mu_2-\mu_3)s & (\mu_4-\mu_2)+t(\mu_2-\mu_3) \\ (\mu_1-\mu_3)r+(\mu_3-\mu_4) & (\mu_1-\mu_3)s & (\mu_1-\mu_3)t+(\mu_4-\mu_1) \end{vmatrix}.$$

Upon calculation, it is found that the determinant is equal to zero. So, the three vectors are perpendicular to a common vector. So the four lines are parallel to a common plane. Now it is enough to prove that each pair of these lines intersect each other. For this purpose let us find a scalar t such that $(1-t)\vec{a}+t\vec{b}$ is satisfied by the plane ABC: $\vec{r}\cdot\left(\vec{b}\times\vec{c}+\vec{c}\times\vec{a}+\vec{a}\times\vec{b}\right)=\left[\vec{a}\ \vec{b}\ \vec{c}\right]$ and plane $A'B'C'$: $\vec{r}\cdot\left(\mu_1\vec{b}\times\vec{c}+\mu_2\vec{c}\times\vec{a}+\mu_3\vec{a}\times\vec{b}\right)=\left[\vec{a}\ \vec{b}\ \vec{c}\right]$. Hence $\frac{1-t}{\lambda_1}+\frac{t}{\lambda_1}=1$, upon calculation. So, $t=\frac{\lambda_2(\lambda_1-1)}{\lambda_1-\lambda_2}$. Hence $(1-t)\vec{a}+t\vec{b}=\frac{\lambda_1(1-\lambda_2)}{\lambda_1-\lambda_2}\vec{a}+\frac{\lambda_2(\lambda_1-1)}{\lambda_1-\lambda_2}\vec{b}$, which is independent of λ_3. Thus this point lies on both the lines of intersection

$\begin{cases} ABC \\ A'B'C' \end{cases}$ and $\begin{cases} ABD \\ A'B'D' \end{cases}$. Hence that each pair of these four lines intersect each other. Proved.

(30): Let $\vec{d^*}$, $\vec{m^*}$ the Plucker vectors of the required line. So, $\vec{d^*}=\frac{\vec{d}\times\vec{d'}}{\left|\vec{d}\times\vec{d'}\right|}$ or, $\frac{\vec{d}\times\vec{d'}}{N}$ where $N\equiv\left|\vec{d}\times\vec{d'}\right|$. Equation of the required line is

$(*)\cdots \begin{cases} (\vec{r}-\vec{d}\times\vec{m})\cdot(\vec{d}\times\overrightarrow{d^*})=0 \\ (\vec{r}-\vec{d'}\times\overrightarrow{m'})\cdot(\vec{d'}\times\overrightarrow{d^*})=0 \end{cases}$. Here $\vec{d}\times\overrightarrow{d^*}=\frac{1}{N}\{(\vec{d}\cdot\vec{d'})\vec{d}-\vec{d'}\}$ or,

$(\vec{d}\times\vec{m})\cdot(\vec{d}\times\overrightarrow{d^*})=-\frac{1}{N}\{(\vec{d}\times\vec{m})\cdot\vec{d'}\}=\frac{1}{N}[\vec{m}\ \vec{d}\ \vec{d'}]$ $\cdots(**)$. Let $B(\vec{b})$ be any point on the line $(*)$. So, $\overrightarrow{m^*}=\vec{d}\times\overrightarrow{d^*}$ where

$\begin{cases} (\vec{b}-\vec{d}\times\vec{m})\cdot(\vec{d}\times\overrightarrow{d^*})=0 \\ (\vec{b}-\vec{d'}\times\overrightarrow{m'})\cdot(\vec{d'}\times\overrightarrow{d^*})=0 \end{cases}$. So,

$$\begin{cases} \vec{b}\cdot(\vec{d}\times\overrightarrow{d^*})=(\vec{d}\times\vec{m})\cdot(\vec{d}\times\overrightarrow{d^*}) \\ \vec{b}\cdot(\vec{d'}\times\overrightarrow{d^*})=(\vec{d'}\times\overrightarrow{m'})\cdot(\vec{d'}\times\overrightarrow{d^*})=0 \end{cases} \quad \ldots(1)$$

Also, $(\vec{d}\times\overrightarrow{d^*})\times(\vec{d'}\times\overrightarrow{d^*})=-\{(\vec{d}\times\vec{d'})\cdot\overrightarrow{d^*}\}\overrightarrow{d^*}=N\overrightarrow{d^*}=\vec{d}\times\vec{d'}$ $\cdots(2)$

So, $\overrightarrow{m^*}=\vec{b}\times\overrightarrow{d^*}=\vec{b}\times\frac{(\vec{d}\times\vec{d'})}{N}=\vec{b}\times\frac{1}{N}\{(\vec{d}\times\overrightarrow{d^*})\times(\vec{d'}\times\overrightarrow{d^*})\}$, from (2). Hence

$\overrightarrow{m^*}=\frac{1}{N}\{-\{\vec{b}\cdot(\vec{d}\times\overrightarrow{d^*})\}(\vec{d'}\times\overrightarrow{d^*})+\{\vec{b}\cdot(\vec{d'}\times\overrightarrow{d^*})\}(\vec{d}\times\overrightarrow{d^*})\}=-\frac{1}{N}\Big\{\{(\vec{d}\times\vec{m})\cdot$

$(\vec{d}\times\overrightarrow{d^*})\}(\vec{d'}\times\overrightarrow{d^*})-\{(\vec{d'}\times\overrightarrow{m'})\cdot(\vec{d'}\times\overrightarrow{d^*})\}(\vec{d}\times\overrightarrow{d^*})\Big\}$, (from (1)) $\cdots(***)$.

$\vec{d'}\times\overrightarrow{d^*}=\frac{1}{N}\vec{d'}\times(\vec{d}\times\vec{d'})=\frac{1}{N}\{\vec{d}-(\vec{d'}\cdot\vec{d})\vec{d'}\}$. So, $\big((\vec{d}\times\vec{m})\cdot(\vec{d}\times\overrightarrow{d^*})\big)(\vec{d'}\times$

$\overrightarrow{d^*})=\frac{1}{N}[\vec{m}\ \vec{d}\ \vec{d'}](\vec{d'}\times\overrightarrow{d^*})$, from $(**)$. Now $(\vec{d'}\times\overrightarrow{m'})\cdot(\vec{d'}\times\overrightarrow{d^*})=$

$(\vec{d'}\times\overrightarrow{m'})\cdot\frac{1}{N}\{\vec{d}-(\vec{d}\cdot\vec{d'})\vec{d'}\}=\frac{1}{N}[\overrightarrow{m'}\ \vec{d}\ \vec{d'}]$ $\cdots(3)$

So, $\{(\vec{d'}\times\overrightarrow{m'})\cdot(\vec{d'}\times\overrightarrow{d^*})\}(\vec{d}\times\overrightarrow{d^*})=\frac{-1}{N}[\overrightarrow{m'}\ \vec{d'}\ \vec{d}](\vec{d}\times\overrightarrow{d^*})$ $\cdots(4)$

So, $\overrightarrow{m^*}=\frac{-1}{N}\left\{\frac{1}{N}[\vec{m}\ \vec{d}\ \vec{d'}]\vec{d'}\times\left(\frac{\vec{d}\times\vec{d'}}{N}\right)+\frac{1}{N}[\overrightarrow{m'}\ \vec{d'}\ \vec{d}]\left(\vec{d}\times\frac{(\vec{d}\times\vec{d'})}{N}\right)\right\}$, from

$(***),(3),(4)$. So, $\overrightarrow{m^*}=\frac{1}{N^3}\{[\vec{m}\ \vec{d}\ \vec{d'}](\vec{d}\times\vec{d'})\times\vec{d'}+[\overrightarrow{m'}\ \vec{d'}\ \vec{d}](\vec{d}\times\vec{d'})\times\vec{d}\}$.

Ans.

(31): This is solved in the text.

(32): Let $\vec{p} = \lambda\vec{u} + \mu\vec{v} + \nu\vec{n}$. So, $\vec{p}\cdot\vec{l} = \vec{p}\cdot(\vec{v}\times\vec{n}) = \lambda[\,\vec{u}\ \vec{v}\ \vec{n}] = \lambda(\vec{u}\times\vec{v})\cdot\vec{n} = \lambda(\sin\theta)\vec{n}\cdot\vec{u} = \lambda\sin\theta$. So, $\lambda = \frac{\vec{p}\cdot\vec{l}}{\sin\theta}$. Also $\vec{p}\cdot\vec{m} = \vec{p}\cdot(\vec{n}\times\vec{u}) = \lambda[\,\vec{v}\ \vec{n}\ \vec{u}] = \mu\sin\theta$. So, $\mu = \frac{\vec{p}\cdot\vec{m}}{\sin\theta}$. Now $\vec{p}\cdot\vec{n} = \vec{p}\cdot\left(\frac{\vec{u}\times\vec{v}}{\sin\theta}\right) = \frac{\nu\vec{n}\cdot(\vec{u}\times\vec{v})}{\sin\theta} = \nu$. So, $\vec{p} = \frac{\vec{p}\cdot\vec{l}}{\sin\theta}\vec{u} + \frac{\vec{p}\cdot\vec{m}}{\sin\theta}\vec{v} + (\vec{p}\cdot\vec{n})\vec{n}$ $\cdots(*)$. Also, $\frac{(\vec{p}\cdot\vec{m})}{\sin\theta}(-\sin\theta)\vec{n} + (\vec{p}\cdot\vec{n})\vec{n}\times\vec{u} = \vec{p}\times\vec{u} = \vec{q}\times\vec{v} = \frac{\vec{q}\cdot\vec{l}}{\sin\theta}(\sin\theta)\vec{n} + (\vec{p}\cdot\vec{n})\vec{n}\times\vec{v}$. This implies $(\vec{q}\cdot\vec{l} + \vec{p}\cdot\vec{m})\vec{n} - (\vec{p}\cdot\vec{n})\vec{m} - (\vec{p}\cdot\vec{n})\vec{l} = 0$. So, $\vec{p}\cdot\vec{n} = 0$ and $\vec{q}\cdot\vec{n} = 0$. So, from $(*)$, $\vec{p}$ is coplanar with $\vec{u}$ and $\vec{v}$. Similarly $\vec{q}$ is coplanar with $\vec{u}$ and $\vec{v}$. Proved. Hence $p\sin\theta = (\vec{p}\cdot\vec{l})\,\vec{u} + (\vec{p}\cdot\vec{m})\,\vec{v}$, from $(*)$. Next let $\vec{p} = \lambda_1\vec{l} + \mu_1\vec{m}$. So, $\vec{p}\cdot\vec{u} = (\lambda_1\vec{v}\times\vec{n} + \mu_1\vec{n}\times\vec{u})\cdot\vec{u} = \lambda_1[\,\vec{u}\ \vec{v}\ \vec{n}] = \lambda_1\sin\theta$. So, $\lambda_1 = \frac{\vec{p}\cdot\vec{u}}{\sin\theta}$. Next, $\vec{p}\cdot\vec{v} = (\lambda_1\vec{v}\times\vec{n} + \mu_1\vec{n}\times\sin\theta\cos\theta\,\vec{u})\cdot\vec{v} = \mu_1[\,\vec{n}\ \vec{u}\ \vec{v}] = \mu_1\sin\theta$. So, $\mu_1 = \frac{\vec{p}\cdot\vec{v}}{\sin\theta}$. So, $(\sin\theta)\vec{p} = (\vec{p}\cdot\vec{u})\,\vec{l} + (\vec{p}\cdot\vec{v})\,\vec{m}$. Proved.

(33): **(i):** $RHS = \vec{a}\times(\vec{l}\times\vec{m}) = \vec{a}\times((\vec{v}\times\vec{n})\times(\vec{n}\times\vec{u})) = \vec{a}\times([\,\vec{v}\ \vec{n}\ \vec{u}]\vec{n}) = \vec{a}\times(\sin\theta)\vec{n} = \vec{a}\times(\vec{u}\times\vec{v})$. Proved.

(ii): Let $\vec{v} \equiv (\cos\theta, \sin\theta)$, $\vec{l} = (\sin\theta, -\cos\theta)$ and $\vec{m} = (0,1)$. $\vec{l}\sin\theta = \sin\theta(\sin\theta, -\cos\theta) = (\sin^2\theta, -\sin\theta\cos\theta)$ and $\vec{u} - \vec{v}\cos\theta = (1,0) - \cos\theta(\cos\theta, \sin\theta) = (1-\cos^2\theta, -) = (\sin^2\theta, -\sin\theta\cos\theta)$. This proves (ii). Also, $\vec{m}\sin\theta = (0,1)\sin\theta = (0, \sin\theta)$ and $\vec{v} - \vec{u}\cos\theta = (\cos\theta, \sin\theta) - (1,0)\cos\theta = (0, \sin\theta)$. Proved.

(iii): $\vec{u}\sin\theta = (1,0)\sin\theta = (\sin\theta, 0)$ and $\vec{l} + \vec{m}\cos\theta = (\sin\theta, -\cos\theta) + (0,1)\cos\theta = (\sin\theta, -\cos\theta + \cos\theta) = (\sin\theta, 0)$. Proved.

Also, $\vec{v}\sin\theta = (\cos\theta, \sin\theta)\sin\theta = (\sin\theta\cos\theta, \sin^2\theta)$ and $\vec{m} + \vec{l}\cos\theta = (0,1) + (\sin\theta, -\cos\theta)\cos\theta = (\sin\theta\cos\theta, 1-\cos^2\theta) = (\sin\theta\cos\theta, \sin^2\theta)$. Proved.

(34): Since $(\vec{a}\cdot\vec{l})(\vec{u}\times\vec{a}) + (\vec{a}\cdot\vec{m})(\vec{v}\times\vec{a}) = \{(\vec{a}\cdot\vec{l})\vec{u} + (\vec{a}\cdot\vec{m})\vec{v}\}\times\vec{a}$, since $\vec{p} = \frac{\vec{p}\cdot\vec{l}}{\sin\theta}\vec{u} + \frac{\vec{p}\cdot\vec{m}}{\sin\theta}\vec{v} + (\vec{p}\cdot\vec{n})\vec{n}$ (see Ex 32, equation $(*)$). So, $(\sin\theta)(\vec{u}\times\vec{a}) +$

$(\vec{a}\cdot\vec{m})(\vec{v}\times\vec{a}) = \{(\sin\theta)\vec{a} - (\sin\theta)(\vec{a}\cdot\vec{n})\vec{n}\}\times\vec{a} = (\vec{a}\cdot\vec{n})(\vec{a}\times\vec{n})\sin\theta =$ $(\vec{a}\cdot\vec{n})\vec{a}\times(\vec{u}\times\vec{v})$. Proved. Also, $(\vec{a}\cdot\vec{u})(\vec{l}\times\vec{a}) + (\vec{a}\cdot\vec{v})(\vec{m}\times\vec{a}) =$ $\sin\theta\left\{\frac{\vec{a}\cdot\vec{u}}{\sin\theta}\vec{l} + \frac{\vec{a}\cdot\vec{v}}{\sin\theta}\vec{m}\right\}\times\vec{a} = (\sin\theta)\{\vec{a} - (\vec{a}\cdot\vec{n})\vec{n}\}\times\vec{a}$ (see Ex. 32, equation $(*)$). So, $(\vec{a}\cdot\vec{u})(\vec{l}\times\vec{a}) + (\vec{a}\cdot\vec{v})(\vec{m}\times\vec{a}) = (\vec{a}\cdot\vec{n})(\vec{a}\times(\sin\theta)\vec{n}) = (\vec{a}\cdot\vec{n})\big(\vec{a}\times(\vec{u}\times\vec{v})\big)$. Proved. $LHS = (\vec{a}\cdot\vec{n})\left(\vec{a}\times\left(\vec{l}\times\vec{m}\right)\right) = (\vec{a}\cdot\vec{n})\big(\vec{a}\times(\vec{u}\times\vec{v})\big)$. (*Reason:* $\vec{l}\times\vec{m} = (\vec{v}\times\vec{n})\times(\vec{n}\times\vec{u}) = [\vec{u}\ \ \vec{v}\ \ \vec{n}]\vec{n} - \vec{0} = (\sin\theta)\vec{n} = \vec{u}\times\vec{v}$). Proved.

5

DIFFERENTIATION AND INTEGRATION OF VECTORS CURVATURE AND TORSION OF CURVES

Examples (page 96): (1): **(i):** $\frac{d}{dt}\left(r^2\vec{r}+(\vec{a}\cdot\vec{r})\vec{b}\right)=\frac{d}{dt}(r^2\vec{r})+\frac{d}{dt}\left((\vec{a}\cdot\vec{r})\vec{b}\right)=\left(\left(\frac{d}{dt}r^2\right)\vec{r}+r^2\frac{d}{dt}\vec{r}\right)+\left\{\frac{d}{dt}(\vec{a}\cdot\vec{r})\right\}\vec{b}=(2r\dot{r})\vec{r}+r^2\dot{\vec{r}}+\left(\vec{a}\cdot\frac{d\vec{r}}{dt}\right)\vec{b}=2r\dot{r}\vec{r}+r^2\dot{\vec{r}}+(\vec{a}\cdot\dot{\vec{r}})\vec{b}$ Ans.

(ii): $\frac{d}{dt}\left(r^3\vec{r}+\vec{a}\times\frac{d\vec{r}}{dt}\right)=\frac{d}{dt}(r^3\vec{r})+\vec{a}\times\frac{d}{dt}\left(\frac{d\vec{r}}{dt}\right)=\left((3r^2\dot{r})\vec{r}+r^3\dot{\vec{r}}\right)+\vec{a}\times\ddot{\vec{r}}$ Ans.

(iii): $\frac{d}{dt}\left(a\vec{r}+r\vec{b}\right)^2=2\left(a\vec{r}+r\vec{b}\right)\cdot\frac{d}{dt}\left(a\vec{r}+r\vec{b}\right)=2\left(a\vec{r}+r\vec{b}\right)\cdot\left(a\dot{\vec{r}}+\dot{r}\vec{b}\right)$ Ans.

(iv): $\frac{d}{dt}\left(\frac{\vec{r}}{r^2}+\frac{r\vec{b}}{\vec{a}\cdot\vec{r}}\right)=\frac{d}{dt}\left(\frac{\vec{r}}{r^2}\right)+\frac{d}{dt}\left(\frac{r\vec{b}}{\vec{a}\cdot\vec{r}}\right)=\frac{r^2\dot{\vec{r}}-(2r\dot{r})\vec{r}}{r^4}+\frac{(\vec{a}\cdot\vec{r})(\dot{r}\vec{b})-(r\vec{b})(\vec{a}\cdot\dot{\vec{r}})}{(\vec{a}\cdot\vec{r})^2}=\frac{\dot{\vec{r}}}{r^2}-\left(\frac{2\dot{r}}{r^3}\right)\vec{r}+\frac{\dot{r}\vec{b}}{\vec{a}\cdot\vec{r}}-\frac{r(\vec{a}\cdot\dot{\vec{r}})\vec{b}}{(\vec{a}\cdot\vec{r})^2}$ Ans.

(v): $\frac{d}{dt}\left(r^2+\frac{1}{r^2}\right)=\frac{d}{dt}r^2+\frac{d}{dt}r^{-2}=2r\dot{r}-\frac{2}{r^3}\dot{r}$ Ans.

(vi): $\frac{d}{dt}\left(\frac{1}{2}m\left(\frac{d\vec{r}}{dt}\right)^2\right)=\frac{1}{2}m\frac{d}{dt}\left(\left(\frac{d\vec{r}}{dt}\right)^2\right)=\frac{1}{2}m\left\{2\left(\frac{d\vec{r}}{dt}\right)\cdot\left\{\frac{d}{dt}\left(\frac{d\vec{r}}{dt}\right)\right\}\right\}=m\dot{\vec{r}}\cdot\ddot{\vec{r}}$ Ans.

(2): $\frac{d}{dt}\left[\vec{r}\ \ \frac{d\vec{r}}{dt}\ \ \frac{d^2\vec{r}}{dt^2}\right]=\frac{d}{dt}\left(\left(\vec{r}\times\frac{d\vec{r}}{dt}\right)\cdot\frac{d^2\vec{r}}{dt^2}\right)=\left(\frac{d}{dt}\left(\vec{r}\times\frac{d\vec{r}}{dt}\right)\right)\cdot\frac{d^2\vec{r}}{dt^2}+\left(\vec{r}\times\frac{d\vec{r}}{dt}\right)\cdot\frac{d^3\vec{r}}{dt^3}=\left(\frac{d\vec{r}}{dt}\times\frac{d\vec{r}}{dt}+\vec{r}\times\frac{d^2\vec{r}}{dt^2}\right)\cdot\frac{d^2\vec{r}}{dt^2}+\left(\vec{r}\times\frac{d\vec{r}}{dt}\right)\cdot\frac{d^3\vec{r}}{dt^3}=\left(\vec{0}+\vec{r}\times\frac{d^2\vec{r}}{dt^2}\right)\cdot\frac{d^2\vec{r}}{dt^2}+\left(\vec{r}\times\frac{d\vec{r}}{dt}\right)\cdot\frac{d^3\vec{r}}{dt^3}=\left[\vec{r}\ \ \frac{d^2\vec{r}}{dt^2}\ \ \frac{d^2\vec{r}}{dt^2}\right]+\left[\vec{r}\ \ \frac{d\vec{r}}{dt}\ \ \frac{d^3\vec{r}}{dt^3}\right]=0+\left[\vec{r}\ \ \frac{d\vec{r}}{dt}\ \ \frac{d^3\vec{r}}{dt^3}\right]=\left[\vec{r}\ \ \frac{d\vec{r}}{dt}\ \ \frac{d^3\vec{r}}{dt^3}\right]$ Ans.

$\frac{d}{dt}\left(\vec{r}\times\left(\frac{d\vec{r}}{dt}\times\frac{d^2\vec{r}}{dt^2}\right)\right)=\frac{d\vec{r}}{dt}\times\left(\frac{d\vec{r}}{dt}\times\frac{d^2\vec{r}}{dt^2}\right)+\vec{r}\times\frac{d}{dt}\left(\frac{d\vec{r}}{dt}\times\frac{d^2\vec{r}}{dt^2}\right)=\frac{d\vec{r}}{dt}\times\left(\frac{d\vec{r}}{dt}\times\frac{d^2\vec{r}}{dt^2}\right)+$ $\vec{r}\times\left(\frac{d^2\vec{r}}{dt^2}\times\frac{d^2\vec{r}}{dt^2}+\frac{d\vec{r}}{dt}\times\frac{d^3\vec{r}}{dt^3}\right)=\frac{d\vec{r}}{dt}\times\left(\frac{d\vec{r}}{dt}\times\frac{d^2\vec{r}}{dt^2}\right)+\vec{r}\times\left(\frac{d\vec{r}}{dt}\times\frac{d^3\vec{r}}{dt^3}\right)$ Ans.

$\frac{d^2}{dt^2}\left[\vec{r}\ \frac{d\vec{r}}{dt}\ \frac{d^2\vec{r}}{dt^2}\right]=\frac{d}{dt}\left[\vec{r}\ \frac{d\vec{r}}{dt}\ \frac{d^3\vec{r}}{dt^3}\right]=\left[\frac{d\vec{r}}{dt}\ \frac{d\vec{r}}{dt}\ \frac{d^3\vec{r}}{dt^3}\right]+\left[\vec{r}\ \frac{d^2\vec{r}}{dt^2}\ \frac{d^3\vec{r}}{dt^3}\right]+\left[\vec{r}\ \frac{d\vec{r}}{dt}\ \frac{d^4\vec{r}}{dt^4}\right]=$ $\left[\vec{r}\ \frac{d^2\vec{r}}{dt^2}\ \frac{d^3\vec{r}}{dt^3}\right]+\left[\vec{r}\ \frac{d\vec{r}}{dt}\ \frac{d^4\vec{r}}{dt^4}\right]$ Ans. $\frac{d^2}{dt^2}\left(\vec{r}\times\left(\frac{d\vec{r}}{dt}\times\frac{d^2\vec{r}}{dt^2}\right)\right)=\frac{d}{dt}\left\{\frac{d\vec{r}}{dt}\times\left(\frac{d\vec{r}}{dt}\times\frac{d^2\vec{r}}{dt^2}\right)+\right.$ $\left.\vec{r}\times\left(\frac{d\vec{r}}{dt}\times\frac{d^3\vec{r}}{dt^3}\right)\right\}=\frac{d}{dt}\left\{\frac{d\vec{r}}{dt}\times\left(\frac{d\vec{r}}{dt}\times\frac{d^2\vec{r}}{dt^2}\right)\right\}+\frac{d}{dt}\left\{\vec{r}\times\left(\frac{d\vec{r}}{dt}\times\frac{d^3\vec{r}}{dt^3}\right)\right\}=\left\{\frac{d^2\vec{r}}{dt^2}\times\right.$ $\left(\frac{d\vec{r}}{dt}\times\frac{d^2\vec{r}}{dt^2}\right)+\frac{d\vec{r}}{dt}\times\left\{\frac{d^2\vec{r}}{dt^2}\times\frac{d^2\vec{r}}{dt^2}+\frac{d\vec{r}}{dt}\times\frac{d^3\vec{r}}{dt^3}\right\}\bigg\}+\left\{\frac{d\vec{r}}{dt}\times\left(\frac{d\vec{r}}{dt}\times\frac{d^3\vec{r}}{dt^3}\right)+\vec{r}\times\left\{\frac{d^2\vec{r}}{dt^2}\times\frac{d^3\vec{r}}{dt^3}+\right.\right.$ $\left.\left.\frac{d\vec{r}}{dt}\times\frac{d^4\vec{r}}{dt^4}\right\}\right\}=\frac{d^2\vec{r}}{dt^2}\times\left(\frac{d\vec{r}}{dt}\times\frac{d^2\vec{r}}{dt^2}\right)+\frac{d\vec{r}}{dt}\times\left(\frac{d\vec{r}}{dt}\times\frac{d^3\vec{r}}{dt^3}\right)+\frac{d\vec{r}}{dt}\times\left(\frac{d\vec{r}}{dt}\times\frac{d^3\vec{r}}{dt^3}\right)+\vec{r}\times$ $\left(\frac{d^2\vec{r}}{dt^2}\times\frac{d^3\vec{r}}{dt^3}\right)+\vec{r}\times\left(\frac{d\vec{r}}{dt}\times\frac{d^4\vec{r}}{dt^4}\right)=\frac{d^2\vec{r}}{dt^2}\times\left(\frac{d\vec{r}}{dt}\times\frac{d^2\vec{r}}{dt^2}\right)+2\frac{d\vec{r}}{dt}\times\left(\frac{d\vec{r}}{dt}\times\frac{d^3\vec{r}}{dt^3}\right)+\vec{r}\times$ $\left(\frac{d^2\vec{r}}{dt^2}\times\frac{d^3\vec{r}}{dt^3}\right)+\vec{r}\times\left(\frac{d\vec{r}}{dt}\times\frac{d^4\vec{r}}{dt^4}\right)$ Ans.

(3): $\frac{d}{dt}\frac{\vec{r}+\vec{a}}{\vec{r}^2+\vec{a}^2}=\frac{\left\{\frac{d}{dt}(\vec{r}+\vec{a})\right\}(\vec{r}^2+\vec{a}^2)-(\vec{r}+\vec{a})\frac{d}{dt}(\vec{r}^2+\vec{a}^2)}{(\vec{r}^2+\vec{a}^2)^2}=\frac{(\vec{r}^2+\vec{a}^2)\dot{\vec{r}}-(2\vec{r}\cdot\dot{\vec{r}})(\vec{r}+\vec{a})}{(\vec{r}^2+\vec{a}^2)^2}$ Ans.

$\frac{d}{dt}\left(\frac{\vec{r}\times\vec{a}}{\vec{r}\cdot\vec{a}}\right)=\frac{(\vec{r}\cdot\vec{a})\frac{d}{dt}(\vec{r}\times\vec{a})-\left\{\frac{d}{dt}(\vec{r}\cdot\vec{a})\right\}(\vec{r}\times\vec{a})}{(\vec{r}\cdot\vec{a})^2}=\frac{(\vec{r}\cdot\vec{a})(\dot{\vec{r}}\times\vec{a})-(\vec{a}\cdot\dot{\vec{r}})(\vec{r}\times\vec{a})}{(\vec{r}\cdot\vec{a})^2}$ Ans.

(4): $LHS=\frac{d^2\vec{r}}{dt^2}+n^2\vec{r}=\frac{d^2}{dt^2}(\cos nt\,\vec{a}+\sin nt\,\vec{b})+n^2(\cos nt\,\vec{a}+\sin nt\,\vec{b})=$ $(-n^2\cos nt\,\vec{a}+(-n^2)\sin nt\,\vec{b})+n^2(\cos nt\,\vec{a}+\sin nt\,\vec{b})=\vec{0}=RHS.$ Next $LHS=\vec{r}\times\frac{d\vec{r}}{dt}=(\cos nt\,\vec{a}+\sin nt\,\vec{b})\times\frac{d}{dt}(\cos nt\,\vec{a}+\sin nt\,\vec{b})=(\cos nt\,\vec{a}+$ $\sin nt\,\vec{b})\times n(-\sin nt\,\vec{a}+\cos nt\,\vec{b})=n(\vec{a}\times\vec{b})=RHS.$ Proved.

(5): (i): Here $\frac{d^2\vec{r}}{dt^2}=\vec{a}$ or, $\frac{d}{dt}\left(\frac{d\vec{r}}{dt}\right)=\vec{a}$ or, $d\left(\frac{d\vec{r}}{dt}\right)=\vec{a}\,dt.$ On integration, $\frac{d\vec{r}}{dt}=\vec{a}t+\vec{c}$ where $\vec{c}$ is a constant vector. Now $d\vec{r}=(\vec{a}t+\vec{c})dt.$ On integration, $\vec{r}=\vec{a}\frac{t^{1+1}}{1+1}+\vec{c}t+\vec{c_1}$, where $\vec{c_1}$ is a constant vector. So; $\vec{r}=\frac{1}{2}t^2\vec{a}+t\vec{c}+\vec{c_1}$ Ans.

(ii): Here $\vec{a}\times\frac{d^2\vec{r}}{dt^2}=\vec{b}$ or $\vec{b}=\frac{d}{dt}\left(\vec{a}\times\frac{d\vec{r}}{dt}\right)$ or, $d\left(\vec{a}\times\frac{d\vec{r}}{dt}\right)=\vec{b}\,dt$. On integration, $\vec{a}\times\frac{d\vec{r}}{dt}=\vec{b}\,dt+\vec{c}$, where $\vec{c}$ is a constant vector. Hence $\vec{a}\times d\vec{r}=(\vec{b}\,dt+\vec{c})dt$. On integration, $\vec{a}\times\vec{r}=\vec{b}\frac{t^2}{2}+\vec{c}t+\vec{d}$ where $\vec{d}$ is a constant vector. Taking cross product with $\vec{a}$ we get $\vec{a}^2\vec{r}-(\vec{r}\cdot\vec{a})\vec{a}=\frac{1}{2}(\vec{b}\times\vec{a})t^2+(\vec{c}\times\vec{a})t+\vec{d}\times\vec{a}$ or, $\vec{r}=\frac{\vec{r}\cdot\vec{a}}{\vec{a}^2}\vec{a}+\frac{1}{2}\frac{\vec{b}\times\vec{a}}{\vec{a}^2}t^2+\frac{\vec{c}\times\vec{a}}{\vec{a}^2}t+\frac{\vec{d}\times\vec{a}}{\vec{a}^2}$ $\cdots(*)$. Let us put $\vec{r}=\lambda\vec{a}+\frac{1}{2}\frac{\vec{b}\times\vec{a}}{\vec{a}^2}t^2+\frac{\vec{c}\times\vec{a}}{\vec{a}^2}t+\frac{\vec{d}\times\vec{a}}{\vec{a}^2}$ in $(*)$, $RHS=\frac{1}{\vec{a}^2}\left(\lambda\vec{a}\cdot\vec{a}+\frac{1}{2}\frac{(\vec{b}\times\vec{a})\cdot\vec{a}}{\vec{a}^2}t^2+\frac{(\vec{c}\times\vec{a})\cdot\vec{a}}{\vec{a}^2}t+\frac{(\vec{d}\times\vec{a})\cdot\vec{a}}{\vec{a}^2}\right)\vec{a}+\frac{1}{2}\frac{\vec{b}\times\vec{a}}{\vec{a}^2}t^2+\frac{\vec{c}\times\vec{a}}{\vec{a}^2}t+\frac{\vec{d}\times\vec{a}}{\vec{a}^2}=\frac{1}{\vec{a}^2}(\lambda\vec{a}^2)\vec{a}+\vec{0}+\vec{0}+\vec{0}+\frac{1}{2}\frac{\vec{b}\times\vec{a}}{\vec{a}^2}t^2+\frac{\vec{c}\times\vec{a}}{\vec{a}^2}t+\frac{\vec{d}\times\vec{a}}{\vec{a}^2}=\lambda\vec{a}+\frac{1}{2}\frac{\vec{b}\times\vec{a}}{\vec{a}^2}t^2+\frac{\vec{c}\times\vec{a}}{\vec{a}^2}t+\frac{\vec{d}\times\vec{a}}{\vec{a}^2}=\vec{r}=LHS$. So, required general solution of the given differential equation is $\vec{r}=\lambda\vec{a}+\frac{1}{2}\frac{\vec{b}\times\vec{a}}{\vec{a}^2}t^2+\vec{c_1}t+\vec{c_2}$ where $\vec{c_1}$, $\vec{c_2}$ are arbitrary constant vectors and λ is a parameter. <u>Ans.</u>

(6): Here $\frac{d^2\vec{r}}{dt^2}=\vec{a}t+\vec{b}$. On integration, $\frac{d\vec{r}}{dt}=\vec{a}\frac{t^2}{2}+\vec{b}t+\vec{c}$ where $\vec{c}$ is an arbitrary vector. Given: $\left[\frac{d\vec{r}}{dt}\right]_{t=0}=\vec{0}$. So, $\vec{0}=\vec{a}\frac{(0)^2}{2}+\vec{b}(0)+\vec{c}$ or, $\vec{c}=\vec{0}$. So, $\frac{d\vec{r}}{dt}=\vec{a}\frac{t^2}{2}+\vec{b}t$. On integration, $\vec{r}=\frac{1}{2}\vec{a}\frac{t^3}{3}+\vec{b}\frac{t^2}{2}+\vec{d}$ where $\vec{d}$ is an arbitrary vector. Given: $[\vec{r}]_{t=0}=\vec{0}$. So, $\vec{0}=\vec{0}+\vec{0}+\vec{d}$ or, $\vec{d}=\vec{0}$. So, $\vec{r}=\vec{a}\frac{t^3}{6}+\vec{b}\frac{t^2}{2}$. <u>Ans.</u>

(7): In the first case, $\frac{d}{ds}|\vec{r}|^2=\frac{d}{ds}\vec{r}^2=2\vec{r}\cdot\frac{d\vec{r}}{ds}=2(0)=0$ so, $|\vec{r}|^2=$ constant or, $|\vec{r}|=$ constant. Hence the curve $\vec{r}=\vec{r}(s)$ represents a circle. In the second case, $\vec{0}=\vec{r}\times\frac{d\vec{r}}{ds}$ so, $\vec{r}\times d\vec{r}=\vec{0}$. Hence $\vec{r}\parallel d\vec{r}$. So, there is no change in the direction of the curve $\vec{r}=\vec{r}(s)$. <u>Ans.</u>

Examples (page 100): (1): If $\kappa=0$ at all points of the curve then $\frac{d}{ds}\hat{t}=\kappa\hat{n}=0\hat{n}=\vec{0}$ so, $\hat{t}=$ (constant at all points of the curve $\vec{r}=\vec{r}(s)$). Hence the curve is a straight line. <u>Proved</u>. If $\tau=0$ at all points of the curve then $\frac{d}{ds}\hat{b}=-\tau\hat{n}=$

$-(0)\hat{n} = \vec{0}$ so, $\hat{t} \times \hat{n} = \hat{b} =$ (constant at all points of the curve $\vec{r} = \vec{r}(s)$). Hence the plane of curvature is fixed. So, the curve is plane. Proved.

(2): Here $LHS = |\vec{r}''| = |(\vec{r}')'| = |\hat{t}'| = |\kappa\hat{n}| = \kappa(1) = \kappa = RHS$. Proved. Again $LHS = \vec{r}' \times \vec{r}'' = \hat{t} \times \hat{t}' = \hat{t} \times \kappa\hat{n} = \kappa\hat{b} = RHS$. Proved. Further $LHS = \vec{r}''' = \hat{t}'' = (\kappa\hat{n})' = \kappa'\hat{n} + \kappa\hat{n}' = \kappa'\hat{n} + \kappa(-\kappa\hat{t} + \lambda\hat{b}) = \kappa(\lambda\hat{b} - \kappa\hat{t}) + \kappa'\hat{n} = RHS$. Proved. For the last equality, we shall use the above results. $LHS = [\vec{r}'\ \vec{r}''\ \vec{r}'''] = (\vec{r}' \times \vec{r}'') \cdot \vec{r}''' = \kappa\hat{b} \cdot \vec{r}''' = \kappa\hat{b} \cdot \{\kappa(\lambda\hat{b} - \kappa\hat{t}) + \kappa'\hat{n}\} = \kappa^2\lambda(1) + 0 + 0 = \kappa^2\lambda = RHS$. Proved.

(3): Here $\hat{t}' = \kappa\hat{n}$, $\hat{t}'' = (\kappa\hat{n})' = \kappa'\hat{n} + \kappa\hat{n}' = \kappa'\hat{n} + \kappa(\lambda\hat{b} - \kappa\hat{t}) = \kappa'\hat{n} + \kappa\lambda\hat{b} - \kappa^2\hat{t}$ and $\hat{t}''' = (\kappa'\hat{n} + \kappa\lambda\hat{b} - \kappa^2\hat{t})' = \{\kappa''\hat{n} + \kappa'(\lambda\hat{b} - \kappa\hat{t})\} + \{(\kappa'\lambda + \kappa\lambda')\hat{b} + \kappa\lambda(-\lambda\hat{n})\} - \{2\kappa\kappa'\hat{t} + \kappa^2(\kappa\hat{n})\} = (-\kappa\kappa' - 2\kappa\kappa')\hat{t} + (\kappa'' - \kappa\lambda^2 - \kappa^3)\hat{n} + (\kappa'\lambda + \kappa'\lambda + \kappa\lambda')\hat{b} = -3\kappa\kappa'\hat{t} + (\kappa'' - \kappa\lambda^2 - \kappa^3)\hat{n} + (2\kappa'\lambda + \kappa\lambda')\hat{b}$. Hence $LHS = [\hat{t}'\ \hat{t}''\ \hat{t}'''] = (\hat{t}' \times \hat{t}'') \cdot \hat{t}''' = \{\kappa\hat{n} \times (\kappa'\hat{n} + \kappa\lambda\hat{b} - \kappa^2\hat{t})\} \cdot \{-3\kappa\kappa'\hat{t} + (\kappa'' - \kappa\lambda^2 - \kappa^3)\hat{n} + (2\kappa'\lambda + \kappa\lambda')\hat{b}\} = (\kappa^2\lambda\hat{t} + \kappa^3\hat{b}) \cdot \{-3\kappa\kappa'\hat{t} + (\kappa'' - \kappa\lambda^2 - \kappa^3)\hat{n} + (2\kappa'\lambda + \kappa\lambda')\hat{b}\} = -3\kappa^3\lambda\kappa' + \kappa^3(2\kappa'\lambda + \kappa\lambda') = \kappa^3(\kappa\lambda' - \kappa'\lambda) = \kappa^5 \frac{d}{ds}\left(\frac{\lambda}{\kappa}\right) = RHS$. Proved. Next $LHS = [\hat{b}'\ \hat{b}''\ \hat{b}'''] = \{-\lambda\hat{n} \times (-\lambda'\hat{n} - \lambda(\lambda\hat{b} - \kappa\hat{t}))\} \cdot \frac{d}{ds}(-\lambda'\hat{n} - \lambda(\lambda\hat{b} - \kappa\hat{t})) = (\lambda^3\hat{t} + \kappa\lambda^2\hat{b}) \cdot \{-\lambda''\hat{n} - \lambda'(\lambda\hat{b} - \kappa\hat{t}) - 2\lambda\lambda'\hat{b} + \lambda^3\hat{n} + (\lambda'\kappa + \kappa'\lambda)\hat{t} + \lambda\kappa^2\hat{n}\} = \lambda^2(\lambda\hat{t} + \kappa\hat{b}) \cdot \{(2\lambda'\kappa + \kappa'\lambda)\hat{t} + (\cdots)\hat{n} + (-3\lambda\lambda')\hat{b}\} = \lambda^2((2\lambda'\lambda\kappa + \kappa'\lambda^2) - 3\kappa\lambda\lambda') = \lambda^2(\kappa'\lambda^2 - \kappa\lambda\lambda') = \lambda^3(\kappa'\lambda - \kappa\lambda') = \lambda^5 \frac{d}{ds}\left(\frac{\kappa}{\lambda}\right) = RHS$. Proved.

(4): Here $LHS = \vec{d} \times \hat{t} = (\lambda\hat{t} + \kappa\hat{b}) \times \hat{t} = \vec{0} + \kappa(\hat{n}) = \hat{t}' = RHS$. Proved. Next $LHS = \vec{d} \times \hat{n} = (\lambda\hat{t} + \kappa\hat{b}) \times \hat{n} = \lambda\hat{b} + \kappa(-\hat{t}) = \hat{n}' = RHS$. Proved. Further $LHS = \vec{d} \times \hat{b} = (\lambda\hat{t} + \kappa\hat{b}) \times \hat{b} = \lambda(-\hat{n}) = \hat{b}' = RHS$. Proved. Also $\hat{n} \times \vec{n}' = \hat{n} \times (\vec{d} \times \hat{n}) = (\hat{n} \cdot \hat{n})\vec{d} - (\hat{n} \cdot \vec{d})\vec{n} = \vec{d} - (\hat{n} \cdot (\lambda\hat{t} + \kappa\hat{b}))\vec{n} = \vec{d} - (0 + 0)\vec{n} = \vec{d} = RHS$. Proved. Next $(\hat{n} \times \vec{d}) \times \hat{n} = (\hat{n} \cdot \hat{n})\vec{d} - (\hat{d} \cdot \hat{n})\vec{n} = (\hat{n} \cdot \hat{n})\vec{d} - (\hat{n} \cdot \vec{d})\vec{n} = \hat{n} \times (\vec{d} \times \hat{n})$ so, $(\hat{n} \times \vec{d}) \times \hat{n} = \hat{n} \times (\vec{d} \times \hat{n})$. Hence, by $\hat{n} \times \vec{d} \times \hat{n}$ we can mean any of $(\hat{n} \times \vec{d}) \times \hat{n}$ and $\hat{n} \times (\vec{d} \times \hat{n})$. Proved.

(5): This is solved in the text.

(6): Here equation to the curve is $\vec{r} = (4a\cos^3\theta, 4a\sin^3\theta, 3c\cos 2\theta)$. Hence $\hat{t} = \vec{r}' = \theta'(4a(3)\cos^2\theta\,(-\sin\theta), 4a(3)\sin^2\theta\,(\cos\theta), 3c(-2)\sin 2\theta) = \theta'(12\sin\theta\cos\theta)(-a\cos\theta, a\sin\theta, -c) = 6\theta'\sin 2\theta\,(-a\cos\theta, a\sin\theta, -c)$. So,

$1 = |\hat{t}| = |6\theta'\sin 2\theta\,(-a\cos\theta, a\sin\theta, -c)| = 6\theta'\sin 2\theta\sqrt{(-a\cos\theta)^2 + (a\sin\theta)^2 + (-c)^2} = 6\theta'\sin 2\theta\sqrt{a^2+c^2}$ or,

$\theta' = \frac{\csc 2\theta}{6\sqrt{a^2+c^2}}$ Proved. Next since $\hat{t}$ is the unit vector in the direction of $(-a\cos\theta, a\sin\theta, -c)$, so

$\hat{t} = \frac{(-a\cos\theta, a\sin\theta, -c)}{|(-a\cos\theta, a\sin\theta, -c)|} = \frac{1}{\sqrt{a^2+c^2}}(-a\cos\theta, a\sin\theta, -c)$. Proved. Now

$\kappa\hat{n} = \hat{t}' = \frac{d}{ds}\left\{\frac{1}{\sqrt{a^2+c^2}}(-a\cos\theta, a\sin\theta, -c)\right\} = \frac{1}{\sqrt{a^2+c^2}}\theta'(a\sin\theta, a\cos\theta, 0) = \frac{1}{\sqrt{a^2+c^2}}\frac{\csc 2\theta}{6\sqrt{a^2+c^2}}(a\sin\theta, a\cos\theta, 0) = \frac{\csc 2\theta}{6(a^2+c^2)}(a\sin\theta, a\cos\theta, 0)$. Hence

$\kappa = |\kappa\hat{n}| = \left|\frac{\csc 2\theta}{6(a^2+c^2)}(a\sin\theta, a\cos\theta, 0)\right| = \frac{\csc 2\theta}{6(a^2+c^2)}(a) = \frac{a\csc 2\theta}{6(a^2+c^2)}$. Proved.

Here $\hat{n}$ is the unit vector in the direction of $(a\sin\theta, a\cos\theta, 0)$ so, $\hat{n} = \frac{(a\sin\theta, a\cos\theta, 0)}{|(a\sin\theta, a\cos\theta, 0)|} = \frac{1}{a}(a\sin\theta, a\cos\theta, 0) = (\sin\theta, \cos\theta, 0)$. Proved. Next

$\hat{b} = \hat{t}\times\hat{n} = \frac{1}{\sqrt{a^2+c^2}}(-a\cos\theta, a\sin\theta, -c)\times(\sin\theta, \cos\theta, 0) = \frac{1}{\sqrt{a^2+c^2}}(0 + c\cos\theta, -c\sin\theta - 0, -a\cos^2\theta - a\sin^2\theta) = \frac{1}{\sqrt{a^2+c^2}}(c\cos\theta, -c\sin\theta, -a)$.

Proved. Also,

$-\lambda\hat{n} = \hat{b}' = \frac{d}{ds}\left\{\frac{1}{\sqrt{a^2+c^2}}(c\cos\theta, -c\sin\theta, -a)\right\} = \frac{1}{\sqrt{a^2+c^2}}\theta'(-c\sin\theta, -c\cos\theta, 0) = \frac{1}{\sqrt{a^2+c^2}}\frac{\csc 2\theta}{6\sqrt{a^2+c^2}}(-c\sin\theta, -c\cos\theta, 0) = \frac{-c\csc 2\theta}{6(a^2+c^2)}(\sin\theta, \cos\theta, 0)$ so, $\lambda = |-\lambda\hat{n}| = \left|\frac{-c\csc 2\theta}{6(a^2+c^2)}(\sin\theta, \cos\theta, 0)\right| = \frac{c\csc 2\theta}{6(a^2+c^2)}(1) = \frac{c\csc 2\theta}{6(a^2+c^2)}$. Proved.

(7): Here equation of the curve is $\vec{r} = \left(\ln\cos\theta, \ln\sin\theta, \theta\sqrt{2}\right)$. So, $\hat{t} = \frac{d}{ds}\hat{r} = \frac{d}{ds}\left(\ln\cos\theta, \ln\sin\theta, \theta\sqrt{2}\right) = \theta'\left(\frac{d}{d\theta}\ln\cos\theta, \frac{d}{d\theta}\ln\sin\theta, \frac{d}{d\theta}\theta\sqrt{2}\right) =$

$\theta'\left(\frac{1}{\cos\theta}(-\sin\theta), \frac{1}{\sin\theta}(\cos\theta), \sqrt{2}\right) = \theta'(-\tan\theta, \cot\theta, \sqrt{2})$, which is a vector in the direction of $\hat{t}$.

Here $1 = |\hat{t}| = |\theta'(-\tan\theta, \cot\theta, \sqrt{2})| = \theta'\sqrt{(-\tan\theta)^2 + (\cot\theta)^2 + (\sqrt{2})^2} =$ $\theta'\sqrt{(\tan\theta + \cot\theta)^2} = \theta'(\tan\theta + \cot\theta) = \theta'\left(\frac{1}{\sin\theta\cos\theta}\right)$ so, $\theta' = \sin\theta\cos\theta$. Proved. Also, $\hat{t} = \frac{(-\tan\theta, \cot\theta, \sqrt{2})}{|(-\tan\theta, \cot\theta, \sqrt{2})|} = \frac{1}{\tan\theta + \cot\theta}(-\tan\theta, \cot\theta, \sqrt{2}) =$ $\sin\theta\cos\theta\,(-\tan\theta, \cot\theta, \sqrt{2}) = (-\sin^2\theta, \cos^2\theta, \sqrt{2}\sin\theta\cos\theta)$. Proved. Further $\kappa\hat{n} = \hat{t}' = \theta'\left(-2\sin\theta\cos\theta, -2\sin\theta\cos\theta, \sqrt{2}(\cos^2\theta - \sin^2\theta)\right) =$ $\theta'(-\sin 2\theta, -\sin 2\theta, \sqrt{2}\cos 2\theta)$, which is a vector in the direction of $\hat{n}$. So, $\kappa = |\kappa\hat{n}| = |\theta'(-\sin 2\theta, -\sin 2\theta, \sqrt{2}\cos 2\theta)| =$ $\theta'\sqrt{(-\sin 2\theta)^2 + (-\sin 2\theta)^2 + (\sqrt{2}\cos 2\theta)^2} = \theta'\sqrt{2} = \sqrt{2}\sin\theta\cos\theta$. Proved.

Further $\hat{n} = \frac{(-\sin 2\theta, -\sin 2\theta, \sqrt{2}\cos 2\theta)}{|(-\sin 2\theta, -\sin 2\theta, \sqrt{2}\cos 2\theta)|} = \frac{1}{\sqrt{2}}(-\sin 2\theta, -\sin 2\theta, \sqrt{2}\cos 2\theta)$.

Proved. Hence

$\hat{b} = \hat{t}\times\hat{n} =$

$(-\sin^2\theta, \cos^2\theta, \sqrt{2}\sin\theta\cos\theta)\times\frac{1}{\sqrt{2}}(-\sin 2\theta, -\sin 2\theta, \sqrt{2}\cos 2\theta) =$

$\frac{1}{\sqrt{2}}(\sqrt{2}\cos 2\theta\cos^2\theta + \sqrt{2}\sin\theta\cos\theta\sin 2\theta, -\sqrt{2}\sin\theta\cos\theta\sin 2\theta +$ $\sqrt{2}\cos 2\theta\sin^2\theta, \sin^2\theta\sin 2\theta + \cos^2\theta\sin 2\theta) =$

$\frac{1}{\sqrt{2}}(\sqrt{2}(\cos^4\theta - \sin^2\theta\cos^2\theta + 2\sin^2\theta\cos^2\theta), \sqrt{2}(-\sin^4\theta + \sin^2\theta\cos^2\theta -$ $2\sin^2\theta\cos^2\theta), \sin 2\theta) = (\cos^2\theta, -\sin^2\theta, \sqrt{2}\sin\theta\cos\theta)$. Proved. Now

$-\lambda\hat{n} = \hat{b}' = \frac{d}{ds}(\cos^2\theta, -\sin^2\theta, \sqrt{2}\sin\theta\cos\theta) =$

$\theta'\left(-2\sin\theta\cos\theta, -2\sin\theta\cos\theta, \sqrt{2}(\cos^2\theta - \sin^2\theta)\right) =$

$\theta'(-\sin 2\theta, -\sin 2\theta, \sqrt{2}\cos 2\theta) = \theta'(\sqrt{2}\hat{n})$. Hence $\lambda = -\sqrt{2}\,\theta' =$

$-\sqrt{2}\sin\theta\cos\theta$. Thus $\kappa = -\lambda$. Proved.

Examples (page 102): (1): Here equation of the curve is $\vec{r} = (x, y, 0) \equiv (x(t), y(t), 0)$. So, $\dot{s}\hat{t} = \dot{\vec{r}} = (\dot{x}, \dot{y}, 0)$, $\ddot{\vec{r}} = (\ddot{x}, \ddot{y}, 0)$. Now $\kappa(\dot{s})^3\hat{b} = \dot{\vec{r}} \times \ddot{\vec{r}} = (\dot{x}, \dot{y}, 0) \times (\ddot{x}, \ddot{y}, 0) = (0,0, \dot{x}\ddot{y} - \dot{y}\ddot{x})$. Hence $\dot{s} = |\dot{s}\hat{t}| = |(\dot{x}, \dot{y}, 0)| = \sqrt{(\dot{x})^2 + (\dot{y})^2}$ and therefore $\kappa((\dot{x})^2 + (\dot{y})^2)^{3/2} = \kappa(\dot{s})^3 = \left|\kappa(\dot{s})^3\hat{b}\right| = |(0,0, \dot{x}\ddot{y} - \dot{y}\ddot{x})| = |\dot{x}\ddot{y} - \dot{y}\ddot{x}|$. Hence $\kappa = \frac{|\dot{x}\ddot{y} - \dot{y}\ddot{x}|}{((\dot{x})^2 + (\dot{y})^2)^{3/2}}$. Proved.

(2): Here equation of the curve is $\vec{r} = a(3t - t^3, 3t^2, 3t + t^3)$. So, $\dot{s}\hat{t} = \dot{\vec{r}} = a(3 - 3t^2, 6t, 3 + 3t^2) = 3a(1 - t^2, 2t, 1 + t^2)$. Hence $\dot{s} = |\dot{s}\hat{t}| = |3a(1 - t^2, 2t, 1 + t^2)| = 3a\sqrt{(1 - t^2)^2 + (2t)^2 + (1 + t^2)^2} = 3a\sqrt{2(1 + t^4) + 4t^2} = 3a\sqrt{2}\sqrt{1 + t^4 + 2t^2} = 3a\sqrt{2}(1 + t^2)$. Next $\ddot{\vec{r}} = 3a(-2t, 2, 2t) = 6a(-t, 1, t)$. So, $\kappa(\dot{s})^3\hat{b} = \dot{\vec{r}} \times \ddot{\vec{r}} = 3a(1 - t^2, 2t, 1 + t^2) \times 6a(-t, 1, t) = 18a^2(2t^2 - 1 - t^2, -t - t^3 - t + t^3, 1 - t^2 + 2t^2) = 18a^2(t^2 - 1, -2t, 1 + t^2)$. Hence $\kappa(\dot{s})^3 = \left|\kappa(\dot{s})^3\hat{b}\right| = |18a^2(t^2 - 1, -2t, 1 + t^2)| = 18a^2\sqrt{(t^2 - 1)^2 + (-2t)^2 + (1 + t^2)^2} = 18a^2\sqrt{2(1 + t^4) + 4t^2} = 18\sqrt{2}a^2(1 + t^2)$. Therefore $\kappa = \frac{18\sqrt{2}a^2(1+t^2)}{\{3a\sqrt{2}(1+t^2)\}^3} = \frac{18\sqrt{2}a^2(1+t^2)}{54\sqrt{2}a^3(1+t^2)^3} = \frac{1}{3a(1+t^2)^2}$. Proved. Next $\dddot{\vec{r}} = 6a(-1,0,1)$, hence $\kappa^2\lambda(\dot{s})^6 = \left[\dot{\vec{r}}\ \ddot{\vec{r}}\ \dddot{\vec{r}}\right] = \left(\dot{\vec{r}} \times \ddot{\vec{r}}\right) \cdot \dddot{\vec{r}} = 18a^2(t^2 - 1, -2t, 1 + t^2) \cdot 6a(-1,0,1) = 108a^3(-t^2 + 1 + 1 + t^2) = 216a^3$. So, $\lambda = \frac{216a^3}{\kappa^2(\dot{s})^6} = \frac{216a^3}{\kappa^2\left(3a\sqrt{2}(1+t^2)\right)^6} = \{3a(1 + t^2)^2\}^2 \frac{216a^3}{\left(3a\sqrt{2}(1+t^2)\right)^6} = \frac{216}{a(1+t^2)^2 81(8)} = \frac{1}{3a(1+t^2)^2} = \kappa$. Proved. Since $\dot{s}\hat{t} = 3a(1 - t^2, 2t, 1 + t^2)$ so,

$\hat{t} = \frac{3a(1-t^2,2t,1+t^2)}{\dot{s}} = \frac{3a(1-t^2,2t,1+t^2)}{3a\sqrt{2}(1+t^2)} = \frac{(1-t^2,2t,1+t^2)}{\sqrt{2}(1+t^2)}$ Ans. Since $\kappa(\dot{s})^3\hat{b} = 18a^2(t^2 - 1, -2t, 1 + t^2)$ so, $\hat{b} = \frac{18a^2(t^2-1,-2t,1+t^2)}{\kappa(\dot{s})^3} = \frac{18a^2(t^2-1,-2t,1+t^2)}{\frac{1}{3a(1+t^2)^2}\left(3a\sqrt{2}(1+t^2)\right)^3} = \frac{(t^2-1,-2t,1+t^2)}{\sqrt{2}(1+t^2)}$. Ans. $\hat{n} = \hat{b} \times \hat{t} = \frac{(t^2-1,-2t,1+t^2)}{\sqrt{2}(1+t^2)} \times \frac{(1-t^2,2t,1+t^2)}{\sqrt{2}(1+t^2)} = \frac{-1}{2(1+t^2)^2}\left(1 - t^2, 2t, -(1 + t^2)\right) \times (1 - t^2, 2t, 1 + t^2) = \frac{-1}{2(1+t^2)^2}(4t(1 + t^2), 2(t^4 - 1), 0) = \frac{1}{1+t^2}(-2t, 1 - t^2, 0)$. Ans.

(3): Here equation of the curve is $\vec{r} = (a\cos t, a\sin t, bt)$. So, $\dot{s}\hat{t} = \dot{\vec{r}} = (-a\sin t, a\cos t, b)$, $\ddot{\vec{r}} = (-a\cos t, -a\sin t, 0)$, $\dddot{\vec{r}} = (a\sin t, -a\cos t, 0)$. Now $\dot{s} = |\dot{s}\hat{t}| = |(-a\sin t, a\cos t, b)| = \sqrt{a^2+b^2}$ Proved. Also $\hat{t} = \frac{(-a\sin t, a\cos t, b)}{|(-a\sin t, a\cos t, b)|} = \frac{1}{\sqrt{a^2+b^2}}(-a\sin t, a\cos t, b)$. Since $\kappa(\dot{s})^3\hat{b} = \dot{\vec{r}} \times \ddot{\vec{r}} = (-a\sin t, a\cos t, b) \times (-a\cos t, -a\sin t, 0) = (ab\sin t, -ab\cos t, a^2) = a(b\sin t, -b\cos t, a)$ <u>Proved</u>. Further

$\kappa(\dot{s})^3 = |\kappa(\dot{s})^3\hat{b}| = |a(b\sin t, -b\cos t, a)| = a\sqrt{b^2+a^2}$ or, $\kappa = \frac{a\sqrt{b^2+a^2}}{(\dot{s})^3} = \frac{a\sqrt{b^2+a^2}}{\left(\sqrt{a^2+b^2}\right)^3} = \frac{a}{a^2+b^2}$. <u>Proved</u>. Next

$\kappa^2\lambda(\dot{s})^6 = [\dot{\vec{r}}\ \ddot{\vec{r}}\ \dddot{\vec{r}}] = (\dot{\vec{r}} \times \ddot{\vec{r}}) \cdot \dddot{\vec{r}} = a(b\sin t, -b\cos t, a) \cdot (a\sin t, -a\cos t, 0) = a(ab\sin^2 t + ab\cos^2 t) = a^2 b$. So, $\lambda = \frac{a^2 b}{\kappa^2(\dot{s})^6} = \frac{a^2 b}{\left(\frac{a}{a^2+b^2}\right)^2\left(\sqrt{a^2+b^2}\right)^6} = \frac{b}{a^2+b^2}$. <u>Proved</u>.

(4): Here equation of the curve is $\vec{r} = \left(\ln\cos\theta, \ln\sin\theta, \theta\sqrt{2}\right)$. So, $\dot{s}\hat{t} = \dot{\vec{r}} = \left(-\tan\theta, \cot\theta, \sqrt{2}\right)$,

$\ddot{\vec{r}} = (-\sec^2\theta, -\csc^2\theta, 0) = -(\sec^2\theta, \csc^2\theta, 0)$ and $\dddot{\vec{r}} = (-2\sec^2\theta\tan\theta, 2\csc^2\theta\cot\theta, 0)$. Now

$\frac{ds}{d\theta} = \dot{s} = |\dot{s}\hat{t}| = \left|\left(-\tan\theta, \cot\theta, \sqrt{2}\right)\right| = \sqrt{\tan^2\theta + \cot^2\theta + 2} = \tan\theta + \cot\theta = \frac{1}{\sin\theta\cos\theta}$ so, $\theta' = \frac{d\theta}{ds} = \sin\theta\cos\theta$. <u>Proved</u>. Since

$\dot{s}\hat{t} = \left(-\tan\theta, \cot\theta, \sqrt{2}\right)$ so, .

$\hat{t} = \frac{1}{\dot{s}}\left(-\tan\theta, \cot\theta, \sqrt{2}\right) = \sin\theta\cos\theta\left(-\tan\theta, \cot\theta, \sqrt{2}\right) = \left(-\sin^2\theta, \cos^2\theta, \sqrt{2}\sin\theta\cos\theta\right)$. <u>Proved</u>. Since $\kappa(\dot{s})^3\hat{b} = \dot{\vec{r}} \times \ddot{\vec{r}} = \left(-\tan\theta, \cot\theta, \sqrt{2}\right) \times (-\sec^2\theta, -\csc^2\theta, 0) = \left(\sqrt{2}\csc^2\theta, -\sqrt{2}\sec^2\theta, \tan\theta\csc^2\theta + \cot\theta\sec^2\theta\right) = \left(\frac{\sqrt{2}}{\sin^2\theta}, \frac{-\sqrt{2}}{\cos^2\theta}, \frac{2}{\sin\theta\cos\theta}\right)$

so,

$\kappa(\dot{s})^3 = |\kappa(\dot{s})^3\hat{b}| = |\dot{\vec{r}} \times \ddot{\vec{r}}| = \left|\left(\frac{\sqrt{2}}{\sin^2\theta}, \frac{-\sqrt{2}}{\cos^2\theta}, \frac{2}{\sin\theta\cos\theta}\right)\right| =$

$\sqrt{\left(\frac{\sqrt{2}}{\sin^2\theta}\right)^2 + \left(\frac{-\sqrt{2}}{\cos^2\theta}\right)^2 + \left(\frac{2}{\sin\theta\cos\theta}\right)^2} = \sqrt{2}\left(\frac{1}{\sin^2\theta} + \frac{1}{\cos^2\theta}\right) = \frac{\sqrt{2}}{\sin^2\theta\cos^2\theta}$ or,

$\kappa = \frac{\sqrt{2}}{(\dot{s})^3\sin^2\theta\cos^2\theta} = \frac{\sqrt{2}}{\left(\frac{1}{\sin\theta\cos\theta}\right)^3\sin^2\theta\cos^2\theta} = \sqrt{2}\sin\theta\cos\theta$. <u>Proved</u>. Next

$\kappa^2\lambda(\dot{s})^6 = [\dot{\vec{r}}\ \ddot{\vec{r}}\ \dddot{\vec{r}}] = (\dot{\vec{r}}\times\ddot{\vec{r}})\cdot\dddot{\vec{r}} =$

$\left(\frac{\sqrt{2}}{\sin^2\theta}, \frac{-\sqrt{2}}{\cos^2\theta}, \frac{2}{\sin\theta\cos\theta}\right)\cdot(-2\sec^2\theta\tan\theta, 2\csc^2\theta\cot\theta, 0) =$

$-2\sqrt{2}\sec^2\theta\csc^2\theta\tan\theta - 2\sqrt{2}\sec^2\theta\csc^2\theta\cot\theta =$

$-2\sqrt{2}\sec^2\theta\csc^2\theta(\tan\theta + \cot\theta) = -2\sqrt{2}\sec^3\theta\csc^3\theta$ so,

$\lambda = \frac{-2\sqrt{2}\sec^3\theta\csc^3\theta}{\kappa^2(\dot{s})^6} = \frac{-2\sqrt{2}\sec^3\theta\csc^3\theta}{(\sqrt{2}\sin\theta\cos\theta)^2\left(\frac{1}{\sin\theta\cos\theta}\right)^6} = -\sqrt{2}\sin\theta\cos\theta = -\kappa$. <u>Proved</u>.

Since $\kappa(\dot{s})^3\hat{b} = \left(\frac{\sqrt{2}}{\sin^2\theta}, \frac{-\sqrt{2}}{\cos^2\theta}, \frac{2}{\sin\theta\cos\theta}\right)$ so, $\hat{b} = \frac{\left(\frac{\sqrt{2}}{\sin^2\theta}, \frac{-\sqrt{2}}{\cos^2\theta}, \frac{2}{\sin\theta\cos\theta}\right)}{\left|\left(\frac{\sqrt{2}}{\sin^2\theta}, \frac{-\sqrt{2}}{\cos^2\theta}, \frac{2}{\sin\theta\cos\theta}\right)\right|} =$

$\frac{\left(\frac{\sqrt{2}}{\sin^2\theta}, \frac{-\sqrt{2}}{\cos^2\theta}, \frac{2}{\sin\theta\cos\theta}\right)}{\frac{\sqrt{2}}{\sin^2\theta\cos^2\theta}} = (\cos^2\theta, -\sin^2\theta, \sqrt{2}\sin\theta\cos\theta)$. <u>Proved</u>. Finally

$\hat{n} = \hat{b}\times\hat{t} =$

$(\cos^2\theta, -\sin^2\theta, \sqrt{2}\sin\theta\cos\theta)\times\sin\theta\cos\theta(-\tan\theta, \cot\theta, \sqrt{2}) =$

$\sin\theta\cos\theta\left(-\sqrt{2}\sin^2\theta - \sqrt{2}\cos^2\theta, -\sqrt{2}\sin^2\theta - \sqrt{2}\cos^2\theta, \frac{\cos^3\theta}{\sin\theta} - \frac{\sin^3\theta}{\cos\theta}\right) =$

$\sin\theta\cos\theta\left(-\sqrt{2}, -\sqrt{2}, \frac{\cos 2\theta}{\sin\theta\cos\theta}\right) = \frac{1}{\sqrt{2}}(-\sin 2\theta, -\sin 2\theta, \sqrt{2}\cos 2\theta)$. <u>Proved</u>.

(5): Here equation of the curve is $\vec{r} = a(\cos\theta, \sin\theta, \ln\sec\theta)$. So, $\dot{s}\hat{t} = \dot{\vec{r}} =$ $a(-\sin\theta, \cos\theta, \tan\theta)$, $\ddot{\vec{r}} = a(-\cos\theta, -\sin\theta, \sec^2\theta)$ and

$\dddot{\vec{r}} = a(\sin\theta, -\cos\theta, 2\sec^2\theta\tan\theta)$. Now

$\dot{s} = |\dot{s}\hat{t}| = |a(-\sin\theta, \cos\theta, \tan\theta)| = a\sqrt{(-\sin\theta)^2 + (\cos\theta)^2 + (\tan\theta)^2} =$ $a\sec\theta$. <u>Proved</u>. Since

$\kappa\dot{s}^3\hat{b} = \dot{\vec{r}}\times\ddot{\vec{r}} = a(-\sin\theta, \cos\theta, \tan\theta)\times a(-\cos\theta, -\sin\theta, \sec^2\theta) =$

$a^2(\sec\theta + \tan\theta\sin\theta, -\sin\theta + \sin\theta\sec^2\theta, \sin^2\theta + \cos^2\theta) =$

$a^2\left(\frac{1+\sin^2\theta}{\cos\theta}, \sin\theta\tan^2\theta, 1\right)$ so,

$\kappa\dot{s}^3 = |\kappa\dot{s}^3\hat{b}| = \left|a^2\left(\frac{1+\sin^2\theta}{\cos\theta}, \sin\theta\tan^2\theta, 1\right)\right| =$

$a^2\sqrt{\left(\frac{1+\sin^2\theta}{\cos\theta}\right)^2+\left(\frac{\sin^3\theta}{\cos^2\theta}\right)^2+1}=$

$a^2\sqrt{\frac{\cos^2\theta(4-4\cos^2\theta+\cos^4\theta)+(1-3\cos^2\theta+3\cos^4\theta-\cos^6\theta)+\cos^4\theta}{\cos^4\theta}}=\frac{a^2}{\cos^2\theta}\sqrt{1+\cos^2\theta}.$

Hence $\kappa=\frac{a^2}{\dot{s}^3\cos^2\theta}\sqrt{1+\cos^2\theta}=\frac{a^2}{(a\sec\theta)^3\cos^2\theta}\sqrt{1+\cos^2\theta}=$ $\frac{\cos\theta}{a}\sqrt{1+\cos^2\theta}$ or, $a\kappa=\cos\theta\sqrt{1+\cos^2\theta}$. Proved. Now $\kappa^2\lambda(\dot{s})^6=$ $[\dot{\vec{r}}\ \ddot{\vec{r}}\ \dddot{\vec{r}}]=(\dot{\vec{r}}\times\ddot{\vec{r}})\cdot\dddot{\vec{r}}=$

$a^2\left(\frac{1+\sin^2\theta}{\cos\theta},\sin\theta\tan^2\theta,1\right)\cdot a(\sin\theta,-\cos\theta,2\sec^2\theta\tan\theta)=$

$a^3\left(\frac{1+\sin^2\theta}{\cos\theta}\sin\theta-\frac{\sin^3\theta}{\cos\theta}+2\frac{\sin\theta}{\cos^3\theta}\right)=a^3\left(\frac{\sin\theta\cos^2\theta+2\sin\theta}{\cos^3\theta}\right)=a^3\sin\theta\left(\frac{\cos^2\theta+2}{\cos^3\theta}\right)$

so, $\lambda=\frac{1}{\kappa^2(\dot{s})^6}a^3\sin\theta\left(\frac{\cos^2\theta+2}{\cos^3\theta}\right)=\frac{1}{\left(\frac{\cos\theta}{a}\sqrt{1+\cos^2\theta}\right)^2(a\sec\theta)^6}a^3\sin\theta\left(\frac{\cos^2\theta+2}{\cos^3\theta}\right)=$

$\frac{\sin\theta\cos\theta(\cos^2\theta+2)}{a(\cos^2\theta+1)}$ or, $a\lambda=\frac{\sin\theta\cos\theta(2+\cos^2\theta)}{(1+\cos^2\theta)}$. Proved.

Exercises on Chapter 5

(1): This is solved in the text.

(2): $LHS=\hat{e}\cdot\hat{t}=\frac{1}{r}\vec{r}\cdot\hat{t}=\frac{1}{r}\vec{r}\cdot\frac{d\vec{r}}{ds}=\frac{1}{2r}\frac{d}{ds}(\vec{r}\cdot\vec{r})=\frac{1}{2r}\frac{d}{ds}(r^2)=\frac{1}{2r}2r\frac{dr}{ds}=\frac{dr}{ds}=$ RHS. Proved. Next $\frac{dp}{ds}=\frac{d}{ds}(-\vec{r}\cdot\hat{n})=-\left(\hat{t}\cdot\hat{n}+\vec{r}\cdot(\lambda\hat{b}-\kappa\hat{t})\right)=-(0+\lambda\vec{r}\cdot\hat{b}-\kappa\vec{r}\cdot\hat{t})=0+\kappa\vec{r}\cdot$ $\hat{t}$, because $\lambda=0$ for plane curves. Hence $\frac{dp}{ds}=\kappa\vec{r}\cdot\hat{t}=\kappa(r\hat{e})\cdot\hat{t}=\kappa r\cos\varphi=$ $\kappa r\frac{dr}{ds}$. Proved.

(3): Equation of the osculating plane at point $P(\vec{r})$ is $(\vec{R}-\vec{r})\cdot\hat{b}=0$. Suppose, this plane passes through a fixed point $A(\vec{a})$. So, $(\vec{a}-\vec{r})\cdot\hat{b}=0$. Hence $0=\frac{d}{ds}(0)=\frac{d}{ds}\{(\vec{a}-\vec{r})\cdot\hat{b}\}=-\frac{d\vec{r}}{ds}\cdot\hat{b}+(\vec{r}-\vec{a})\cdot\lambda\hat{n}=-\hat{t}\cdot\hat{b}+$ $\lambda((\vec{r}-\vec{a})\cdot\hat{n})=-0+\lambda((\vec{r}-\vec{a})\cdot\hat{n})=\lambda((\vec{r}-\vec{a})\cdot\hat{n})$. Hence $\lambda=0$, and therefore the curve is plane. Proved.

(4): Here $\overrightarrow{m_1}' = \frac{d}{ds}\overrightarrow{m_1} = \frac{d}{ds}(\vec{r}\times\hat{t}) = \hat{t}\times\hat{t} + \vec{r}\times(\kappa\vec{n}) = \vec{0} + \kappa(\vec{r}\times\vec{n}) = \kappa\overrightarrow{m_2}$, $\overrightarrow{m_2}' = \frac{d}{ds}\overrightarrow{m_2} = \frac{d}{ds}(\vec{r}\times\hat{n}) = \hat{t}\times\hat{n} + \vec{r}\times\left(\lambda\hat{b} - \kappa\hat{t}\right) = \hat{b} + \lambda\vec{r}\times\hat{b} - \kappa\vec{r}\times\hat{t} =$ $\hat{b} + \lambda\overrightarrow{m_2} - \kappa\overrightarrow{m_1} = \hat{b} - \kappa\overrightarrow{m_1} + \lambda\overrightarrow{m_2}$ and $\overrightarrow{m_3}' = \frac{d}{ds}\overrightarrow{m_3} = \frac{d}{ds}\left(\vec{r}\times\hat{b}\right) = \hat{t}\times\hat{b} +$ $\vec{r}\times(-\lambda\hat{n}) = -\hat{n} - \lambda\vec{r}\times\hat{n} = -\hat{n} - \lambda\overrightarrow{m_2}$. Proved.

(5): Suppose $\hat{a}$ is the unit vector in the fixed direction. Given that $\hat{a}\cdot\hat{t} = \cos\theta$ and that $\hat{a}\cdot\hat{b} = \cos\varphi$. Hence, on differentiation, $\hat{a}\cdot\kappa\hat{n} = (-\sin\theta)\theta' \quad \cdots(1)$ and $\hat{a}\cdot(-\lambda\hat{n}) = (-\sin\varphi)\varphi' \quad \cdots(2)$. Now (1)÷(2) gives $\frac{-\kappa}{\lambda} = \frac{\sin\theta\ \theta'}{\sin\varphi\ \varphi\prime} = \frac{\sin\theta\ d\theta}{\sin\varphi\ d\varphi}$. Proved.

(6): Here equation of the curve is $\vec{r} = \left(a\cos t, a\sin t, c\cosh\frac{at}{c}\right)$. So, $\dot{s}\hat{t} = \dot{\vec{r}} = \left(-a\sin t, a\cos t, a\sinh\frac{at}{c}\right)$, $\ddot{\vec{r}} = \left(-a\cos t, -a\sin t, \frac{a^2}{c}\cosh\frac{at}{c}\right)$, $\dddot{\vec{r}} = \left(a\sin t, -a\cos t, \frac{a^3}{c^2}\sinh\frac{at}{c}\right)$. Hence

$$\dot{s} = |\dot{s}\hat{t}| = \left|\left(-a\sin t, a\cos t, a\sinh\frac{at}{c}\right)\right| = \sqrt{a^2 + \left(a\sinh\frac{at}{c}\right)^2} = a\cosh\frac{at}{c}.$$

Also,

$$\kappa\dot{s}^3\hat{b} = \dot{\vec{r}}\times\ddot{\vec{r}} = a\left(-\sin t, \cos t, \sinh\frac{at}{c}\right)\times a\left(-\cos t, -\sin t, \frac{a}{c}\cosh\frac{at}{c}\right) =$$
$$a^2\left(\frac{a}{c}\cos t\cosh\frac{at}{c} + \sin t\sinh\frac{at}{c}, -\cos t\sinh\frac{at}{c} + \frac{a}{c}\sin t.\cosh\frac{at}{c}, 1\right).$$ So,

$$\kappa\dot{s}^3 = |\kappa\dot{s}^3\hat{b}| = \left|\dot{\vec{r}}\times\ddot{\vec{r}}\right| = \left|a^2\left(\frac{a}{c}\cos t\cosh\frac{at}{c} + \sin t\sinh\frac{at}{c}, -\cos t\sinh\frac{at}{c} + \frac{a}{c}\sin t\cosh\frac{at}{c}, 1\right)\right| =$$
$$a^2\sqrt{\left(\frac{a}{c}\cos t\cosh\frac{at}{c}\right)^2 + \left(\sin t\sinh\frac{at}{c}\right)^2 + \left(\cos t\sinh\frac{at}{c}\right)^2 + \left(\frac{a}{c}\sin t\cosh\frac{at}{c}\right)^2 + 1}$$
$$= a^2\sqrt{\left(\frac{a}{c}\cosh\frac{at}{c}\right)^2 + \left(\sinh\frac{at}{c}\right)^2 + 1} = a^2\sqrt{\left(\frac{a}{c}\cosh\frac{at}{c}\right)^2 + \left(\cosh\frac{at}{c}\right)^2} = \frac{a^2\sqrt{a^2+c^2}}{c}\cosh\frac{at}{c}.$$ Proved.

Hence $\kappa = \frac{a^2\sqrt{a^2+c^2}}{c\dot{s}^3}\cosh\frac{at}{c} = \frac{a^2\sqrt{a^2+c^2}}{c\left(a\cosh\frac{at}{c}\right)^3}\cosh\frac{at}{c} = \frac{\sqrt{a^2+c^2}}{ca}\operatorname{sech}^2\frac{at}{c}$. Proved.

Further

$$\kappa^2\lambda(\dot{s})^6 = \left[\dot{\vec{r}}\ \ddot{\vec{r}}\ \dddot{\vec{r}}\right] = \left(\dot{\vec{r}}\times\ddot{\vec{r}}\right)\cdot\dddot{\vec{r}} =$$
$$a^2\left(\frac{a}{c}\cos t\cosh\frac{at}{c} + \sin t\sinh\frac{at}{c}, -\cos t\sinh\frac{at}{c} + \frac{a}{c}\sin t\cosh\frac{at}{c}, 1\right)\cdot$$

$\left(a\sin t\,,-a\cos t\,,\frac{a^3}{c^2}\sinh\frac{at}{c}\right) = a^2\left(\frac{a^2}{c}\sin t\cos t\cosh\frac{at}{c} + a\sin^2 t\sinh\frac{at}{c} + a\cos^2 t\sinh\frac{at}{c} - \frac{a^2}{c}\sin t\cos t\cosh\frac{at}{c} + \frac{a^3}{c^2}\sinh\frac{at}{c}\right) = a^2\left(a\sinh\frac{at}{c} + \frac{a^3}{c^2}\sinh\frac{at}{c}\right) = a^3\left(\frac{a^2+c^2}{c^2}\right)\sinh\frac{at}{c}$. Proved. Hence

$\lambda = \frac{1}{\kappa^2(\dot{s})^6}a^3\left(\frac{a^2+c^2}{c^2}\right)\sinh\frac{at}{c} = \frac{1}{\left(\frac{\sqrt{a^2+c^2}}{ca}\operatorname{sech}^2\frac{at}{c}\right)^2\left(a\cosh\frac{at}{c}\right)^6}a^3\left(\frac{a^2+c^2}{c^2}\right)\sinh\frac{at}{c} = \frac{1}{a}\operatorname{sech}^2\frac{at}{c}\sinh\frac{at}{c}$. Proved.

(7): Here equation of the curve is $\vec{r} = (2a\cos t\,,2a\sin t\,,bt^2)$. So, $\dot{s}\hat{t} = \dot{\vec{r}} = (-2a\sin t\,,2a\cos t\,,2bt) = 2(-a\sin t\,,a\cos t\,,bt)$, $\ddot{\vec{r}} = (-2a\cos t\,,-2a\sin t\,,2b) = -2(a\cos t\,,a\sin t\,,-b)$ and $\dddot{\vec{r}} = -2(-a\sin t\,,a\cos t\,,0)$. Hence $\dot{s} = |\dot{s}\hat{t}| = |(-2a\sin t\,,2a\cos t\,,2bt)| = 2\sqrt{a^2+b^2t^2}$. Since

$\kappa\dot{s}^3\hat{b} = \dot{\vec{r}}\times\ddot{\vec{r}} = 2(-a\sin t\,,a\cos t\,,bt)\times(-2)(a\cos t\,,a\sin t\,,-b) = -4(-ab\cos t - abt\sin t\,,abt\cos t - ab\sin t\,,-a^2) = 4a(b(\cos t + t\sin t),b(\sin t - t\cos t),a)$ so, $\kappa\dot{s}^3 = \left|\kappa\dot{s}^3\hat{b}\right| = \left|\dot{\vec{r}}\times\ddot{\vec{r}}\right| = |4a(b(\cos t + t\sin t),b(\sin t - t\cos t),a)| = 4a\sqrt{b^2+b^2t^2+a^2}$. Proved.

Hence $\kappa = \frac{1}{\dot{s}^3}4a\sqrt{a^2+b^2+b^2t^2} = \frac{1}{\left(2\sqrt{a^2+b^2t^2}\right)^3}4a\sqrt{a^2+b^2+b^2t^2} = \frac{a\sqrt{a^2+b^2+b^2t^2}}{2\left(\sqrt{a^2+b^2t^2}\right)^3}$. Proved. Since $\kappa^2\lambda(\dot{s})^6 = \left[\dot{\vec{r}}\ \ddot{\vec{r}}\ \dddot{\vec{r}}\right] = \left(\dot{\vec{r}}\times\ddot{\vec{r}}\right)\cdot\dddot{\vec{r}} = 4a(b(\cos t + t\sin t),b(\sin t - t\cos t),a)\cdot 2a(\sin t\,,-\cos t\,,0) = 8a^2\big(b\sin t\,(\cos t + t\sin t) - b\cos t\,(\sin t - t\cos t)\big) = 8a^2bt$. Proved. Now $\lambda = \frac{1}{\kappa^2(\dot{s})^6}8a^2bt = \frac{8a^2bt}{\left(\frac{a\sqrt{a^2+b^2+b^2t^2}}{2\left(\sqrt{a^2+b^2t^2}\right)^3}\right)^2\left(2\sqrt{a^2+b^2t^2}\right)^6} \doteq \frac{bt}{2(a^2+b^2+b^2t^2)}$. Proved.

(8): Here equation of the curve is $\vec{r} = (3t,3t^2,2t^3)$. So, $\dot{s}\hat{t} = \dot{\vec{r}} = (3,6t,6t^2) = 3(1,2t,2t^2)$, $\ddot{\vec{r}} = 3(0,2,4t) = 6(0,1,2t)$ and $\dddot{\vec{r}} = 6(0,0,2) = 12(0,0,1)$. Now $\dot{\vec{r}}\times\ddot{\vec{r}} = 3(1,2t,2t^2)\times 6(0,1,2t) = 18(4t^2-2t^2,-2t,1) =$

$18(2t^2, -2t, 1)$. Proved. Next $[\dot{\vec{r}}\ \ddot{\vec{r}}\ \dddot{\vec{r}}] = (\dot{\vec{r}} \times \ddot{\vec{r}}) \cdot \dddot{\vec{r}} = 18(2t^2, -2t, 1) \cdot 12(0,0,1) = 216$. Proved.

Here $\dot{s} = |\dot{s}\hat{t}| = |3(1,2t,2t^2)| = 3\sqrt{1+4t^2+4t^4} = 3(1+2t^2)$. Since $\kappa\dot{s}^3\hat{b} = \dot{\vec{r}} \times \ddot{\vec{r}} = 18(2t^2, -2t, 1)$ so,
$\kappa\dot{s}^3 = |\kappa\dot{s}^3\hat{b}| = |\dot{\vec{r}} \times \ddot{\vec{r}}| = |18(2t^2, -2t, 1)| = 18\sqrt{4t^4+4t^2+1} = 18(2t^2+1)$
and hence $\kappa = \frac{1}{\dot{s}^3}\left(18(2t^2+1)\right) = \frac{1}{(3(1+2t^2))^3}\left(18(2t^2+1)\right) = \frac{2}{3(1+2t^2)^2}$.

Proved. Since $\kappa^2\lambda(\dot{s})^6 = [\dot{\vec{r}}\ \ddot{\vec{r}}\ \dddot{\vec{r}}] = 216$ so,
$\lambda = \frac{216}{\kappa^2(\dot{s})^6} = \frac{216}{\left(\frac{2}{3(1+2t^2)^2}\right)^2(3(1+2t^2))^6} = \frac{2}{3(1+2t^2)^2}$. Proved. Since $\dot{s}\hat{t} = 3(1,2t,2t^2)$,

so $\hat{t} = \frac{1}{\dot{s}}3(1,2t,2t^2) = \frac{1}{3(1+2t^2)}3(1,2t,2t^2) = \frac{1}{1+2t^2}(1,2t,2t^2)$. Ans.

Since $\kappa\dot{s}^3\hat{b} = \dot{\vec{r}} \times \ddot{\vec{r}}$ so, $\hat{b} = \frac{\dot{\vec{r}}\times\ddot{\vec{r}}}{|\dot{\vec{r}}\times\ddot{\vec{r}}|} = \frac{18(2t^2,-2t,1)}{18(2t^2+1)} = \frac{(2t^2,-2t,1)}{2t^2+1}$. Ans. Finally
$\hat{n} = \hat{b} \times \hat{t} = \frac{(2t^2,-2t,1)}{2t^2+1} \times \frac{1}{1+2t^2}(1,2t,2t^2) = \frac{1}{(1+2t^2)^2}(-4t^3-2t, 1-4t^4, 4t^3+2t) = \frac{1}{1+2t^2}(-2t, 1-2t^2, 2t)$. Ans.

(9): Here equation of the curve is $\vec{r} = (e^t, e^{-t}, \sqrt{2}t)$. So, $\dot{s}\hat{t} = \dot{\vec{r}} = (e^t, -e^{-t}, \sqrt{2})$, $\ddot{\vec{r}} = (e^t, e^{-t}, 0)$ and $\dddot{\vec{r}} = (e^t, -e^{-t}, 0)$. Hence
$\dot{s} = |(e^t, -e^{-t}, \sqrt{2})| = \sqrt{e^{2t}+e^{-2t}+2} = e^t + e^{-t} = 2\cosh t$. Proved. Next
$\hat{t} = \frac{1}{\dot{s}}(e^t, -e^{-t}, \sqrt{2}) = \frac{1}{2\cosh t}(e^t, -e^{-t}, \sqrt{2})$. Ans. Now $\kappa\dot{s}^3\hat{b} = \dot{\vec{r}} \times \ddot{\vec{r}} = (e^t, -e^{-t}, \sqrt{2}) \times (e^t, e^{-t}, 0) = (-\sqrt{2}e^{-t}, \sqrt{2}e^t, 2)$ so, $\kappa\dot{s}^3 = |\kappa\dot{s}^3\hat{b}| = |\dot{\vec{r}} \times \ddot{\vec{r}}| = |(-\sqrt{2}e^{-t}, \sqrt{2}e^t, 2)| = \sqrt{2e^{-2t}+2e^{2t}+4} = \sqrt{2}(e^t+e^{-t}) = 2\sqrt{2}\cosh t$. Proved. Further $\kappa = \frac{1}{\dot{s}^3}2\sqrt{2}\cosh t = \frac{1}{(2\cosh t)^3}2\sqrt{2}\cosh t = \frac{\sqrt{2}}{4}\operatorname{sech}^2 t$. Proved.

Also, $\hat{b} = \frac{1}{\kappa\dot{s}^3}(-\sqrt{2}e^{-t}, \sqrt{2}e^t, 2) = \frac{1}{\left(\frac{\sqrt{2}}{4}\operatorname{sech}^2 t\right)(2\cosh t)^3}(-\sqrt{2}e^{-t}, \sqrt{2}e^t, 2) = \frac{\operatorname{sech} t}{2\sqrt{2}}(-\sqrt{2}e^{-t}, \sqrt{2}e^t, 2) = \frac{\operatorname{sech} t}{2}(-e^{-t}, e^t, \sqrt{2})$. Ans. Hence $\hat{n} = \hat{b} \times \hat{t} =$

$\frac{\operatorname{sech} t}{2}(-e^{-t}, e^t, \sqrt{2}) \times \frac{1}{2\cosh t}(e^t, -e^{-t}, \sqrt{2}) = \frac{\operatorname{sech}^2 t}{4}(\sqrt{2}e^t + \sqrt{2}e^{-t}, \sqrt{2}e^t + \sqrt{2}e^{-t}, e^{-2t} - e^{2t}) = \frac{\operatorname{sech}^2 t}{4}(\sqrt{2}(e^t + e^{-t}), \sqrt{2}(e^t + e^{-t}), -2\sinh 2t) = \frac{\operatorname{sech}^2 t}{2}(\sqrt{2}\cosh t, \sqrt{2}\cosh t, -\sinh 2t) =$. Ans. Since $\kappa^2\lambda(\dot{s})^6 = [\dot{\vec{r}}\ \ddot{\vec{r}}\ \dddot{\vec{r}}] = (\dot{\vec{r}} \times \ddot{\vec{r}}) \cdot \dddot{\vec{r}} = (-\sqrt{2}e^{-t}, \sqrt{2}e^t, 2) \cdot (e^t, -e^{-t}, 0) = -\sqrt{2} - \sqrt{2} = -2\sqrt{2}$ so,

$\lambda = \frac{1}{\kappa^2(\dot{s})^6}(-2\sqrt{2}) = \frac{1}{\left(\frac{\sqrt{2}}{4}\operatorname{sech}^2 t\right)^2 (2\cosh t)^6}(-2\sqrt{2}) = -\frac{\sqrt{2}}{4}\operatorname{sech}^2 t.$ Proved.

(10): This is solved in the text.

(11): Here required vector area $= \sum \frac{1}{2}(\vec{r} \times (\vec{r} + \delta\vec{r})) = \sum \frac{1}{2}(\vec{r} \times \vec{r} + \vec{r} \times \delta\vec{r}) = \sum \frac{1}{2}(\vec{0} + \vec{r} \times \delta\vec{r}) = \frac{1}{2}\sum(\vec{r} \times \delta\vec{r})$, where summation is taken over the curve from A to B. For smooth curve, required vector area can be written as $\frac{1}{2}\int_A^B \vec{r} \times d\vec{r} = \frac{1}{2}\int_A^B \vec{r} \times (\hat{t}\, ds) = \frac{1}{2}\int_A^B (\vec{r} \times \hat{t}) ds.$ Proved.

(12): Let us imagine that the space inside the closed surface is sliced by three sets of parallel planes in large number such that each piece is a small-sized cube.

For each such cube, it is easy to see that $\sum \hat{n}\, dS = 0$, where $\hat{n}$ denotes unit outward normal to the plane surface of the cube. Further we observe that for two adjacent cubes, $\hat{n}\, dS$ for the common wall contributed by these cubes cancel each other. So, the sum of all $\sum \hat{n}\, dS$ for all cubes is equal to $\sum \hat{n}\, dS$ over the outer surface of the given closed surface. Hence, for smooth closed surface we can write $\int \hat{n}\, dS = 0$; that is, vector area of a closed surface is zero. Proved.

(13):

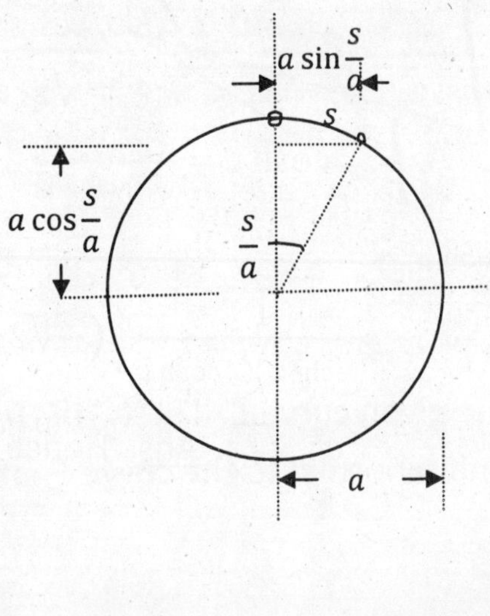

It is clear from the figure that equation of the given curve can be written as $\vec{r}(x,y,z) = \left(b\cos\left\{\frac{1}{b}\left(a\sin\frac{s}{a}\right)\right\}, b\sin\left\{\frac{1}{b}\left(a\sin\frac{s}{a}\right)\right\}, a\cos\frac{s}{a}\right)$. So, $\hat{t} = \vec{r}\,' = \frac{d}{ds}\vec{r} = \frac{d}{ds}\left(b\cos\left\{\frac{1}{b}\left(a\sin\frac{s}{a}\right)\right\}, b\sin\left\{\frac{1}{b}\left(a\sin\frac{s}{a}\right)\right\}, a\cos\frac{s}{a}\right) =$
$\left(-b\sin\left\{\frac{1}{b}\left(a\sin\frac{s}{a}\right)\right\}\frac{1}{b}\cos\frac{s}{a}, b\cos\left\{\frac{1}{b}\left(a\sin\frac{s}{a}\right)\right\}\frac{1}{b}\cos\frac{s}{a}, -\sin\frac{s}{a}\right) =$
$\left(-\sin\left\{\frac{1}{b}\left(a\sin\frac{s}{a}\right)\right\}\cos\frac{s}{a}, \cos\left\{\frac{1}{b}\left(a\sin\frac{s}{a}\right)\right\}\cos\frac{s}{a}, -\sin\frac{s}{a}\right) =$
$\left(-\sin\left(\frac{a}{b}\sin\frac{s}{a}\right)\cos\frac{s}{a}, \cos\left(\frac{a}{b}\sin\frac{s}{a}\right)\cos\frac{s}{a}, -\sin\frac{s}{a}\right)$ so, $\kappa\hat{n} = \frac{d}{ds}\hat{t} =$
$\left(\left\{-\cos\left(\frac{a}{b}\sin\frac{s}{a}\right)\frac{1}{b}\cos\frac{s}{a}\right\}\cos\frac{s}{a} + \left\{-\sin\left(\frac{a}{b}\sin\frac{s}{a}\right)\right\}\left(-\frac{1}{a}\sin\frac{s}{a}\right), \left\{-\sin\left(\frac{a}{b}\sin\frac{s}{a}\right)\frac{1}{b}\cos\frac{s}{a}\right\}\cos\frac{s}{a} + \left\{\cos\left(\frac{a}{b}\sin\frac{s}{a}\right)\right\}\left(-\frac{1}{a}\sin\frac{s}{a}\right), -\frac{1}{a}\cos\frac{s}{a}\right) = \left(\left\{-\frac{1}{b}\cos\left(\frac{a}{b}\sin\frac{s}{a}\right)\cos^2\frac{s}{a}\right\} + \left\{\frac{1}{a}\sin\left(\frac{a}{b}\sin\frac{s}{a}\right)\right\}\sin\frac{s}{a}, \left\{-\frac{1}{b}\sin\left(\frac{a}{b}\sin\frac{s}{a}\right)\cos^2\frac{s}{a}\right\} - \left\{\frac{1}{a}\cos\left(\frac{a}{b}\sin\frac{s}{a}\right)\right\}\sin\frac{s}{a}, -\frac{1}{a}\cos\frac{s}{a}\right)$.
Hence $\kappa^2 = \frac{1}{b^2}\cos^4\frac{s}{a} + \frac{1}{a^2}\sin^2\frac{s}{a} + \frac{1}{a^2}\cos^2\frac{s}{a} = \frac{1}{a^2} + \frac{1}{b^2}\cos^4\frac{s}{a}$. Proved.

(14):

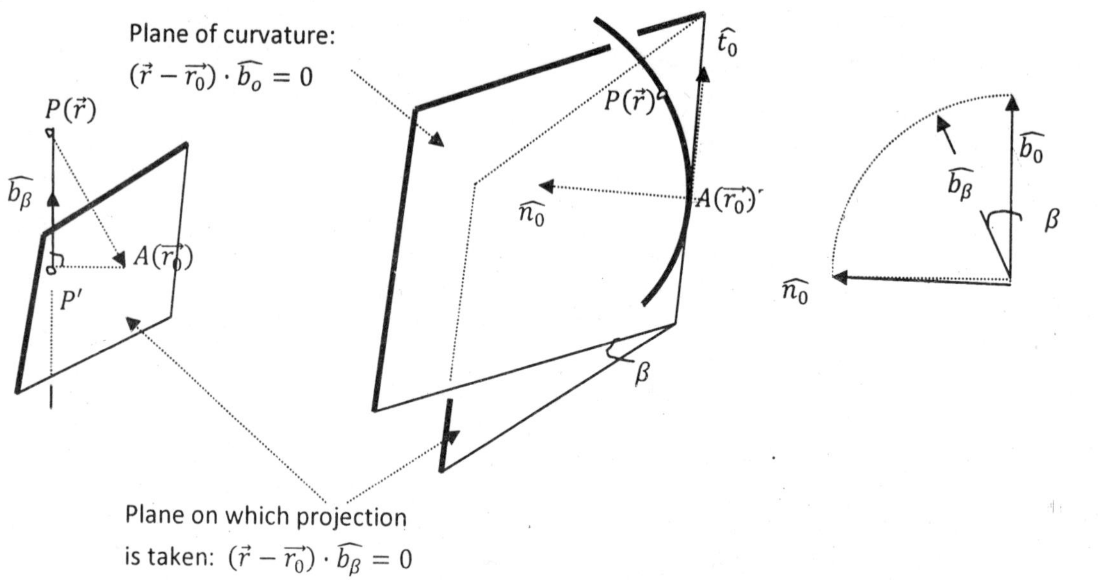

Let $A(\vec{r_0})$ be a fixed point on the given curve Γ. Let $\widehat{t_0}, \widehat{n_0}, \widehat{b_0}$ be , respectively, the unit tangent, unit normal, unit binormal to the curve Γ at the point $A(\vec{r_0})$. Let

$P(\vec{r})$ be any point on the given curve. Let Π be the plane through A and parallel to $\widehat{t_0}$. Let $P'(\vec{R})$ be the point of projection of P on Π. Let $\widehat{b_\beta}$ be the unit vector normal to the plane Π. Clearly length of PP' is $(\vec{r_0}-\vec{r})\cdot(-\widehat{b_\beta})$ and hence $\vec{R}=\vec{r}+\{(\vec{r_0}-\vec{r})\cdot(-\widehat{b_\beta})\}(-\widehat{b_\beta})$ or, $\vec{R}=\vec{r}+\{(\vec{r_0}-\vec{r})\cdot(\widehat{b_\beta})\}(\widehat{b_\beta})$. Let s denotes the arc-length of the curve Γ and let S denotes the arc-length of the projected curve Γ' of Γ on plane Π. All other notations will be self-explanatory.

Now $\hat{T}\dfrac{dS}{ds}=\dfrac{d\vec{R}}{dS}\dfrac{dS}{ds}=\hat{t}-(\hat{t}\cdot\widehat{b_\beta})\widehat{b_\beta}$ $\cdots(1)$

Hence $\left(\dfrac{dS}{ds}\right)^2=1+(\hat{t}\cdot\widehat{b_\beta})^2-2\hat{t}\cdot(\hat{t}\cdot\widehat{b_\beta})\widehat{b_\beta}=1-(\hat{t}\cdot\widehat{b_\beta})^2$. Hence $\left[\dfrac{dS}{ds}\right]^2_{\vec{r}=\vec{r_0}}=$ $1-(\widehat{t_0}\cdot\widehat{b_\beta})^2=1-0$, because $\widehat{b_\beta}=(\cos\beta)\widehat{b_0}+(\sin\beta)\widehat{n_0}$ in the first situation of the question. So, $\left[\dfrac{dS}{ds}\right]_{\vec{r}=\vec{r_0}}=1$. Further, on differentiating (1), we get

$\mathrm{K}\hat{N}\left(\dfrac{dS}{ds}\right)^2+\hat{T}\dfrac{d^2S}{ds^2}=\kappa\hat{n}-(\kappa\hat{n}\cdot\widehat{b_\beta})\widehat{b_\beta}$ $\cdots(2)$

Since $\left(\dfrac{dS}{ds}\right)^2=1-(\hat{t}\cdot\widehat{b_\beta})^2$ so, upon differentiation, we get $2\dfrac{dS}{ds}\dfrac{d^2S}{ds^2}=$ $-2(\hat{t}\cdot\widehat{b_\beta})(\kappa\hat{n}\cdot\widehat{b_\beta})$. At $\vec{r}=\vec{r_0}$, $2(1)\left[\dfrac{d^2S}{ds^2}\right]_{\vec{r}=\vec{r_0}}=-2(\widehat{t_0}\cdot\widehat{b_\beta})(\kappa\widehat{n_0}\cdot\widehat{b_\beta})=$ $-2(0)(\kappa\widehat{n_0}\cdot\widehat{b_\beta})=0$. So, $\left[\dfrac{d^2S}{ds^2}\right]_{\vec{r}=\vec{r_0}}=0$. At $\vec{r}=\vec{r_0}$, (2) gives $\mathrm{K}_0\widehat{N_0}(1)^2+\vec{0}=$ $\kappa_0\widehat{n_0}-(\kappa_0\widehat{n_0}\cdot\widehat{b_\beta})\widehat{b_\beta}=\kappa_0\widehat{n_0}-(\kappa_0\sin\beta)\widehat{b_\beta}$ and hence, on taking dot product with itself, we get $(\mathrm{K}_0)^2=(\kappa_0)^2\{1-2\sin\beta\,(\widehat{n_0}\cdot\widehat{b_\beta})+\sin^2\beta\}=(\kappa_0)^2\{1-$ $2\sin\beta\,(\sin\beta)+\sin^2\beta\}=(\kappa_0)^2\cos^2\beta$ or, $\mathrm{K}_0=\kappa_0\cos\beta$. Proved. In the second situation our plane, on which the curve Γ is to be projected, passes through $A(\vec{r_0})$ and is parallel to $\widehat{n_0}$. So, equation of this plane is $\vec{R}=\vec{r}+$ $\{(\vec{r_0}-\vec{r})\cdot(-\widehat{b_\beta})\}(-\widehat{b_\beta})$ where $\widehat{b_\beta}\equiv(\cos\beta)\widehat{b_0}+(\sin\beta)\widehat{t_0}$. So, as above, $\left[\dfrac{dS}{ds}\right]^2_{\vec{r}=\vec{r_0}}=1-(\widehat{t_0}\cdot\widehat{b_\beta})^2=1-\sin^2\beta=\cos^2\beta$ or, $\left[\dfrac{dS}{ds}\right]_{\vec{r}=\vec{r_0}}=\cos\beta$. Further, as above, we get: $2(\cos\beta)\left[\dfrac{d^2S}{ds^2}\right]_{\vec{r}=\vec{r_0}}=-2(\widehat{t_0}\cdot\widehat{b_\beta})(\kappa\widehat{n_0}\cdot\widehat{b_\beta})=-2(\widehat{t_0}\cdot\widehat{b_\beta})0=$

0, hence $\left[\frac{d^2S}{ds^2}\right]_{\vec{r}=\vec{r_0}} = 0$. As above, we also get: $\mathrm{K}_0\widehat{N_0}(\cos\beta)^2 + \vec{0} = \kappa_0\widehat{n_0} - (\kappa_0\widehat{n_0}\cdot\widehat{b_\beta})\widehat{b_\beta} = \kappa_0\widehat{n_0} - \vec{0} = \kappa_0\widehat{n_0}$. Hence $\mathrm{K}_0(\cos\beta)^2 = \kappa_0$ or, $\mathrm{K}_0 = \kappa_0\sec^2\beta$. Proved.

(15): Let $\kappa(s) = K$, $\lambda(s) = \Lambda$ for every value of s. We have to prove that the curve $\vec{r} = (x(s), y(s), z(s))$ is a circular helix. Here $\frac{d}{ds}(\Lambda\hat{t} + K\hat{b}) = \Lambda\hat{t}' + K\hat{b}' = \Lambda(K\hat{n}) + K(-\Lambda\hat{n}) = \vec{0}$, so $\Lambda\hat{t} + K\hat{b}$ is a constant vector, say, $\vec{a}$. Now $(1)|\vec{a}|\cos(\measuredangle\hat{t},\vec{a}) = \hat{t}\cdot\vec{a} = \hat{t}\cdot(\Lambda\hat{t} + K\hat{b}) = \Lambda$, a constant. Hence $\hat{t}$ always makes a constant angle with the constant direction $\vec{a}$. So the curve is a helix. If we take the direction of $\vec{a}$ as the direction of z-axis, then $\frac{dz}{ds} = z' = (x', y', z')\cdot\frac{\vec{a}}{|\vec{a}|} = \hat{t}\cdot\frac{\vec{a}}{|\vec{a}|} = \frac{\Lambda}{|\vec{a}|}$. So, we can suppose that $z = (\cos\beta)s$ for some β. Hence $\vec{r} = (x(s), y(s), (\cos\beta)s)$ is the equation of the curve (the helix). It remains to prove that this is a circular helix, Here $\hat{t} = \frac{d\vec{r}}{ds} = (x', y', \cos\beta)$, $1 = |\hat{t}|^2 = (x')^2 + (y')^2 + \cos^2\beta$ or, $(x')^2 + (y')^2 = \sin^2\beta$. So, we can suppose that $x' = (\sin\beta)\cos\theta$ and $y' = (\sin\beta)\sin\theta$ $\cdots(*)$. Further $K\hat{n} = \hat{t}' = (x'', y'', 0)$ so, $K^2 = (x'')^2 + (y'')^2$. So, from $(*)$, $K^2 = \{-\theta'\sin\theta\sin\beta\}^2 + \{\theta'\cos\theta\sin\beta\}^2$ or, $K^2 = (\sin^2\beta)(\theta')^2$ or, $\frac{d\theta}{ds} = \theta' = \frac{K}{\sin\beta}$, a constant. So, we can suppose that $\theta = \alpha s$ where α is a constant. Hence, from $(*)$, $\frac{dx}{ds} = x' = \sin\beta\cos(\alpha s)$. On integration, we can write $x = (\sin\beta)\left\{\frac{1}{\alpha}\sin(\alpha s)\right\}$. Similarly $y = (\sin\beta)\left\{\frac{-1}{\alpha}\cos(\alpha s)\right\}$. So equation of the curve is $\vec{r} = \left(\frac{\sin\beta}{\alpha}\sin(\alpha s), -\frac{\sin\beta}{\alpha}\cos(\alpha s), (\cos\beta)s\right)$ which represents a circular helix. Proved.

6

KINEMATICS AND DYNAMICS OF A PARTICLE

Examples (page 113): (3): At time t, let $Q(q\cos\omega t, q\sin\omega t)$. Hence $P(q\cos\omega t + p\cos(-\omega t), q\sin\omega t + p\sin(-\omega t))$ i.e., $P((q+p)\cos\omega t, (q-p)\sin\omega t)$. Let $P(x,y)$. So, $x=(q+p)\cos\omega t$ and $y=(q-p)\sin\omega t$. Hence $\frac{x^2}{(q+p)^2}+\frac{y^2}{(q-p)^2}=1$ is the equation of path of P. This is an equation of ellipse. Here eccentric angle of P is ωt, so the eccentric angle of D is $\frac{\pi}{2}+\omega t$. So, $D\left((q+p)\cos\left(\frac{\pi}{2}+\omega t\right),(q-p)\sin\left(\frac{\pi}{2}+\omega t\right)\right)$ or $D(-(q+p)\sin\omega t,(q-p)\cos\omega t)$. Hence $OD=\sqrt{(q+p)^2\sin^2\omega t+(q-p)^2\cos^2\omega t}=\sqrt{q^2(\sin^2\omega t+\cos^2\omega t)+p^2(\sin^2\omega t+\cos^2\omega t)-2pq(\cos^2\omega t-\sin^2\omega t)}=\sqrt{q^2+p^2-2pq\cos 2\omega t}\ \cdots(1)$. Here angle between $\overrightarrow{OQ}$ and $\overrightarrow{QP}$ is $\omega t-((-\omega)t)\ (=2\omega t)$. So, $|\hat{k}\times(\mathbf{q}-\mathbf{p})|=|\hat{k}||\mathbf{q}-\mathbf{p}|\sin 90°=|\mathbf{q}-\mathbf{p}|=\sqrt{q^2+p^2-2\mathbf{p}\cdot\mathbf{q}}=\sqrt{q^2+p^2-2pq\cos\omega t}=OD$, from (1). Hence semi-conjugate diameter OD is equal to $|\hat{k}\times(\mathbf{q}-\mathbf{p})|$. Finally, we have to prove that $OP^2+OD^2=(q+p)^2+(q-p)^2$. Here, $LHS=[\{(q+p)\cos\omega t\}^2+\{(q-p)\sin\omega t\}^2]+[(q+p)^2\sin^2\omega t+(q-p)^2\cos^2\omega t]=(q+p)^2(\cos^2\omega t+\sin^2\omega t)+(q-p)^2(\sin^2\omega t+\cos^2\omega t)=(q+p)^2+(q-p)^2=RHS$. Proved.

(4): Let A be a fixed point on the circle of radius a. Let its centre be the origin and OA be the positive direction of the x-axis. Let tangent at Q makes an angle θ with the tangent at A. Let $\hat{r}$ be the unit vector along $\overrightarrow{OQ}$ and $\hat{\theta}$ be the unit vector perpendicular to OQ. Clearly $\hat{r} = (\cos\theta)\hat{\imath} + (\sin\theta)\hat{\jmath}$ and $\hat{\theta} = \left(\cos\left(\theta + \frac{\pi}{2}\right)\right)\hat{\imath} + \left(\sin\left(\theta + \frac{\pi}{2}\right)\right)\hat{\jmath} = -(\sin\theta)\hat{\imath} + (\sin\theta)\hat{\jmath}$.Further $\dot{\hat{r}} = (-\sin\theta)\dot{\theta}\hat{\imath} + (\cos\theta)\hat{\jmath} = \dot{\theta}\hat{\theta}$ and $\dot{\hat{\theta}} = -(\cos\theta)\dot{\theta}\hat{\imath} - (\sin\theta)\dot{\theta}\hat{\jmath} = -\dot{\theta}\hat{r}$. Also, $\hat{r}\cos\theta - \hat{\theta}\sin\theta = \cos\theta\big((\cos\theta)\hat{\imath} + (\sin\theta)\hat{\jmath}\big) - \sin\theta\big((-\sin\theta)\hat{\imath} + (\cos\theta)\hat{\jmath}\big) = (\cos^2\theta + \sin^2\theta)\hat{\imath} = \hat{\imath}$. Thus $\hat{\imath} = \hat{r}\cos\theta - \hat{\theta}\sin\theta$. Similarly $\hat{\jmath} = \hat{r}\cos\theta + \hat{\theta}\cos\theta$. Here $\overrightarrow{OP} = \overrightarrow{OQ} + \overrightarrow{QP} = \big((a\cos\theta)\hat{\imath} + (a\sin\theta)\hat{\jmath}\big) + \left(\left(r\cos\left(\frac{\pi}{2} + \theta\right)\right)\hat{\imath} + \left(r\sin\left(\frac{\pi}{2} + \theta\right)\right)\hat{\jmath}\right) = (a\cos\theta - r\sin\theta)\hat{\imath} + (a\sin\theta + r\cos\theta)\hat{\jmath} = (a\cos\theta - r\sin\theta)(\hat{r}\cos\theta - \hat{\theta}\sin\theta) + (a\sin\theta + r\cos\theta) = \hat{r}(a\cos^2\theta - r\sin\theta\cos\theta + a\sin^2\theta + r\sin\theta\cos\theta) + \hat{\theta}(-a\sin\theta\cos\theta + r\sin^2\theta + a\sin\theta\cos\theta + r\cos^2\theta) = a\hat{r} + r\hat{\theta}$. Thus $\overrightarrow{OP} = a\hat{r} + r\hat{\theta}$. Hence velocity of P at time t is $\frac{d}{dt}(a\hat{r} + r\hat{\theta}) = a\dot{\hat{r}} + \left(\dot{r}\hat{\theta} + r\dot{\hat{\theta}}\right) = a(\dot{\theta}\hat{\theta}) + \dot{r}\hat{\theta} + r(-\dot{\theta}\hat{r}) = (a\dot{\theta} + \dot{r})\hat{\theta} - r\dot{\theta}\hat{r}$. Hence acceleration of P at time t is $\frac{d^2}{dt^2}\overrightarrow{OP} = \frac{d}{dt}\left((a\dot{\theta} + \dot{r})\hat{\theta} - r\dot{\theta}\hat{r}\right) = \{(a\ddot{\theta} + \ddot{r})\hat{\theta} + (a\dot{\theta} + \dot{r})(-\dot{\theta}\hat{r})\} - \{\dot{r}\dot{\theta}\hat{r} + r\ddot{\theta}\hat{r} + r\dot{\theta}(\dot{\theta}\hat{\theta})\} = (\ddot{r} + a\ddot{\theta} - r\dot{\theta}^2)\hat{\theta} + (-a\dot{\theta}^2 - \dot{r}\dot{\theta} - \dot{r}\dot{\theta} - r\ddot{\theta})\hat{r}$. So resolute of acceleration along $\overrightarrow{QP}$ is $\ddot{r} + a\ddot{\theta} - r\dot{\theta}^2$ and resolute of acceleration perpendicular to $\overrightarrow{QP}$ is $2\dot{r}\dot{\theta} + r\ddot{\theta} + a\dot{\theta}^2$ $\left(= \frac{1}{r}\frac{d}{dt}(r^2\dot{\theta}) + a\dot{\theta}^2\right)$. Proved.

Exercises on Chapter 6

(1): Here (rate of change of vectorial area) $= \frac{d}{dt}\left(\frac{1}{2}(\vec{r_2} - \vec{r_1}) \times (\vec{r_3} - \vec{r_1})\right) = \frac{1}{2}\frac{d}{dt}(\vec{r_2} \times \vec{r_3} + \vec{r_1} \times \vec{r_2} + \vec{r_3} \times \vec{r_1}) = \frac{1}{2}(\dot{\vec{r_2}} \times \vec{r_3} - \dot{\vec{r_3}} \times \vec{r_2} + \dot{\vec{r_1}} \times \vec{r_2} - \dot{\vec{r_2}} \times \vec{r_1} + \dot{\vec{r_3}} \times \vec{r_1} - \dot{\vec{r_1}} \times \vec{r_3}) = \frac{1}{2}(\vec{v_2} \times \vec{r_3} - \vec{v_3} \times \vec{r_2} + \vec{v_1} \times \vec{r_2} - \vec{v_2} \times \vec{r_1} + \vec{v_3} \times \vec{r_1} - \vec{v_1} \times \vec{r_3}) = -\frac{1}{2}[\vec{r_1} \times (\vec{v_2} - \vec{v_3}) + \vec{r_2} \times (\vec{v_3} - \vec{v_1}) + \vec{r_3} \times (\vec{v_1} - \vec{v_2})]$. Proved.

(2): Before developing the answer to this problem, let us note the following facts:

(i): Suppose, a particle of mass m is moving under a central force $-F(r)\hat{r}$. Then $\ddot{\vec{r}} = -\frac{F(r)\vec{r}}{mr}$. Taking dot product with $2\dot{\vec{r}}$

we get $\frac{d}{dt}(v^2) = \frac{d}{dt}\left(\dot{\vec{r}}^2\right) = \quad 2\dot{\vec{r}} \cdot \ddot{\vec{r}} = -\frac{F(r)}{mr}\left(2\vec{r} \cdot \dot{\vec{r}}\right) = -\frac{F(r)}{mr}\frac{d}{dt}(\vec{r}^2) = -\frac{F(r)}{mr}\frac{d}{dt}(r^2)$. On integration, $v^2 = C - \frac{1}{m}\int \frac{F(r)}{r}(2r)dr$ or $v^2 = C - \frac{2}{m}\int F(r)dr$ $\cdots (1)$

where C is a constant. Also, $\hat{k}pv = \hat{k}(r\sin\varphi)v = \vec{r} \times \vec{v} = r\hat{r} \times \left(\dot{r}\hat{r} + r\dot{\theta}\hat{\theta}\right) = r^2\dot{\theta}\hat{k}$, that is, $\hat{k}pv = \hat{k}r^2\dot{\theta}$ $\cdots (2)$

where symbols have their usual meanings. Here force is central, so, $0 =$ transversal acceleration $= \frac{1}{r}\frac{d}{dt}\left(r^2\dot{\theta}\right)$, and hence $r^2\dot{\theta} = \text{constant, say } h$. By $\vec{h}$ we mean $h\hat{k}$. From (2), $\vec{r} \times \vec{v} = \vec{h}$ and $pv = h$ (These are important results). Also, from (1), we get $\frac{h^2}{p^2} = C - \frac{2}{m}\int F(r)dr$. On differentiation, we get

$$-\frac{2}{m}F(r) = h^2(-2)p^{-2-1}\frac{dp}{dr} \text{ or, } F(r) = \frac{mh^2}{p^3}\frac{dp}{dr} \qquad \cdots (3)$$

This equation gives the law of force for a motion under central forces, provided we are given with the pedal equation of the path of the motion.

(ii): Let us find the pedal equation (that is, *p,r* equation) of ellipse $\frac{x^2}{a^2} + \frac{y^2}{b^2} = 1$ referred to a focus $(-ae, 0)$. Here equation of tangent at $(a\cos\theta, b\sin\theta)$ is $\frac{\cos\theta}{a}x + \frac{\sin\theta}{b}y - 1 = 0$. So, $p = \frac{\left|\frac{\cos\theta}{a}(-ae) + \frac{\sin\theta}{b}(0) - 1\right|}{\sqrt{\left(\frac{\cos\theta}{a}\right)^2 + \left(\frac{\sin\theta}{b}\right)^2}}$ and hence $\frac{1}{p^2} = \frac{\left(\frac{\cos^2\theta}{a^2} + \frac{\sin^2\theta}{b^2}\right)}{(e\cos\theta + 1)^2} = \frac{b^2\cos^2\theta + a^2\sin^2\theta}{a^2b^2(e\cos\theta + 1)^2} = \frac{a^2(1-e^2)\cos^2\theta + a^2\sin^2\theta}{a^2b^2(e\cos\theta + 1)^2} = \frac{1 - e^2\cos^2\theta}{b^2(e\cos\theta + 1)^2} = \frac{1 - e\cos\theta}{b^2(1 + e\cos\theta)}$. Thus $\frac{1}{p^2} = \frac{1 - e\cos\theta}{b^2(1 + e\cos\theta)}$ $\cdots (1)$

Also $r^2 = (a\cos\theta - (-ae))^2 + (b\sin\theta - 0)^2 = a^2(\cos^2\theta + 2e\cos\theta + e^2 + \sin^2\theta - e^2\sin^2\theta) = a^2(1 + 2e\cos\theta + e^2(1 - \sin^2\theta)) = a^2(1 + 2e\cos\theta + e^2\cos^2\theta) = a^2(1 + e\cos\theta)^2$ and hence $r = a(1 + e\cos\theta)$ ...(2)

Now from (1) and (2), $\frac{1}{p^2} = \frac{1}{b^2}\left(\frac{1-\left(\frac{r}{a}-1\right)}{1+\left(\frac{r}{a}-1\right)}\right) = \frac{1}{b^2}\left(\frac{2-\frac{r}{a}}{\frac{r}{a}}\right) = \frac{1}{b^2}\left(\frac{2a}{r} - 1\right)$ or $\frac{1}{p^2} = \frac{1}{b^2}\left(\frac{2a}{r} - 1\right)$ ...(3)

This is the required pedal equation of the ellipse reffered to focus $(-ae, 0)$.

(iii): If a particle is moving on an ellipse under a central force towards the focus O, then from the above discussion, $F(r) = \frac{mh^2}{p^3}\frac{dp}{dr} = \frac{mh^2}{-2}\frac{d}{dr}\left(\frac{1}{p^2}\right) = -\frac{mh^2}{2}\frac{d}{dr}\left(\frac{a}{b^2}\left(\frac{2}{r} - \frac{1}{a}\right)\right) = -\frac{1}{2}mh^2\frac{a}{b^2}\frac{(-2)}{r^2} = \left(\frac{mh^2 a}{b^2}\right)\frac{1}{r} \propto \frac{1}{r^2}$. Thus the particle will follow the inverse square law.

(iv): Let P be any point on the ellipse, whose focus is the pole O and major axis is the initial line. Let (r,θ) be the polar coordinates of the point P, l be the semi-latus rectum of the ellipse and M be the foot of perpendicular drawn from P to the directrix of the ellipse. Here $\frac{OP}{PM} = e$ (called eccentricity of the ellipse) and $0 < e < 1$. Hence $r = e(PM) = e\left(r\cos\theta + \frac{l}{r}\right)$ or $\frac{l}{r} = 1 - e\cos\theta$.

Answer to the main problem (2): Here $\ddot{\vec{r}} = -\frac{\mu}{r^2}\hat{r}$ for some positive constant μ. Taking cross product with $\vec{h}$, we get $\frac{d}{dt}(\vec{h}\times\vec{v}) = \vec{h}\times\frac{d}{dt}\vec{v} = \vec{h}\times\ddot{\vec{r}} = \frac{\mu}{r^2}\hat{r}\times\vec{h} = \frac{\mu}{r^2}\hat{r}\times(\vec{r}\times\vec{v}) = \frac{\mu}{r^2}\hat{r}\times\left(r\hat{r}\times\left(\dot{r}\hat{r} + r\dot{\theta}\hat{\theta}\right)\right) = \frac{\mu}{r^2}\hat{r}\times\left(r^2\dot{\theta}(\hat{r}\times\hat{\theta})\right) = \mu\dot{\theta}\left(\hat{r}\times(\hat{r}\times\hat{\theta})\right) = \mu\dot{\theta}\left((\hat{r}\cdot\hat{\theta})\hat{r} - (\hat{r}\cdot\hat{r})\hat{\theta}\right) = \mu\dot{\theta}(0\hat{r} - 1\hat{\theta}) = -\mu\dot{\theta}\hat{\theta} = \mu(-\dot{\theta}\hat{\theta}) = -\mu(\hat{r})^{\cdot} = \frac{d}{dt}(-\mu\hat{r})$. Hence, on integration, $\vec{h}\times\vec{v} = -\mu\hat{r} + \vec{a}$ ...(1)

where $\vec{a}$ is a constant. On taking cross product with $\vec{h}$, we get $-h^2\vec{v} = \vec{0} - h^2\vec{v} = (\vec{h}\cdot\vec{v})\vec{h} - h^2\vec{v} = \vec{h}\times(\vec{h}\times\vec{v}) = \vec{h}\times(-\mu\hat{r} + \vec{a}) = \mu\hat{r}\times\vec{h} + \vec{h}\times\vec{a} = \mu(\hat{r}\times h\hat{k}) + h\hat{k}\times\vec{a}$, so, $\vec{v} = \frac{\mu}{h}(\hat{k}\times\hat{r}) + \left(\frac{1}{h}\right)(\vec{a}\times\hat{k}) = \frac{\mu}{h}(\hat{\theta}) + \frac{1}{h}(\vec{a}\times\hat{k})$ ...(2)

Here direction of $\hat{\theta}$ is perpendicular to $\overrightarrow{OP}$. On taking dot product with $\vec{r}$, (1) becomes $-h^2 = \vec{h}\cdot(-\vec{h}) = \vec{h}\cdot(\vec{v}\times\vec{r}) = (\vec{h}\times\vec{v})\cdot\vec{r} = (-\mu\hat{r}+\vec{a})\cdot\vec{r} = -\mu r + \vec{a}\cdot\vec{r} = -\mu r + ar\cos\theta$ where θ is the angle which $\vec{r}$ makes with the constant vector $\vec{a}$. So, $\frac{(h^2)}{r} = 1 - \left(\frac{a}{\mu}\right)\cos\theta$ $\cdots(3)$

which is the polar equation of the ellipse. Here θ is the angle which $\vec{r}$ makes with the major axis. Hence the direction of $\vec{a}$ is along the major axis, and therefore the direction of $\vec{a}\times\hat{k}$ is perpendicular to major axis. So from (2), required components of $\vec{v}$ are $\frac{\mu}{h}$ and $\frac{1}{h}$ which are constants. Proved.

NOTE (1): From (3), we find that the eccentricity e of the ellipse is given by $e = \frac{a}{\mu}$. Further from (1), $(\mu e)^2 = a^2 = \left(\mu\hat{r} + \vec{h}\times\vec{v}\right)^2 = \mu^2 + 2\mu\hat{r}\cdot\left(\vec{h}\times\vec{v}\right) + \left(\vec{h}\times\vec{v}\right)^2 = \mu^2 - 2\frac{\mu}{r}\vec{h}\cdot(\vec{r}\times\vec{v}) + (hv\sin 90^\circ)^2 = \mu^2 - 2\frac{\mu}{r}\vec{h}\cdot\vec{h} + h^2v^2 = \mu^2 - 2\frac{\mu}{r}h^2 + h^2v^2$ and hence $e^2 = 1 - 2\frac{h^2}{\mu r} + \left(\frac{hv}{\mu}\right)^2$ $\cdots(4)$

Further, from (3), semi-latus rectum l is given by $l = \frac{h^2}{\mu}$.

For ellipse $\frac{h^2}{\mu} = l = \frac{b^2}{a}$ $\cdots(5)$

NOTE (2): For a particle, describing an ellipse under a central force to the focus, equation of motion $\ddot{\vec{r}} = -\frac{\mu}{r^2}$. Taking dot product with $2\dot{\vec{r}}$, we get $\frac{d}{dt}(v^2) = \frac{d}{dt}\left(\dot{\vec{r}}\right)^2 = 2\dot{\vec{r}}\cdot\ddot{\vec{r}} = -\frac{2\mu}{r^3}\dot{\vec{r}}\cdot\vec{r} = -\frac{\mu}{r^3}\frac{d}{dt}(\vec{r}^2) = -\frac{\mu}{r^3}2r\frac{dr}{dt} = -\frac{2\mu}{r^2}$ and hence $d(v^2) = -2\mu r^{-2}dr$. On integration, $v^2 = -2\mu\frac{r^{-2+1}}{-2+1} + C$ or $v^2 = \frac{2\mu}{r} + C$ $\cdots(6)$

where C is a constant. Considering the velocity $\frac{h}{b}$ at one end of the minor axis, we get $\frac{\mu}{a} = \left(\frac{h}{b}\right)^2 = \frac{2\mu}{a} + C$ (from (5)), so, $C = -\frac{\mu}{a}$. So, from (6), $v^2 = \frac{2\mu}{r} + \frac{-2\mu}{a}$ or $v^2 = \mu\left(\frac{2}{r} - \frac{1}{a}\right)$ $\cdots(7)$

(3): Here $(x,y)=\vec{r}=(\cos nt)\vec{a}+(\sin nt)\vec{b}=(\cos nt)(a_1,a_2)+(\sin nt)(b_1,b_2)$ where $\vec{a}\equiv(a_1,a_2)$, $\vec{b}\equiv(b_1,b_2)$. So, $x=a_1\cos nt+b_1\sin nt$ and $y=a_2\cos nt+b_2\sin nt$. So, $a_1\cos nt+b_1\sin nt-x=0$ and $a_2\cos nt+b_2\sin nt-y=0$. So, $\frac{\cos nt}{yb_1-xb_2}=\frac{\sin nt}{xa_2-ya_1}=\frac{-1}{a_1b_2-a_2b_1}$ or $1=\cos^2 nt+\sin^2 nt=\left(\frac{-(yb_1-xb_2)}{a_1b_2-a_2b_1}\right)^2+\left(\frac{-(xa_2-ya_1)}{a_1b_2-a_2b_1}\right)^2$ or $(a_1b_2-a_2b_1)^2={b_2}^2x^2-2b_1b_2xy+{b_1}^2y^2+({a_1}^2y^2-2a_1a_2xy+{a_2}^2x^2)$ or $({a_2}^2+{b_2}^2)x^2-2(a_1a_2+b_1b_2)xy+({a_1}^2+{b_1}^2)y^2=(a_1b_2-a_2b_1)^2$ $\cdots(1)$

Here $\left(\frac{1}{2}\text{coefficient of } xy\right)^2-(\text{coefficient of } x^2)(\text{coefficient of } y^2)=(a_1a_2+b_1b_2)^2-({a_2}^2+{b_2}^2)({a_1}^2+{b_1}^2)=-(a_1b_2-a_2b_1)^2$, which is negative. So, the equation of path (1) is an ellipse. Further $\dot{\vec{r}}=(-\sin nt)n\vec{a}+(\cos nt)n\vec{b}=n\left((-\sin nt)\vec{a}+(\cos nt)\vec{b}\right)$ and $\ddot{\vec{r}}=n\left((-\cos nt)n\vec{a}+(-\sin nt)n\vec{b}\right)=-n^2\left((\cos nt)\vec{a}+(\sin nt)\vec{b}\right)=-n^2\vec{r}$. Hence force on a particle of mass m is $m(n^2)\overrightarrow{PO}$. <u>Proved</u>.

(4): Put $\vec{r}=(x,y,z)$. So, $\hat{t}=\frac{d\vec{r}}{ds}=\left(\frac{dx}{ds},\frac{dy}{ds},\frac{dz}{ds}\right)$ or $\hat{t}=(x',y',z')$ where x' denotes derivative of x with respect to arc-length s, etc. Hence $\hat{n}=\frac{1}{\kappa}(x'',y'',z'')$ and $\hat{h}=\hat{t}\times\hat{n}=\frac{1}{\kappa}(y'z''-z'y'',z'x''-x'z'',x'y''-y'x'')$.

$\frac{1}{m}(\text{net force on } m)=\frac{1}{m}\left[\{mg(0,0,-1)\cdot(x',y',z')\}\hat{t}+\left\{mg(0,0,-1)\cdot\left(\frac{1}{\kappa}(x'',y'',z'')\right)\right\}\hat{n}+\{mg(0,0,-1)\cdot(y'z''-z'y'',z'x''-x'z'',x'y''-y'x'')\}\hat{b}+N_2\hat{n}+N_3\hat{b}-\mu\sqrt{{N_2}^2+{N_3}^2}\hat{t}\right]$, where symbols have their usual meanings. So, $\ddot{s}\hat{t}+(\dot{s})^2\kappa\hat{n}=-gz'\hat{t}-\frac{gz'}{\kappa}\hat{n}-g(x'y''-y'x'')\hat{b}+N_2\hat{n}+N_3\hat{b}$ (since $\mu=0$) or $v\frac{dv}{ds}=\ddot{s}=-gz'=-g\frac{dz}{ds}$ or $vdv=-gdz$. On integration $\frac{v^2}{2}=-gz+C$. Let, initially $z=z_0$ and $v=u$. So, $\frac{v^2}{2}-\frac{u^2}{2}=$

$g(z_0 - z)$ or, $v^2 = u^2 + 2gh$ where
$h \equiv z_0 - z = (\text{depth of } Q(z) \text{ from } P(z_0))$. Proved.

(5): Already solved in the text.

(6): Let $x = a\cos\omega t$, $y = a\sin\omega t$ be the parametric equation of the circular path of the particle, where a, ω are positive constant and t denotes the time. So, $x^2 + y^2 = a^2$. This is the equation of circular path, with centre at the origin and radius a. So, $\vec{r} = a(\cos\omega t, \sin\omega t)$. Hence

$\vec{v} = \dot{\vec{r}} = a\big((-\sin\omega t)\omega, (\cos\omega t)\omega\big) = a\omega(-\sin\omega t, \cos\omega t)$. So, equation of hodograph is $\vec{R} = a\omega(-\sin\omega t, \cos\omega t)$. Clearly this represents a circle with centre at origin and radius $a\omega$. Further $\dot{\vec{R}} = a\omega(-(\cos\omega t)\omega, (-\sin\omega t)\omega) = -a\omega^2(\cos\omega t, \sin\omega t)$. So, $\left|\dot{\vec{R}}\right| = a\omega^2$. Hence hodograph is described uniformity. Proved.

(7): Let O be the centre of the vertical smooth circular ring and let a be its radius. Let us choose O as the pole and upward vertical direction as the initial line for the polar coordinate system. Here $\vec{v} = a\dot{\theta}\hat{\theta}$ and acceleration $=$ $a\left(\ddot{\theta}\hat{\theta} + \dot{\theta}(-\dot{\theta}\hat{r})\right) = (a\ddot{\theta})\hat{\theta} - a(\dot{\theta})^2\hat{r}$. By Newton's second law, $a\ddot{\theta} = \frac{1}{m}(mg\sin\theta)$. So, $\ddot{\theta} = \frac{g}{a}\sin\theta$. Hence $\frac{d}{dt}(\dot{\theta}^2) = 2\dot{\theta}\ddot{\theta} = 2\left(\frac{d\theta}{dt}\right)\left(\frac{g}{a}\sin\theta\right)$ or $d(\dot{\theta}^2) = \frac{2g}{a}\sin\theta\, d\theta$. On integration, $\dot{\theta}^2 = -\frac{2g}{a}\cos\theta + C$. Given that $\dot{\theta} = 0$ when $\theta = 0$. So, $C = \frac{2g}{a}$. Hence $\dot{\theta}^2 = \frac{2g}{a}(1 - \cos\theta) = \frac{2g}{a} 2\sin^2\frac{\theta}{2}$ or $\dot{\theta} = 2\sqrt{\frac{g}{a}}\sin\frac{\theta}{2}$. Hence $\vec{v} = a\left(2\sqrt{\frac{g}{a}}\sin\frac{\theta}{2}\right)\hat{r}$. So, required equation of hodograph is $\vec{R} = \left(2\sqrt{ag}\sin\frac{\theta}{2}\right)\hat{r}$ or $R = 2\sqrt{ag}\sin\frac{\theta}{2}$ which is of the form $r = c\sin\frac{\theta}{2}$. Proved.

(8): Already solved in text.

(9):

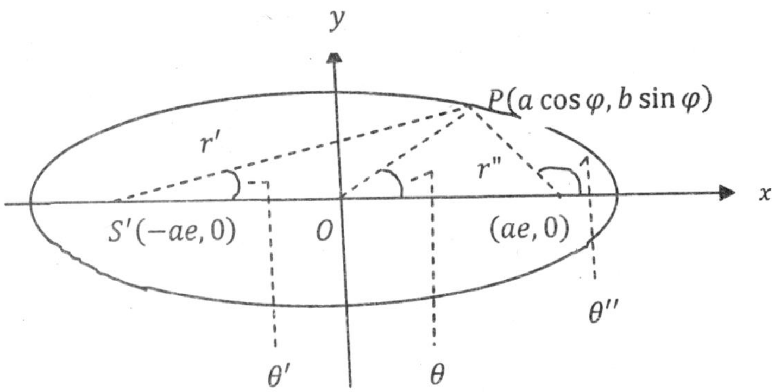

Here

$r' = \sqrt{(a\cos\varphi - (-ae))^2 + (b\sin\varphi)^2} =$
$\sqrt{a^2(\cos\varphi + e)^2 + a^2(1-e^2)\sin^2\varphi} =$

$a\sqrt{\cos^2\varphi + 2e\cos\varphi + e^2 + \sin^2\varphi - e^2\sin^2\varphi} =$
$a\sqrt{1 + 2e\cos\varphi + e^2\cos^2\varphi} = a(1+e\cos\varphi)$ Similarly $r'' = a(1 - e\cos\varphi)$.

Further $\tan\theta' =$ (slope of line $S'P$) $= \frac{b\sin\varphi - 0}{a\cos\varphi - (-ae)} = \frac{b\sin\varphi}{a(\cos\varphi + e)}$ and

$\tan\theta'' = \frac{b\sin\varphi}{a(\cos\varphi - e)}$. Hence $r'\dot{\theta}' = \frac{a(1+e\cos\varphi)}{\sec^2\theta'}\left[\frac{b}{a}\frac{(\cos\varphi+e) - \sin\varphi(-\sin\varphi)}{(e+\cos\varphi)^2}\dot{\varphi}\right] =$

$\left[\frac{b(1+e\cos\varphi)}{(e+\cos\varphi)^2}(1+e\cos\varphi)\dot{\varphi}\right]\left[\frac{a^2(\cos\varphi+e)^2}{(b\sin\varphi)^2 + \left(a(\cos\varphi+e)\right)^2}\right] =$

$\frac{b(1+e\cos\varphi)^2}{(e+\cos\varphi)^2}\dot{\varphi}\frac{a^2(\cos\varphi+e)^2}{a^2[(1-e^2)\sin^2\varphi + \cos^2\varphi + 2e\cos\varphi + e^2]} =$

$b(1+e\cos\varphi)^2\dot{\varphi}\frac{1}{1+2e\cos\varphi + e^2\cos^2\varphi}$. Thus $r'\dot{\theta}' = b\dot{\varphi}$. Now $r''\frac{d}{dt}\theta'' =$

$\frac{a(1-e\cos\varphi)}{\sec^2\theta''}\frac{b}{a}\frac{[\cos\varphi(\cos\varphi - e) - \sin\varphi(-\sin\varphi)]\dot{\varphi}}{(\cos\varphi - e)^2} = \frac{b(1-e\cos\varphi)}{\sec^2\theta''}\frac{(1-e\cos\varphi)\dot{\varphi}}{(\cos\varphi - e)^2} =$

$\frac{b(1-e\cos\varphi)^2\dot{\varphi}}{(\cos\varphi - e)^2}\left[\frac{a^2(\cos\varphi - e)^2}{b^2\sin^2\varphi + a^2(\cos\varphi - e)^2}\right] =$

$b(1-e\cos\varphi)^2\dot{\varphi}\frac{a^2}{a^2(1-e^2)\sin^2\varphi + a^2(\cos\varphi - e)^2} = \frac{b(1-e\cos\varphi)^2\dot{\varphi}}{[1 - 2e\cos\varphi + e^2\cos^2\varphi]} = \frac{b}{a}\dot{\varphi}$. Also

$h^2 = (OP)^2\dot{\theta} = \{(a\cos\varphi)^2 + (b\sin\varphi)^2\}\frac{d}{dt}\left(\tan^{-1}\frac{b\sin\varphi}{a\cos\varphi}\right) = (a^2\cos^2\varphi +$

$b^2\sin^2\varphi)\frac{1}{1+\left(\frac{b}{a}\tan\varphi\right)^2}\left(\frac{b}{a}\sec^2\varphi\right)\dot{\varphi} = (a^2\cos^2\varphi)\left(1 + \frac{b^2}{a^2}\tan^2\varphi\right)\frac{\left(\frac{b}{a}\sec^2\varphi\right)\dot{\varphi}}{1+\left(\frac{b}{a}\tan\varphi\right)^2} =$

$ab\dot{\varphi}$. Thus, $\dot{\varphi} = \frac{h^2}{ab}$. Hence $\dot{\theta}' = \frac{b\dot{\varphi}}{r'} = \frac{\left(h^2/a\right)}{r'}$. So, $\dot{\theta}' \propto \frac{1}{r'}$. Similarly $\dot{\theta}'' \propto \frac{1}{r''}$. Proved.

Also, $\frac{1}{\dot{\theta}'} + \frac{1}{\dot{\theta}''} = \frac{r'}{h^2/a} + \frac{r''}{h^2/a} = \frac{a}{h^2}(r' + r'') = \frac{a}{h^2}(2a) = \frac{2a^2}{h^2}$ which is a constant. Proved.

(10):

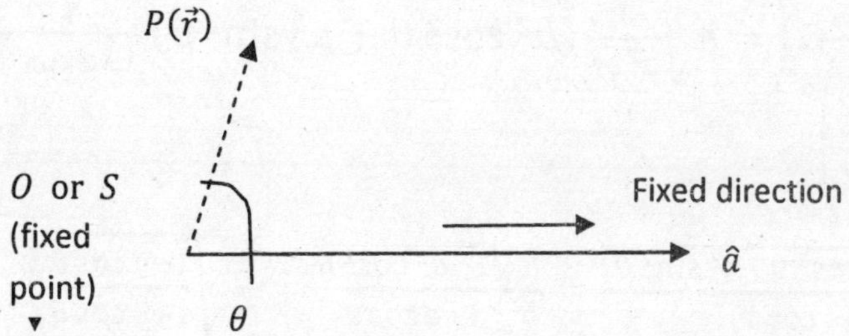

Here $\dot{\vec{r}}$ or $\vec{v} = u_0\hat{a} + v_0\hat{\theta}$ so, $\ddot{\vec{r}} = 0 + v_0(-\dot{\theta}\hat{r}) = -v_0\left(\frac{h}{r^2}\right)\hat{r}$ (Here S is assumed to be the centre of force). So, $\ddot{\vec{r}} = -\frac{\mu}{r^2}\hat{r}$ where $\mu = v_0$ $\cdots(1)$

Taking cross product with $\vec{h}$, $\vec{h} \times \ddot{\vec{r}} = \frac{\mu}{r^2} \times \vec{h} = \frac{\mu}{r^2}\hat{r} \times (\vec{r} \times \dot{\vec{r}}) = \frac{\mu}{r^2}\hat{r} \times (r^2\dot{\theta}\hat{r} \times \hat{\theta})\varphi = \mu\dot{\theta}(-\hat{\theta}) = -\mu\dot{\hat{r}}$. Now on integraton, $\vec{h} \times \dot{\vec{r}} = -\mu\hat{r} + \vec{c}$ where $\vec{c}$ is a constant vector. Here $\dot{\vec{r}}$ lies in the plane of orbit, so, $\vec{h} \times \dot{\vec{r}}$ lies in the plane of orbit. Hence $\vec{c}\ (= \vec{h} \times \dot{\vec{r}} + \mu\hat{r})$ lies in the plane of orbit. On taking dot product with $\vec{r}$ we get $(\vec{h} \times \dot{\vec{r}}) \cdot \vec{r} = -\mu\hat{r} \cdot \vec{r} + \vec{c} \cdot \vec{r}$ or, $-h^2 = \vec{h} \cdot (-\vec{h}) = \vec{h} \cdot (\dot{\vec{r}} \times \vec{r}) = -\mu r + cr\cos\varphi$ where φ denotes the angle between $\overrightarrow{OP}$ and $\vec{c}$. Hence, $\frac{h^2/\mu}{r} = 1 - \left(\frac{c}{\mu}\right)\cos\varphi$, which is a conic having focus O (or S) and eccentricity $\frac{c}{\mu}$.

Further, from the initial condition, $\vec{c} = \vec{h} \times (u_0\hat{a} + v_0\hat{\theta}) + \mu\hat{r} = u_0h\hat{k} \times \vec{a} + v_0h(\hat{k} \times \hat{\theta}) + \mu\hat{r} = u_0h\hat{k} \times \hat{a} + v_0h(-\hat{r}) + \mu\hat{r} = u_0h(\hat{k} \times \hat{a})$, from (1). Hence

$c = u_0h|\hat{k} \times \hat{a}| = u_0h(1) = u_0h$. So, required eccentricity $= \frac{c}{\mu} = \frac{u_0h}{v_0h} = \frac{u_0}{v_0}$. Proved.

(11): Let $\frac{x^2}{a^2}+\frac{y^2}{b^2}=1$ be the equation of ellipse on which the given moves. Let $D(a\cos\theta, b\sin\theta)$ and $D'(-a\cos\theta,-b\sin\theta)$ be the two ends of a diameter.

Here equation of tangent at D is $\frac{\left|\frac{\cos\theta}{a}(-ae)+\frac{\sin\theta}{b}(0)-1\right|}{\sqrt{\left(\frac{\cos\theta}{a}\right)^2+\left(\frac{\sin\theta}{b}\right)^2}}\left(=\frac{1+e\cos\theta}{\sqrt{\frac{\cos^2\theta}{a^2}+\frac{\sin^2\theta}{b^2}}}\right)$.

So, speed of the particle at D is

$\frac{h}{\left(\frac{1+e\cos\theta}{\sqrt{\frac{\cos^2\theta}{b^2}+\frac{\sin^2\theta}{b^2}}}\right)}\left(=h\sqrt{\frac{1}{a^2b^2}(b^2\cos^2\theta+a^2\sin^2\theta)}\frac{1}{1+e\cos\theta}=\right.$

$\left.\frac{h\sqrt{\{a^2(1-e^2)\cos^2\theta+a^2\sin^2\theta\}}}{ab(1+e\cos\theta)}=\frac{h}{b}\frac{\sqrt{1-e^2\cos^2\theta}}{1+e\cos\theta}=\frac{h}{b}\sqrt{\frac{1-e\cos\theta}{1+e\cos\theta}}\right)$. Similarly speed of the

particle at D is $\frac{h}{b}\sqrt{\frac{1+e\cos\theta}{1-e\cos\theta}}$. Also, if we put $\theta=\frac{\pi}{2}$, we get that speed of the particle at ond of the minor axis is $\frac{h}{b}$. Further the geometric mean of the speeds at D and D' is $\sqrt{\left(\frac{h}{b}\sqrt{\frac{1-e\cos\theta}{1+e\cos\theta}}\right)\left(\frac{h}{b}\sqrt{\frac{1+e\cos\theta}{1-e\cos\theta}}\right)}\left(=\frac{h}{b}\right)$. Proved.

(12): Let equation of the path of particle be $\frac{x^2}{a^2}+\frac{y^2}{b^2}=1$. Let $P(a\cos\chi, b\sin\chi)$ be the position of the particle at time t. Here equation of tangent at P is $\frac{\cos\chi}{a}x+\frac{\sin\chi}{b}y-1=0$. So, $\tan\psi=\frac{-\frac{\cos\chi}{a}}{\frac{\sin\chi}{b}}=-\frac{b}{a}\cot\chi$, where $\chi\equiv$ angle between tangent at P with the positive direction of the x-axis. Also, equation of normal at $P(a\cos\chi, b\sin\chi)$ is $\frac{\sin\chi}{b}(x-a\cos\chi)-\frac{\cos\chi}{a}(y-b\sin\chi)=0$ or $\frac{\sin\chi}{b}x-\frac{\cos\chi}{a}y=\sin\chi\cos\chi\left(\frac{a}{b}-\frac{b}{a}\right)=(\sin\chi\cos\chi)\frac{a^2-b^2}{ab}=(\sin\chi\cos\chi)$
$\frac{a^2-a^2(1-e^2)}{ab}=\frac{ae^2}{b}\sin\chi\cos\chi$. Let normal at P intersect x-axis at N.
So, $N(ae^2\cos\chi, 0)$. Hence length of normal
$(=NP)=\sqrt{(ae^2\cos\chi-a\cos\chi)^2+(b\sin\chi)^2}=$

$\sqrt{(a^2\cos^2\chi)(e^2-1)^2+b^2\sin^2\chi}=\sqrt{b^2(1-e^2)\cos^2\chi+b^2\sin^2\chi}=$
$b\sqrt{1-e^2\cos^2\chi}$ ... (1)

We have seen that $\tan\psi=-\frac{b}{a}\cot\chi$. On differentiation, $(\sec^2\chi)\dot{\chi}=$ $\left(-\frac{b}{a}\right)(-\csc^2\chi)\dot{\chi}=\left(\frac{b}{a}\csc^2\chi\right)\dot{\chi}$. So, $\dot{\psi}=\frac{1}{1+\frac{b^2}{a^2}\cot^2\chi}\left(\frac{b}{a}\csc^2\chi\right)\dot{\chi}=$ $\frac{a^2\sin^2\chi}{a^2\sin^2\chi+b^2\cos^2\chi}\left(\frac{b}{a}\csc^2\chi\right)\dot{\chi}=\frac{ab\dot{\chi}}{a^2\sin^2\chi+b^2\cos^2\chi}$. Here let $SP=r$ and $\angle XSP=\theta$ where $S(-ae,0)$. Since S is the centre of force, so $r^2\dot{\theta}=h$ where h is a constant. Further $r^2=(a\cos\chi+ae)^2+(b\sin\chi)^2=a^2(\cos^2\chi+e^2+2e\cos\chi)+a^2(1-e^2)\sin^2\chi=a^2(1+e\cos\chi)^2$. So,
$r=a(1+e\cos\chi)$ ... (2)

Also, $\tan\theta=\frac{b\sin\chi}{a\cos\chi+ae}=\frac{b}{a}\frac{\sin\chi}{e+\cos\chi}$. So, $(\sec^2\theta)\dot{\theta}=\frac{b}{a}\frac{\cos\chi(e+\cos\chi)-\sin\chi(-\sin\chi)}{(e+\cos\chi)^2}$
$\dot{\chi}=\frac{b\dot{\chi}}{a}\frac{(1+e\cos\psi)}{(e+\cos\chi)^2}$ and hence

$$\dot{\theta}=\frac{1}{1+\left(\frac{b\sin\chi}{ae+\cos\chi}\right)^2}\frac{b\dot{\chi}}{a}\frac{(1+e\cos\chi)}{(e+\cos\chi)^2}=\frac{b\dot{\chi}}{a}\frac{(1+e\cos\chi)a^2}{a^2(e+\cos\chi)^2+b^2\sin^2\chi}=$$

$$\frac{b\dot{\chi}}{a}\frac{(1+e\cos\chi)}{[(e+\cos\chi)^2+(1-e^2)\sin^2\chi]}=\frac{b\dot{\chi}}{a}\frac{(1+e\cos\chi)}{[e^2+2e\cos\chi+\cos^2\chi+\sin^2\chi-e^2\sin^2\chi]}=\frac{b\dot{\chi}(1+e\cos\chi)}{a(1+e\cos\chi)^2}.$$

Thus $\dot{\theta}=\frac{b\dot{\chi}}{a(1+e\cos\chi)}$ ... (3)

Since $r^2\dot{\theta}=h$ so, from (2) and (3), $h=a^2(1+e\cos\chi)^2\frac{b\dot{\chi}}{a(1+e\cos\chi)}=$
$ab(1+e\cos\chi)\dot{\chi}$ ... (4)

Let us denote the angle $XS'P$ by θ' where $S'(ae,0)$. We have to prove that $(\dot{\theta}')(NP)^2=$ constant. Here $\tan\theta'=\frac{b\sin\chi}{a\cos\chi-ae}=\frac{b}{a}\frac{\sin\chi}{\cos\chi-e}\Rightarrow(\sec^2\theta')\dot{\theta}'=$ $\frac{b}{a}\frac{\cos\chi(\cos\chi-e)-\sin\chi(-\sin\chi)}{(\cos\chi-e)^2}\dot{\chi}=\frac{ab(1-e\cos\chi)\dot{\chi}}{a^2[(\cos\chi-e)^2+(1-e^2)\sin^2\chi]}=\frac{b}{a}\frac{(1-e\cos\chi)\dot{\chi}}{1-2e\cos\chi+e^2\cos^2\chi}=$ $\frac{b(1-e\cos\chi)}{a(1-e\cos\chi)^2}\dot{\chi}=\frac{b\dot{\chi}}{a(1-e\cos\chi)}=\frac{b}{a(1-e\cos\chi)}\frac{h}{ab(1+e\cos\chi)}$ from (1).
Hence $\dot{\theta}'=\frac{h}{a^2(1-e^2\cos^2\chi)}$. So, $(\dot{\theta}')(NP)^2=\frac{h}{a^2(1-e^2\cos^2\chi)}b^2(1-e^2\cos^2\chi)$ (from (1)). Hence, $(\dot{\theta}')(NP)^2=\frac{hb^2}{a^2}$, which is a constant. Proved.

(13): The particle is describing an ellipse under a central force towards the focus, so by the remark (iii) of question (2), the particle will follow the inverse square law. So, let equation of motion of the particle be $\ddot{\vec{r}} = -\mu\frac{1}{r^2}\hat{r}$, where symbols have their usual meanings. Since $vp = h$, a constant, so from question,

$$v_B b = h \quad \cdots (1)$$

where v_B denotes the speed of the particle at the upper end B of the minor axis of the ellipse. Hence $v_n p_n = h_n$ will give $h_n = v_B b$, where subscript n represents "new". Thus $h_n = h$

$$\cdots (2)$$

from (1). Our new equation of motion is $\ddot{\vec{r}}_n = -\left(\frac{2\mu}{3}\right)\frac{1}{r^2}\hat{r}_n$. On using the formula (iv) of question (2), we get $e_n{}^2 = 1 - \frac{2h_n{}^2}{\frac{2\mu}{3}\sqrt{(ae)^2+b^2}} + \left(\frac{h_n v_B}{\frac{2\mu}{3}}\right)^2 = 1 - \frac{2h^2}{\frac{2\mu}{3}\sqrt{a^2e^2+a^2(1-e^2)}} + \left(\frac{h\frac{h}{b}}{\frac{2\mu}{3}}\right)^2$, from (1) and (2). Hence, $e_n{}^2 = 1 - \frac{3h^2}{a\mu} + \frac{9}{4b^2}\left(\frac{h^2}{\mu}\right)^2 = 1 - \frac{3}{a}\frac{b^2}{a} + \frac{9}{4b^2}\left(\frac{b^2}{a}\right)^2$, from (v) of question (2). Hence $e_n{}^2 = 1 - \frac{3}{4}\frac{b^2}{a^2} = 1 - \frac{3}{4}(1-e^2) = \frac{1}{4}(1+3e^2)$. So, $e_n = \frac{1}{2}\sqrt{1+3e^2}$. Since $0 < e < 1$, so $0 < e^2 < 1$ or $0 < 3e^2 < 3$ or $1 < 1+3e^2 < 4$ or $\frac{1}{2} < \frac{1}{2}\sqrt{1+3e^2} < 1$. Hence $\frac{1}{2} < e_n < 1$. So, new path is again an ellipse. From, formula (vi) of question (2),

$v_n{}^2 = \frac{2\left(\frac{2\mu}{3}\right)}{r_n} + C$, where C is a constant. At B $\frac{\mu}{a} = \in \frac{\left(\mu\frac{b^2}{a}\right)}{b^2} = \frac{h^2}{b^2} = \frac{4\mu}{3a} + C$, so $C = -\frac{\mu}{3a}$. Hence $v_n{}^2 = \frac{4\mu}{3r_n} - \frac{\mu}{3a} = \frac{\mu}{a}\left(\frac{4}{r_n} - \frac{1}{a}\right)$

$$\cdots (3)$$

Further, from formula (vii) of question (2), $v_n{}^2 = \left(\frac{2\mu}{3}\right)\left(\frac{2}{r_n} - \frac{1}{a_n}\right)$. Hence, from (3), $\frac{\mu}{3}\left(\frac{4}{r_n} - \frac{1}{a}\right) = \frac{2\mu}{3}\left(\frac{2}{r_n} - \frac{1}{a_n}\right)$, and hence $\frac{\mu}{3a} = \frac{2\mu}{3a_n}$ or. Further, $b_n{}^2 = a_n{}^2(1 - e_n{}^2)$ $= (2a)^2\left(1 - \frac{1+3e^2}{4}\right) = 4a^2\left(\frac{3(1-e^2)}{4}\right) = 3b^2$ or

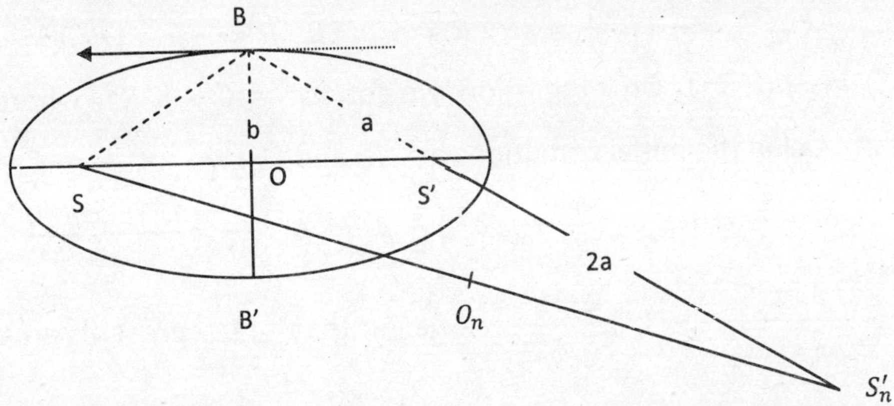

So $SB + BS_n' = 2a_n = 2(2a) = 4a$. Since $SB = BS' = a$, so $\boxed{S'S_n' = 2a}$.

So, clearly $S_n'(3ae, -2b)$. Hence the new centre is $O_n\left(\frac{-ae+3ae}{2}, \frac{0-2b}{2}\right)$ or $O_n(ae, -b)$. The mid-point of SO_n is $\left(\frac{-ae+ae}{2}, \frac{0-b}{2}\right)$ i.e., $(0, -b)$. So, SO_n is bisected by the semi-minor axis OB'. Proved.

(14): Before answering this question, let us obtain the pedal equation of ellipse $\frac{x^2}{a^2} + \frac{y^2}{b^2} = 1$ referred to centre O. Here equation of tangent at $(a\cos\theta, b\sin\theta)$ is $\frac{\cos\theta}{a}x + \frac{\sin\theta}{b}y - 1 = 0$. So, $p =$ (length perpendicular to the tangent at P drawn from origin O)$= \frac{1}{\sqrt{\left(\frac{\cos\theta}{a}\right)^2 + \left(\frac{\sin\theta}{b}\right)^2}}$ or $\frac{1}{p^2} = \frac{\cos^2\theta}{a^2} + \frac{\sin^2\theta}{b^2}$ or, $b^2\cos^2\theta + a^2\sin^2\theta - \frac{a^2b^2}{p^2} = 0$. Also, $a^2\cos^2\theta + b^2\sin^2\theta - r^2 = 0$. Hence $\frac{\cos^2\theta}{-a^2 + \frac{a^2b^4}{p^2}} = \frac{\sin^2\theta}{-\frac{a^4b^2}{p^2} + b^2r^2} = \frac{1}{b^4 - a^4}$. This implies $1 = \cos^2\theta + \sin^2\theta = \frac{(b^2-a^2)r^2 + \frac{a^2b^2}{p^2}(b^2-a^2)}{b^4-a^4} = \frac{r^2 + \frac{a^2b^2}{p^2}}{b^2+a^2}$ or $a^2 + b^2 = r^2 + \frac{a^2b^2}{p^2}$ or $\boxed{\frac{1}{p^2} = \frac{a^2+b^2-r^2}{a^2b^2}}$

Answer to the main question: $\frac{h}{b} = \text{velocity} = \frac{h_n}{b} \Rightarrow h_n = h$ $\quad \cdots (1)$

where subscript n represents "new path". Here equation of the new path is given by $\ddot{\vec{R}} = -\mu'\vec{R}$ where $\mu' a = \mu'(SB) \equiv \frac{\mu}{(SB)^2} = \frac{\mu}{a^2}$ $\quad \cdots (2)$

Hence $\frac{d}{dt}(v_n{}^2) = \frac{d}{dt}\left(\dot{\vec{R}}^2\right) = 2\dot{\vec{R}} \cdot \ddot{\vec{R}} = -\mu' 2\vec{R} \cdot \dot{\vec{R}} = -\mu\frac{d}{dt}(\vec{R}^2) = -\mu\left(\frac{d}{dt}R^2\right)$ or, $d(v_n{}^2) = -\mu d(R^2)$. On integration, we get $v_n{}^2 = C - \mu' R^2$, where C is a constant. Using the initial conditions at B and (1), we get $\frac{h^2}{b^2} = C - \mu' a^2$. Hence $\frac{h^2}{p^2} = v_n{}^2 = \left(\frac{h^2}{b^2} + \mu' a^2\right) - \mu' R^2$ or, $\frac{1}{p^2} = \frac{h^2+\mu' a^2 b^2 - \mu' b^2 R^2}{b^2 h^2}$ or, $\frac{1}{p^2} = \frac{\frac{h^2+\mu' a^2 b^2}{\mu' b^2} - R^2}{\frac{h^2}{\mu'}}$ $\cdots(*)$. $\frac{a^2+b^2-r^2}{a^2b^2}$, we get that $(*)$ represents an ellipse with S as centre of the ellipse. Further, $\left[a_n{}^2 + b_n{}^2 = \frac{h^2+\mu' a^2 b^2}{\mu' b^2} = \frac{h^2}{\mu' b^2} = \frac{h^2}{\mu' b^2} + a^2 = \frac{h^2}{\left(\frac{\mu}{a^3}\right)b^2} + a^2 \text{ (from (2)) or, } a_n{}^2 + b_n{}^2 = \frac{h^2}{\mu}\frac{a^3}{b^2} + a^2 = \frac{b^2}{a}\frac{a^3}{b^2} + a^2 = 2a^2\right]$ and $\left[a_n{}^2 b_n{}^2 = \frac{h^2}{\mu'} = \frac{h^2}{(\mu/a^3)} = \frac{h^2}{\mu}a^3 = \frac{b^2}{a}a^3 = a^2b^2 \ or.\ a_n b_n = ab \quad \cdots(**)\right]$. Hence, the new time period $= \frac{2\pi a_n b_n}{\left(\frac{1}{2}h_n\right)} = \frac{2\pi(ab)}{\frac{1}{2}h}$ (from (1), (**)) = old time period.

Proved.

Further, $(a_n \pm b^n)^2 = \left(a_n{}^2 + b_n{}^2\right) \pm 2a_n b_n = 2a^2 \pm 2(ab) = 2a(a \pm b)$. Hence $\frac{2a_n+2b_n}{2a_n-2b_n} = \frac{\sqrt{2a}\sqrt{a+b}}{\sqrt{2a}\sqrt{a-b}} = \frac{a+b}{a-b} = \frac{a+b}{\sqrt{a^2-a^2(1-e^2)}} = \frac{a+b}{ae} = \frac{2a+2b}{2ae} = \frac{\text{sum of old axes}}{\text{distance between old foci}}$. Proved.

(15): Let the elliptical path of the particle , in pedal form, be $\frac{1}{p^2} = \frac{a}{b^2}\left(\frac{2}{r} - \frac{1}{a}\right)$ $\cdots(*)$ (see remark (ii) of question (2)). Hence $-\frac{2}{p^3}\frac{dp}{dr} = \frac{2a}{b^2}\frac{-1}{r^2}$ or $\frac{dr}{dp} = \frac{b^2r^2}{ap^3}$ $\cdots(1)$

But from calculus, we know that $\rho =$ (radius of curvature at an arbitrary point) $= r\frac{dr}{dp}$ and (curvature at a point) $= \frac{1}{\rho}$, hence from (1),

$\rho = \frac{b^2r^3}{ap^3}$ $\cdots(2)$

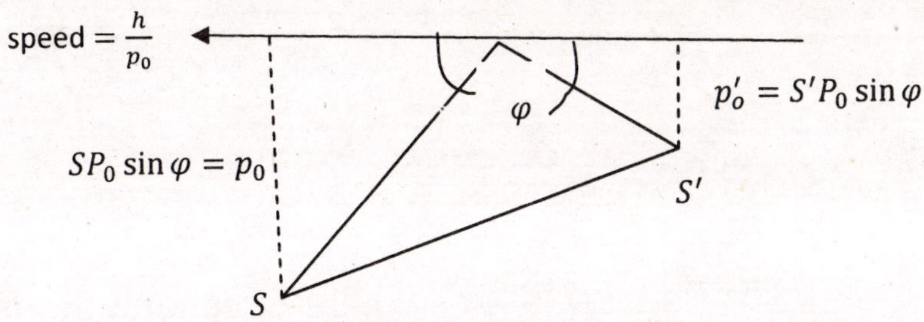

Here let equation of motion of the particle, before alteration, be $\ddot{\vec{r}} = -\mu\frac{1}{r^2}\hat{r}$.

Hence, $\left(\frac{h}{p}\right)^2 = v^2 = \mu\left(\frac{2}{r} - \frac{1}{a}\right)$ (see equation (vii) of question (2)). On comparing with $(*)$, we get $\frac{a}{b^2} = \frac{\mu}{h^2}$ or, $\mu = \frac{ah^2}{b^2}$. Similarly $\mu_n = \frac{a_n h_n^{\,2}}{b_n^{\,2}}$, where subscript n represents "new path". Since magnitude of the law of force has not changed, we have $\mu = \mu_n$. So,

$\frac{ah^2}{b^2} = \frac{a_n h_n^{\,2}}{b_n^{\,2}} = \frac{a_n}{b_n^{\,2}}\left((\text{speed at } P_0)(p_0')\right)^2 = \frac{a_n}{b^2}\left(\frac{h}{p_0}\frac{b^2}{p_0}\right)^2$ (since $p_0 p_0' = b^2$ from geometry) and therefore $\frac{a_n}{b_n^{\,2}} = \frac{ah^2}{b^2}\frac{p_0^{\,4}}{h^2 b^4} = \frac{ap_0^{\,4}}{b^6}$ $\cdots (**)$. On using (2) we get $\kappa : \kappa' = \left[\frac{\rho'}{\rho}\right]_{\text{at } P_0} = \frac{b_n^{\,2}}{a_n}\frac{(S'P_0)^3}{(p_0')^3} \Big/ \frac{b^2}{a}\frac{(SP_0)^3}{p_0^{\,3}} = \frac{b^6}{ap_0^{\,4}}\frac{ap_0^{\,3}}{(p_0')^3 b^2}\left(\frac{S'P_0}{SP_0}\right)^3$, from $(**)$.

Hence $\kappa : \kappa' = \frac{b^4}{p_0(p_0')^3}\left(\frac{S'P_0}{SP_0}\right)^3 = \frac{(p_0 p_0')^2}{p_0(p_0')^3}\left(\frac{S'P_0}{SP_0}\right)^3 = \frac{p_0}{p_0'}\left(\frac{S'P_0}{SP_0}\right)^3 = \frac{SP_0 \sin\varphi}{S'P_0 \sin\varphi}\left(\frac{S'P_0}{SP_0}\right)^3 = (S'P_0)^2 : (SP_0)^2$. Proved.

(16): Let equation of the path of particle be $\frac{x^2}{a^2} + \frac{y^2}{b^2} = 1$. Here centre of force is the focus $S(-ae, 0)$. Let $P(a\cos\chi, b\sin\chi)$ be the position of the particle at time t. Here $\tan\psi =$ (slope of tangent at P) $\frac{b\cos\chi}{-a\sin\chi} = -\frac{b}{a}\cot\chi$ or, $\psi = \tan^{-1}\left(-\frac{b}{a}\cot\chi\right)$, where tangent at P makes with positive direction of the fixed line x-axis. Here we have, $\psi = \tan^{-1}\left(-\frac{b}{a}\cot\chi\right)$, so

$\dot{\psi} = \frac{1}{1+\left(-\frac{b}{a}\cot\chi\right)^2}\left(-\frac{b}{a}\right)(-\csc^2\chi)\dot{\chi} = \frac{b}{a}\dot{\chi}\frac{a^2\sin^2\chi\csc^2\chi}{a^2\sin^2\chi+b^2\cos^2\chi} \Rightarrow \dot{\psi} =$

$$\frac{ab\dot{\chi}}{a^2(\sin^2\chi+(1-e^2)\cos^2\chi)} = \frac{b}{a}\frac{\dot{\chi}}{(1-e^2\cos^2\chi)} \qquad \cdots (1)$$

Also $\tan\theta = \frac{b\sin\psi-0}{a\cos\psi-(-ae)} = \frac{b\sin\psi}{a(\cos\psi+e)}$ where θ denotes the angle which SP makes with the positive direction of x-axis (that is, the initial line of the polar coordinate system). Hence $\theta = \tan^{-1}\left(\frac{b}{a}\frac{\sin\chi}{e+\cos\chi}\right)$. This gives $\frac{h}{(SP)^2} = \frac{h}{r^2} = \dot{\theta} =$
$\frac{1}{1+\left(\frac{b}{a}\frac{\sin\chi}{e+\cos\chi}\right)^2}\frac{b}{a}\frac{\cos\psi(e+\cos\chi)-\sin\chi(0-\sin\chi)}{(e+\cos\chi)^2}\dot{\chi} = \frac{b}{a}\frac{a^2(e+\cos\chi)^2}{a^2(e+\cos\chi)^2+b^2\sin^2\chi}\frac{(1+e\cos\chi)\dot{\chi}}{(e+\cos\chi)^2} =$
$\frac{b}{a}\frac{a^2(1+e\cos\chi)\dot{\chi}}{a^2[e^2+2e\cos\chi+\cos^2\chi+(1-e^2)\sin^2\chi]} = \frac{b}{a}\frac{(1+e\cos\chi)\dot{\chi}}{(1+e\cos\chi)^2} = \frac{b\dot{\chi}}{a(1+e\cos\chi)} =$
$\frac{b}{a(1+e\cos\chi)}\frac{\dot{\psi}a(1-e^2\cos^2\chi)}{b}$, from (1).

So, $\frac{h}{(SP)^2} = (1-e\cos\chi)\dot{\chi}$, and hence
$\dot{\psi} = \frac{h}{(1-e\cos\chi)(SP)^2} = \frac{h}{(1-e\cos\psi)\{(a\cos\chi+ae)^2+(b\sin\chi)^2\}} =$
$\frac{h}{(1-e\cos\chi)a^2[\cos^2\chi+2e\cos\chi+e^2+(1-e^2)\sin^2\chi]}$. Therefore
$\dot{\psi}(\chi) = \frac{h}{a^2(1-e\cos\chi)(1+e\cos\chi)^2}$. Hence, (rate of rotation at $(a\cos\chi, b\sin\chi)$) $=$
$\dot{\psi}(\chi) = \frac{h}{a^2(1-e\cos\chi)(1+e\cos\chi)^2}$.

To prove: (i) $0 = \left[\frac{d}{d\chi}(\text{rate of rotation})\right]_{\chi=0}$

(ii) $\left[\frac{d^2}{dx^2}(\text{rate of rotation})\right]_{\chi=0}$ is negative implies $\frac{1}{3} < e$

(iii) $\left[\frac{d^2}{dx^2}(\text{rate of rotation})\right]_{\chi=0}$ is positive implies $e < \frac{1}{3}$.

Here

$\frac{d}{d\chi}(\text{rate of rotation}) = \frac{-h[(e\sin\chi)(1+e\cos\chi)^2+2(1+e\cos\chi)(-e\sin\chi)(1-e\cos\chi)]}{a^2\{(1-e\cos\chi)^2(1+e\cos\chi)^4\}} =$

$-\frac{h}{a^2}\frac{(e\sin\chi)(1+e\cos\chi)(1+e\cos\chi-2+2e\cos\chi)}{(1-e\cos\chi)^2(1+e\cos\chi)^4}=-\frac{eh}{a^2}\frac{(\sin\chi)(3e\cos\chi)}{(1-e\cos\chi)^2(1+e\cos\chi)^3}$. This clearly proves (i).

Now

$\frac{d}{d\chi^2}$(rate of rotation) =

$$-\frac{eh}{a^2}\frac{\{\cos\chi(3e\cos\chi-1)+\sin\chi(-3e\sin\chi)\}\{(1-e\cos\chi)^2(1+e\cos\chi)^3\}-\sin\chi(3e\cos\chi-1)\frac{d}{d\chi}\{(1-e\cos\chi)^2(1+e\cos\chi)^3\}}{(1-e\cos\chi)^4(1+e\cos\chi)^6}.$$

So, $\left[\frac{d}{d\chi^2}(\text{rate of rotation})\right]_{\chi=0}=-\frac{eh}{a^2}\left(\frac{(3e-1(1-e)^2(1+e)^3)}{(1-e)^4(1+e)^6}\right)=\frac{-eh(3e-1)}{a^2(1-e)^2(1+e)^3}$.

Therefore, $\left[\frac{d}{d\chi^2}(\text{rate of rotation})\right]_{\chi=0}$ is negative $\Leftrightarrow$ $(3e-1$ is positive$)$ $\Leftrightarrow$ $\left(\frac{1}{3}<e\right)$. This completes the proof of (ii), (iii). Proved.

(17): Since path elliptical and force is central towards focus, so the force follows inverse square law (see remark (iii) of question (2)). Hence the "new path" will

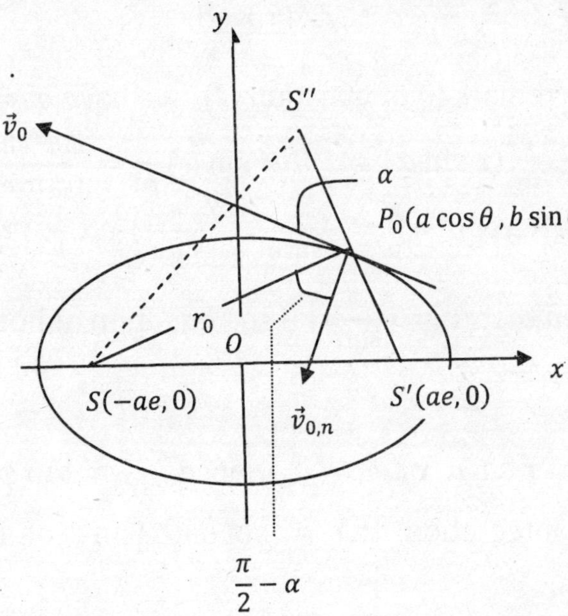

also be elliptical with one focus at $S(-ae'0)$. Let us try to find the another focus S''. Here, $v_0{}^2=\mu\left(\frac{2}{r_0}-\frac{1}{a}\right)$ and $v_0{}^2=\mu\left(\frac{2}{r_0}-\frac{1}{a_n}\right)$, where subscript n represents the "new path". So, $a_n=a$ ... (1)

So, $SP_0 + P_0S'' = 2a_n = 2a = SP_0 + P_0S'$. It follows that $P_0S'' = P_0S'$. Further since angle between P_0S and velocity $\vec{v}_{0,n}$ is $\frac{\pi}{2} - \alpha$ (from question), so, angle between P_0S'' and $-\vec{v}_{0,n}$ is also $\frac{\pi}{2} - \alpha$. It follows that S', P, S'' lie on the same line. Further since $P_0S'' = P_0S'$ so, from figure, P_0 is the mid-point of $S'S''$. So, $S''(2(a\cos\theta - ae, 2b\sin\theta))$. Hence

$e_n = \frac{SS''}{2a_n} = \frac{\sqrt{(2a\cos\theta)^2+(2b\sin\theta)^2}}{2a} = \frac{1}{a}\sqrt{(a\cos\theta)^2 + (b\sin\theta)^2} \propto$ (distance of P from centre O). Proved.

(18): Already solved in the text. See also the remarks of question (2).

(19): Let equation of ellipse in pedal form be $\frac{1}{p^2} = \frac{a^2+b^2-r^2}{a^2b^2}$ (see the initial remark of question (14)). Now, on using the remark (i) of question (2), we have $\ddot{\vec{r}} = -\frac{F(r)}{m}\hat{r}$ where $F(r) = \frac{mh^2}{p^3}\frac{dp}{dr} = \frac{mh^2}{-2}\frac{d}{dr}\left(\frac{1}{p^2}\right) = -\frac{1}{2}mh^2\frac{d}{dr}\left(\frac{1}{a^2b^2}(a^2 + b^2 - r^2)\right) = -\frac{1}{2}mh^2\frac{-1}{a^2b^2}2r = \frac{mh^2}{a^2b^2}r \propto r$. Proved.

(20): On using the remark (i) of question (2), we have $\ddot{\vec{r}} = -\frac{F(r)}{m}\hat{r}$ where $F(r) = \frac{mh^2}{p^3}\frac{dp}{dr} = \frac{mh^2}{p^3}\frac{d}{dr}(r\sin\alpha) = (mh^2\sin\alpha)\frac{1}{p^3} = \frac{mh^2\sin\alpha}{(r\sin\alpha)^3} = \left(\frac{mh^2}{\sin^2\alpha}\right)\frac{1}{r^3}$. This proves the first part. Here $v = \frac{h}{p} = \frac{h}{r\sin\alpha} = \left(\frac{h}{\sin\alpha}\right)\frac{1}{r} \propto \frac{1}{r}$. Proved.

(21): We have seen that $v = \left(\frac{h}{\sin\alpha}\right)\frac{1}{r}$ so, equation of hodograph is

$$R = \left(\frac{h}{\sin\alpha}\right)\frac{1}{r} \quad \cdots(1)$$

where $r\sin\varphi = p = r\sin\alpha$ or $\varphi = \alpha$. Hence $\frac{rd\theta}{dr} = \tan\varphi = \tan\alpha$ or, $\frac{dr}{r} = (\cot\alpha)d\theta$. On integration, $\ln r = (\cot\alpha)\theta + \ln c$ or $\ln\frac{r}{c} = (\cot\alpha)\theta$ or,

$$r = ce^{(\cot\alpha)\theta} \quad \cdots(2)$$

where c is a constant. This is the polar equation of equiangular spiral. Using (1), polar equation of hodograph is $\left(\frac{h}{\sin\alpha}\right)\frac{1}{R} = ce^{(\cot\alpha)\theta}$ or, $R = \left(\frac{h}{c\sin\alpha}\right)e^{(-\cot\alpha)\theta}$

$$\text{or } R = Ae^{\cot(\pi-\alpha)\theta} \quad \cdots(3)$$

where $A \equiv \frac{h}{c \sin\alpha}$, a constant. Here (3) is of the form (2), so (3) is also an equiangular spiral. Proved.

(22): Let O be the centre of force, lying on the cicumference of a circle with diameter d. Let us take the diameter through O as the initial line. Let $P(r, \theta)$ be any point on the circle. Hence $r \sec\theta = d$ or, $\cos\theta = \frac{r}{d}$ or, $\theta = \cos^{-1}\frac{r}{d}$.

Since $p = r\sin\varphi$, so $\frac{p}{\sqrt{r^2-p^2}} = \tan\varphi = r\frac{d\theta}{dr} = r\frac{d}{dr}\left(\cos^{-1}\frac{r}{d}\right) = r\left(\frac{-1}{\sqrt{1-\left(\frac{r}{d}\right)^2}}\frac{1}{d}\right) =$

$\frac{-r}{\sqrt{d^2-r^2}}$ or, $p^2(d^2-r^2) = r^2(r^2-p^2)$ or, $p^2d^2 = r^4$ or, $\frac{1}{p^2} = \frac{d^2}{r^4}$. On using the remark (i) of question (2), we have $\ddot{\vec{r}} = -\frac{F(r)}{m}\hat{r}$ where $F(r) = \frac{mh^2}{p^3}\frac{dp}{dr} =$

$\frac{mh^2}{-2}\frac{d}{dr}\left(\frac{d^2}{r^4}\right) = -\frac{mh^2}{2}d^2(-4)\frac{1}{r^5} = (2mh^2d^2)\frac{1}{r^5} \propto \frac{1}{r^5}$. This proves the first part.

Also $speed = v = \frac{h}{p} = \frac{hd}{r^2} \propto \frac{1}{r^2}$. Proved.

(23): From $v = \frac{hd}{r^2}$, equation of the hodograph is $R = \frac{hd}{r^2} = \frac{hd}{(d\cos\theta)^2} =$

$$\frac{2h}{d(1+\cos 2\theta)} \text{ or, } \frac{\left(\frac{2h}{d}\right)}{R} = 1+\cos 2\theta \text{ or, } \frac{\left(\frac{2h}{d}\right)}{R} = 1+\cos\theta' \quad \cdots(1)$$

where $\theta' \equiv 2\theta$. Here (1) represents a parabola (see remark (iv) of question (2)). Proved.

(24): Let the equation of the parabolic path be $\frac{2a}{r} = 1-\cos\theta \quad \cdots(1)$

On taking natural log, we get $\ln 2a - \ln r = \ln(1-\cos\theta)$. On differentiation w.r.to θ we get $0 - \frac{1}{r}\frac{dr}{d\theta} = \left(\frac{1}{1-\cos\theta}\right)\sin\theta$ or, $\tan\frac{\varphi}{2} = r\frac{d\theta}{dr} = -\frac{1-\cos\theta}{\sin\theta} =$

$-\frac{2\sin^2\frac{\theta}{2}}{2\sin\frac{\theta}{2}\cos\frac{\theta}{2}} = -\tan\frac{\theta}{2} = \tan\left(\pi-\frac{\theta}{2}\right)$ or, $\boxed{\varphi = \pi - \frac{\theta}{2}}$. So, $p = r\sin\varphi =$

$r\sin\left(\pi-\frac{\theta}{2}\right) = r\sin\frac{\theta}{2} = r\sqrt{\frac{1-\cos\theta}{2}} = r\sqrt{\frac{a}{r}}$, from (1). So, $\boxed{p = \sqrt{ar}}$ is the pedal equation of the parabola. Here equation of motion is given by $\ddot{\vec{r}} = -F(r)\hat{r}$ where $F(r) = \frac{h^2}{p^3}\frac{dp}{dr}$ (see remark (i) of question (2)). So, $F(r) = \frac{h^2}{(\sqrt{ar})^3}\frac{d}{dr}\sqrt{ar} = \frac{h^2}{ar\sqrt{ar}}\sqrt{a}\frac{1}{2\sqrt{r}} = \left(\frac{h^2}{2a}\right)\frac{1}{r^2} = \frac{\mu}{r^2}$ where $\boxed{\mu \equiv \frac{h^2}{2a}}$, a positive constant. So, inverse square law is valid here. After T duration, inverse square law is again valid (according to the question). So the new path is a conic. Let the "new" equation of motion be $\ddot{\vec{R}} = -\frac{\mu}{R^2}\hat{R}$. But we know that $\ddot{\vec{R}} = (\ddot{R} - R\dot{\Theta}^2)\hat{R} + \left(\frac{1}{R}\frac{d}{dt}(R^2\dot{\Theta})\right)\hat{\Theta}$, so, $\frac{1}{R}\frac{d}{dt}(R^2\dot{\Theta}) = 0$ and hence $R^2\dot{\Theta} = \text{constant, say } h'$. Put $\boxed{\vec{h'} \equiv h'\hat{k}}$. So, $\boxed{\vec{h'} = R^2\dot{\Theta}\hat{k}}$. Further $\vec{R}\times\vec{V} = \vec{R}\times(\dot{R}\hat{R} + R\dot{\Theta}\hat{\Theta}) = R\hat{R}\times(\dot{R}\hat{R} + R\dot{\Theta}\hat{\Theta}) = \vec{0} + R^2\dot{\Theta}\hat{R}\times\hat{\Theta} = R^2\dot{\Theta}\hat{k} = \vec{h'}$. Hence $\boxed{\vec{R}\times\vec{V} = \vec{h'}}$, where $\vec{V} \equiv \dot{\vec{R}}$. Now $\frac{d}{dt}(\vec{h'}\times\dot{\vec{R}}) = \vec{h'}\times\ddot{\vec{R}} = \vec{h'}\times\left(-\frac{\mu}{R^2}\hat{R}\right) = \frac{\mu}{R^2}(\hat{R}\times\vec{h'}) = \frac{\mu}{R^2}\hat{R}\times(R^2\dot{\Theta}\vec{k}) = \mu\dot{\Theta}(-\hat{\Theta}) = -\mu\left(\frac{d}{dt}\hat{R}\right) = \frac{d}{dt}(-\mu\hat{R})$. On integration, $\vec{h'}\times\vec{V} = \vec{h'}\times\dot{\vec{R}} = -\mu\hat{R} + \mu e\hat{e}$ $\cdots(2)$

where $\mu e\hat{e}$ is a constant vector. On taking dot product with $\hat{R}$ we get $-\frac{h'^2}{R} = -(\vec{R}\times\vec{V})\cdot\frac{\vec{h'}}{R} = (\vec{h'}\times\vec{V})\cdot\frac{\vec{R}}{R} = -\mu + \mu e\cos\theta$ where θ is the angle which SP makes with the fixed direction of $\hat{e}$. Hence, $\frac{(h'^2/\mu)}{R} = 1 - e\cos\theta$, which is an equation of conic with pole as focus and e is the eccentricity. Further, semi –latus rectum l is given by $\boxed{l = (h'^2/\mu)}$. . From (2), $e^2 = (\mu\hat{R} + \vec{h'}\times\vec{V})^2 = \mu^2 + 2\mu\hat{R}\cdot(\vec{h'}\times\vec{V}) + (\vec{h'}\times\vec{V})^2 = \mu^2 - 2\mu\vec{h'}\cdot\left(\frac{1}{R}\vec{R}\times\vec{V}\right) + (h'V\sin 90^\circ)^2$ or, $\mu^2e^2 = \mu^2 - 2\mu\frac{h'^2}{R} + h'^2V^2$. Hence $e^2 = 1 - \frac{2h'^2}{\mu R} + \frac{h'^2}{\mu}\left(\frac{V^2}{\mu}\right) = 1 - 2(l)\left(\frac{1}{R}\right) + (l)\left(\frac{V^2}{\mu}\right) = 1 - 2\frac{l}{R_i} + (l)\left(\frac{v^2}{\mu}\right)$ where $R_i \equiv R(t=0)$ and $v \equiv \frac{h}{p} = \frac{h}{\sqrt{ar}} = \frac{\sqrt{2a\mu}}{\sqrt{ar}} = \sqrt{\frac{2\mu}{r}}$ $\cdots(A)$. So, $e^2 = 1 - \frac{2l}{R_i} + (l)\left(\frac{2}{r}\right)$. New path is an ellipse means $e^2 < 1$, that is, $1 - \frac{2l}{R_i} + \frac{2l}{r} < 1$ or, $\frac{2l}{r} < \frac{2l}{R_i}$ or, $\boxed{R_i < r}$.

Hence, for ellipse, $(\text{maximum } T) = \frac{2r\cos\frac{\theta}{2}}{v} = 2r\sqrt{\frac{1+\cos\theta}{2}} \Big/ \sqrt{\frac{2\mu}{r}}$, from (*A*).

Then $(\max T \text{ for ellipse}) = r\sqrt{\frac{r}{\mu}}\sqrt{1+\left(1-\frac{2a}{r}\right)}$, from (1). Thus

$(\max T \text{ for ellipse}) = 2r\sqrt{\frac{1}{2\mu}(r-a)}$. <u>Proved</u>

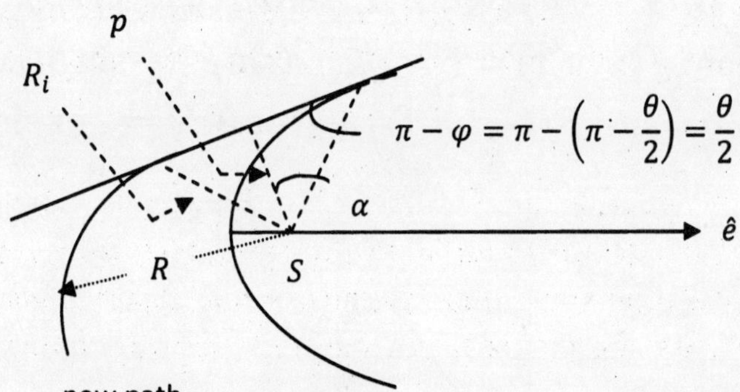

(25): Let equation of helix be $(x, y, z) \equiv \vec{r} = a(\cos\theta, \sin\theta, \theta\cot\beta)$. So,

$\hat{t} = \vec{r'} = a\theta'(-\sin\theta, \cos\theta, \cot\beta)$ where $\theta' \equiv \frac{d\theta}{ds}$. Now $1 = |\hat{t}| =$

$a\theta'\sqrt{(-\sin\theta)^2 + (\cos\theta)^2 + (\cot\beta)^2} = a\theta' \csc\beta$ and hence $\boxed{\frac{d\theta}{ds} = \theta' = \frac{\sin\beta}{a}}$.

Hence $\boxed{\hat{t} = (\sin\beta)(-\sin\theta, \cos\theta, \cot\beta)}$. So,

$\kappa\hat{n} = \frac{d\hat{t}}{ds} = (\sin\beta)\theta'(-\cos\theta, -\sin\theta, 0) = \frac{\sin^2\beta}{a}(-\cos\theta, -\sin\theta, 0)$. Hence

$\boxed{\kappa = \frac{\sin^2\beta}{a}}$ and $\boxed{\hat{n} = (-\cos\theta, -\sin\theta, 0)}$. It follows that normal is horizontal.

Also $\hat{t}\cdot\hat{k} = (\sin\beta)(0 + 0 + \cot\beta) = \cos\beta$. So, tangent vector makes a constant angle β with the initial line. Now equation of motion of the smooth bead is $\ddot{s}\hat{t} + \kappa(\dot{s})^2\hat{n} = \ddot{\vec{r}} = \frac{1}{m}\left(mg(-\hat{k}) + N_1\hat{n} + N_2\hat{b}\right)$ where $\hat{b} \equiv \hat{t}\times\hat{n}$. On taking dot product with $\hat{t}$, we get $\boxed{\ddot{s} = -g\hat{k}\cdot\hat{t} = -g\cos\beta}$, a constant.

So required duration $= \sqrt{\frac{2s}{(g\cos\beta)}}$ where $z = a\theta\cot\beta$. Hence $\frac{dz}{ds} = (a\cot\beta)\theta'$

$= (a\cot\beta)\frac{\sin\beta}{a} = \cos\beta$ and therefore $\boxed{s = \frac{1}{\cos\beta}z}$. So, required duration

$= \sqrt{\frac{2\left(\frac{1}{\cos\beta}z\right)}{g\cos\beta}} = \sec\beta\sqrt{\frac{2z}{g}}$. Proved.

(26): Here $|\vec{v}|$ or $|v| = \sqrt{2(g\cos\beta)s} = \sqrt{2(g\cos\beta)\left(\frac{1}{\cos\beta}z\right)} = \sqrt{2gz}$ and $\vec{v}$ is along $\hat{t}$, so $\vec{v} = \sqrt{2gz}\hat{t} = \sqrt{2gz}(\sin\beta)(-\sin\theta, \cos\theta, \cot\beta)$. Hence required equation of hodograph is $(x, y, z) = \vec{r} = \sqrt{2gz}(\sin\beta)(-\sin\theta, \cos\theta, \cot\beta)$. Hence $x^2 + y^2 = (2gz)\sin^2\beta\,((-\sin\theta)^2 + (\cos\theta)^2) = 2gz\sin^2\beta$ and

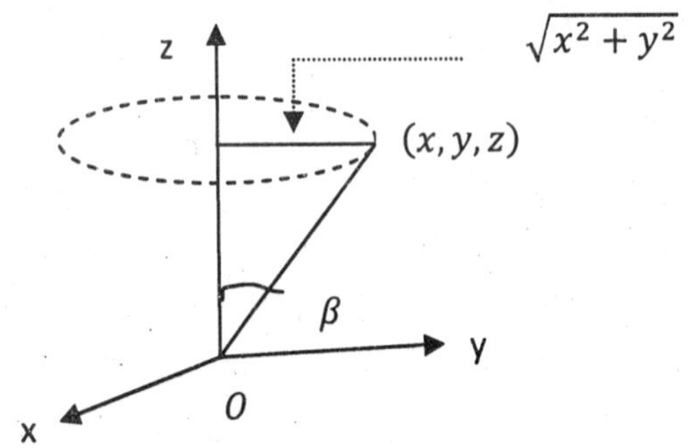

$z^2 = (2gz\sin^2\beta)\cot^2\beta$. Hence $x^2 + y^2 = (\tan^2\beta)z^2$ which is a cone with semi-vertical angle β. Proved.

(27): Here equation of motion is $\ddot{s}\hat{t} + \kappa(\dot{s})^2\hat{n} = \ddot{\vec{r}} = -\mu\overrightarrow{OP} + N_1\hat{n} + N_2\hat{b} = -\mu a(\cos\theta, \sin\theta, \theta\cot\beta) + N_1\hat{n} + N_2\hat{b}$ $\cdots(1)$

On taking dot product with $\hat{n}$, we get $\frac{\sin^2\beta}{a}(\dot{s})^2 = -\mu a(\cos\theta, \sin\theta, \theta\cot\beta)\cdot(-\cos\theta, -\sin\theta\,'0) + \frac{N_1}{m} = -\mu a(-\cos^2\theta - \sin^2\theta + 0) + \frac{N_1}{m} = \mu a + \frac{N_1}{m}$ or

$(\dot{s})^2 = \frac{a}{\sin^2\beta}\left(\mu a + \frac{N_1}{m}\right)$ $\cdots(2)$

Further, on taking dot product with $\hat{b}$ in (1), we get $0 = N_2$. Thus, if action vanishes, that is, $\sqrt{{N_1}^2 + {N_2}^2} = 0$ then $0 = N_1$, and hence, from (2), speed $=$

$\dot{s} = \sqrt{\frac{a}{\sin^2\beta}(\mu a + 0)} = a\sqrt{\mu}\csc\beta$. Proved.

(28):

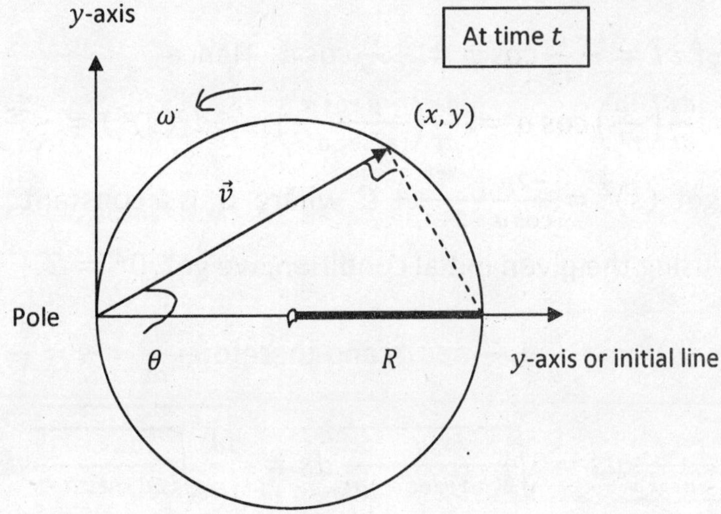

Here $v = 2R\cos\theta$, so $\dot{x} = v\cos\theta = (2R\cos\theta)\cos\theta = R(2\cos^2\theta) = R(1+\cos 2\theta)$ and $\dot{y} = v\sin\theta = (2R\cos\theta)\sin\theta = R(\sin 2\theta)$. Here $\frac{d\theta}{dt} \equiv \omega$, a constant (given), so $dx = \dot{x}\left(\frac{d\theta}{\omega}\right) = \frac{1}{\omega}R(1+\cos 2\theta)d\theta$ and $dy = \dot{y}\left(\frac{d\theta}{\omega}\right) = \frac{1}{\omega}R(\sin 2\theta)d\theta$. On integration, we get $x = \frac{R}{\omega}\left(\theta + \frac{\sin 2\theta}{2}\right) + C_1$ and $y = \frac{R}{\omega}\left(\frac{-\cos 2\theta}{2}\right) + C_2$. On using the initial conditions that $x(t=0)$ and $y(t=0)$ we get $C_1 = 0$ and $C_2 = \frac{R}{2\omega}$. Hence $x = \left(\frac{R}{2\omega}\right)(2\theta + \sin 2\theta)$ and $y = \left(\frac{R}{2\omega}\right)(1-\cos 2\theta)$. If we put $\theta' \equiv 2\theta$, then equation of path is $x = \left(\frac{R}{2\omega}\right)(\theta' + \sin\theta')$ and $y = \left(\frac{R}{2\omega}\right)(1-\cos\theta')$ which represents a cycloid of radius $\frac{R}{2\omega}$. <u>Proved</u>.

(29): Let equation of the equiangular spiral of angle α be $r = ae^{(\cot\alpha)\theta}$. On taking natural logarithm, we get $\ln r = \ln a + (\cot\alpha)\theta$. Hence $\boxed{\frac{1}{r}dr = (\cot\alpha)d\theta}$ that is, $\tan\varphi = r\frac{d\theta}{dr} = \tan\alpha$. So, $\boxed{\varphi = \alpha}$. Now equation of motion of the particle can be written as: $\ddot{s}\hat{t} + (\dot{s})^2\kappa\hat{n} = -\frac{\mu}{r^2}\hat{r} + N_1\hat{n} + N_2\hat{b}$ $\quad\cdots(1)$

On taking dot product with $\hat{t}$

we get $\ddot{s} = -\frac{\mu}{r^2}\hat{r}\cdot\hat{t} = -\frac{\mu}{r^2}\cos\varphi = -\frac{\mu}{r^2}\cos\alpha$. Hence $\frac{d}{dt}(\dot{s})^2 = 2\dot{s}\ddot{s} = 2\frac{ds}{dt}\left(\frac{-\mu}{r^2}\right)\cos\alpha = 2\frac{ds}{dt}\left(\frac{-\mu\cos\alpha}{(s/\sec\alpha)^2}\right)$ or $d((\dot{s})^2) = \frac{-2\mu}{\cos\alpha}s^{-2}\,ds$. On integration, we get $(\dot{s})^2 = \frac{-2\mu}{\cos\alpha}\frac{s^{-2+1}}{-2+1} + C$ where C is a constant. So, $(\dot{s})^2 = C + \frac{2\mu\sec\alpha}{s}$. On using the given initial condition, we get $0^2 = C + \frac{2\mu\sec\alpha}{(\sec\alpha)(2d)}$. So, $C = -\frac{\mu}{d}$. Hence $(\dot{s})^2 = -\frac{\mu}{d} + \frac{2\mu}{s}\sec\alpha$ and therefore $\frac{ds}{dt} = \dot{s} = -\sqrt{-\frac{\mu}{d} + \frac{2\mu\sec\alpha}{s}}$.

So, $-dt = \frac{1}{\sqrt{-\frac{\mu}{d}+\frac{2\mu\sec\alpha}{s}}}ds = \sqrt{\frac{sd}{2(\mu d)\sec\alpha - \mu s}}\,ds = \sqrt{\frac{d}{\mu}}\sqrt{\frac{s}{(2d)\sec\alpha - s}}\,ds$. On putting $s = (2d)(\sec\alpha)\cos^2\chi$, we get $-dt = \sqrt{\frac{d}{\mu}}(2d)(\sec\alpha)(2\cos\chi)(-\sin\chi)d\chi$ or $dt = \sqrt{\frac{d}{\mu}}(4d)\sec\alpha\cos^2\chi\,d\chi = \sqrt{\frac{d}{\mu}}(2d)(\sec\alpha)(1+\cos 2\chi)d\chi$. So, required duration $= \int_{s=(2d)\sec\alpha}^{s=0}\sqrt{\frac{d}{\mu}}(2d)(\sec\alpha)(1+\cos 2\chi)d\chi = \sqrt{\frac{d}{\mu}}(2d)(\sec\alpha)\left[\left(\frac{\pi}{2} + 0\right) - (0+0)\right] = \pi\sqrt{\frac{d^3}{\mu}}\sec\alpha$. Proved.

(30):

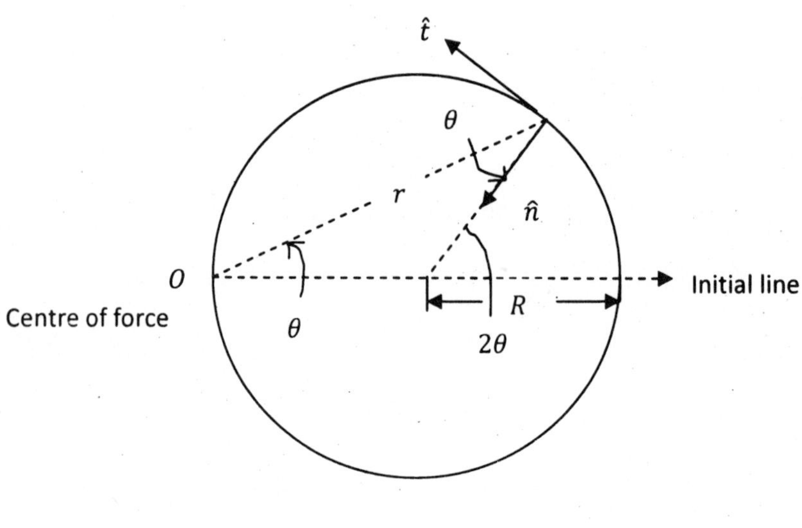

Here equation of motion is $\ddot{s}\hat{t}+\frac{(\dot{s})^2}{R}\hat{n}=\ddot{\vec{r}}=\left(\frac{\mu}{r^5}\sin\left(\frac{1}{2}\frac{3}{R}\right)\right)\hat{t}+\left(\frac{\mu}{r^5}\cos\left(\frac{1}{2}\frac{s}{R}\right)\right)\hat{n}\pm$ $N_1\hat{n}$. Hence $\ddot{s}=\frac{\mu\left(\sin\frac{s}{2R}\right)}{\left(2R\cos\frac{S}{2R}\right)^5}$ ... (1)

and $\frac{(\dot{s})^2}{R}=\frac{\mu\cos\frac{s}{2R}}{\left(2R\cos\frac{s}{2R}\right)^5}\pm N_1$ ... (2)

From (1), $\frac{d}{dt}(\dot{s})^2=2\dot{s}\ddot{s}=2\frac{ds}{dt}\frac{\mu\left(\sin\frac{s}{2R}\right)}{\left(2R\cos\frac{s}{2R}\right)^5}$ or, $d(\dot{s})^2=-2\mu\left(2R\cos\frac{s}{2R}\right)^{-5}$ $d\left(2R\cos\frac{s}{2R}\right)$. On integration, we get $(\dot{s})^2=-2\mu\frac{\left(2R\cos\frac{s}{2R}\right)^{-4}}{-4}+C$, so, from (2), $R\left(\frac{\mu\cos\frac{s}{2R}}{\left(2R\cos\frac{s}{2R}\right)^5}\pm N_1\right)=\frac{\mu}{2}\left(2R\cos\frac{s}{2R}\right)^{-4}+C$ or, $\frac{R\mu\cos\frac{s}{2R}}{2R\cos\frac{s}{2R}}\left(2R\cos\frac{s}{2R}\right)^{-4}\pm RN_1=$ $\frac{\mu}{2}\left(2R\cos\frac{s}{2R}\right)^{-4}+C\Rightarrow N_1=\frac{C}{\pm R}$, a constant. <u>Proved.</u>

7

DYNAMICS OF A SYSTEM OF PARTICLES AND OF A RIGID BODY

Example (page 135): This is solved in the text.

Example (page 137): (the $A.M.$ of the system about O') $= \sum \overrightarrow{O'P} \times m\frac{d}{dt}(\overrightarrow{OP}) = \sum \overrightarrow{O'P} \times m\frac{d}{dt}\left(\overrightarrow{OO'} + \overrightarrow{O'P}\right) = \sum \overrightarrow{O'P} \times m\left(\dot{\overrightarrow{OO'}} + \dot{\overrightarrow{O'P}}\right) = \sum\left(\overrightarrow{O'P} \times m\dot{\overrightarrow{OO'}} + \overrightarrow{O'P} \times m\dot{\overrightarrow{O'P}}\right) = \left(\sum m\overrightarrow{O'P}\right) \times \dot{\overrightarrow{OO'}} + \sum \overrightarrow{O'P} \times \left(m\dot{\overrightarrow{O'P}}\right) = \left(M\overrightarrow{O'G}\right) \times \dot{\overrightarrow{OO'}} + \sum \overrightarrow{O'P} \times \left(m\dot{\overrightarrow{O'P}}\right) =$ ($A.M.$ about O' of a particle of mass M at G moving with the velocity of O') + ($A.M.$ about O' of the motion relative to O'). Proved.

Example (page 140): This is solved in the text.

Examples (page 151): (1): This is solved in the text.

(2): This is solved in the text.

(3): Our Euler's dynamical equations become $A\dot{\omega}_1 - (A - C)\omega_2\omega_3 = 0 \quad \cdots(1)$ and $A\dot{\omega}_2 - (C - A)\omega_3\omega_1 = 0 \quad \cdots(2)$

Now $\omega_1(1) + \omega_2(2)$ gives $A(\omega_1\dot{\omega}_1 + \omega_2\dot{\omega}_2) = 0$ or $\frac{d}{dt}({\omega_1}^2 + {\omega_2}^2) = 0,$ that is, ${\omega_1}^2 + {\omega_2}^2 =$ constant.

Hence the vector $\omega_1\hat{\imath} + \omega_2\hat{\jmath}$ revolves in the xy-plane and in a circular path with centre at the origin. Now, let its angular velocities be $\beta\hat{k}$. It remains to prove that $\beta = \frac{(C-A)\omega_3}{A}$. Here $(-\omega_2\beta, \omega_1\beta, 0) = (0,0,\beta) \times (\omega_1, \omega_2, 0) = (\dot{\omega}_1, \dot{\omega}_2, 0) = \left(\frac{A-C}{A}\omega_2\omega_3, \frac{C-A}{A}\omega_3\omega_1, 0\right)$. So, $-\omega_2\beta = \frac{A-C}{A}\omega_2\omega_3$ or, $\beta = \frac{C-A}{A}\omega_3$. Proved.

Examples (page 153): **(1):** Here $Av^2 \cot\theta = Cv\omega_3 - Mgh\sin\theta$, where $v = \omega' \sin\theta$. So, $A(\sin\theta\cos\theta)(\omega')^2 - (C\omega_3\sin\theta)\omega' + Mgh\sin\theta = 0$ or, $(A\cos\theta)(\omega')^2 - (C\omega_3)\omega' + Mgh = 0$. So, $\omega' = \frac{C\omega_3 \pm \sqrt{C^2\omega_3^2 - 4(A\cos\theta)(Mgh)}}{2(A\cos\theta)}$. If $4AMgh\cos\theta \ll C^2\omega_3^2$ then one of the values of ω' is approximately $\frac{C\omega_3 + \sqrt{C^2\omega_3^2}}{2(A\cos\theta)}$ that is, $\frac{C\omega_3}{A\cos\theta}$, and hence the other value of ω' is approximately $\frac{1}{\left(\frac{C\omega_3}{A\cos\theta}\right)}$(product of two values of ω') $= \left(\frac{A\cos\theta}{C\omega_3}\right)\left(\frac{Mgh}{A\cos\theta}\right)$, that is, $\frac{Mgh}{C\omega_3}$. Proved.

(2): Since $Av^2\cot\theta = Cv\omega_3 - Mgh\sin\theta$ and g is nearly zero, then $Av^2\cot\theta \approx Cv\omega_3 - 0$ or, $Av\cot\theta \approx C\omega_3$ or, $\tan\theta \approx \frac{Av}{C\omega_3}$. Proved.

Exercises on Chapter 7

(1):

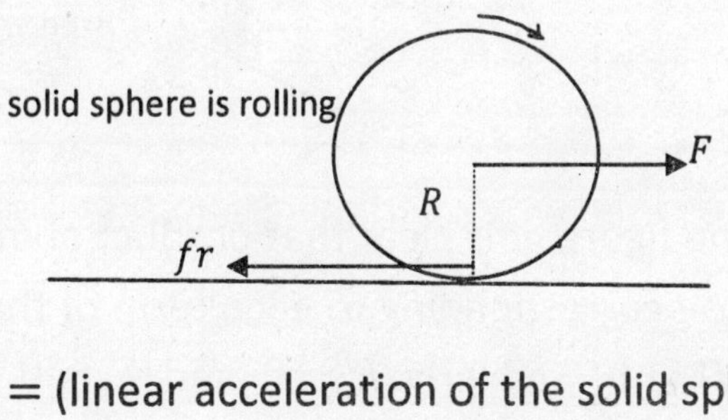

Here, we have $\left(\frac{R(fr)}{I}\right)R =$ (linear acceleration of the solid sphere)

$$= \frac{F - fr}{M} \qquad \cdots (1)$$

where symbols have their usual meanings. Further, we know that

$$I = \frac{2}{5}MR^2 \qquad \cdots (2)$$

From (1) and (2), $fr = \frac{2}{7}F$. So, required linear acceleration $= \frac{1}{M}(F - fr)$ $= \frac{1}{M}\left(F - \frac{2}{7}F\right) = \frac{\frac{5}{7}F}{M} =$ (acceleration of the sphere, where all the forces were reduced to $\frac{5}{7}$th of the former value). Proved.

(2): In order to fix the edge AB (see the figure), external impulses $\vec{J_1}$, $\vec{J_2}$ should be applied at A and B. Since these impulses have no angular momentum about AB, we have $(A.M.)_{\text{before}} \cdot \frac{\overrightarrow{AB}}{AB} + 0 = (A.M.)_{\text{after}} \cdot \frac{\overrightarrow{AB}}{AB}$.

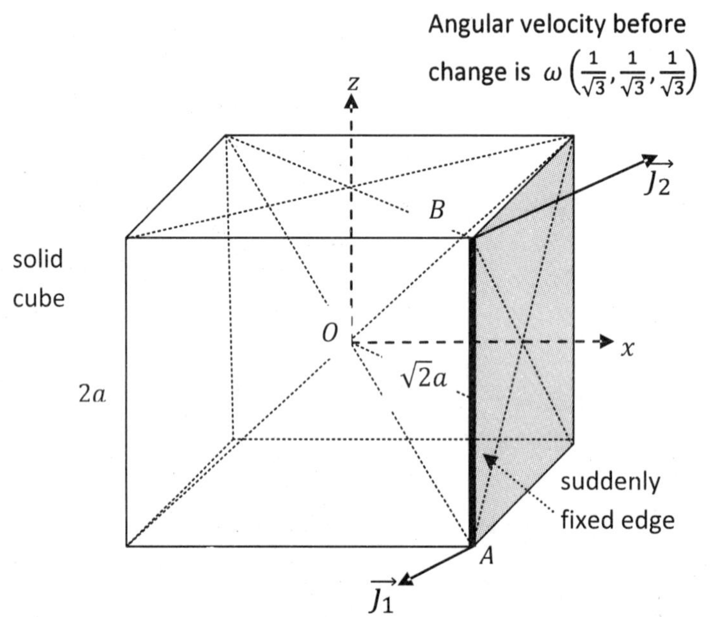

From symmetry of the figure, it is clear that product of inertia relative to y and z-axes is zero, because corresponding to each term of the type myz there is a $m(-y)z$ term. So, $D = 0$. Similarly $E = 0$ and $F = 0$. From symmetry of the figure, $A = B = C$. So, x, y, z-axes, shown in the figure, are the principal axes of inertia of the solid cube. Further, (moment of inertia of the cube about diagonal) $= A\left(\frac{1}{\sqrt{3}}\right)^2 + B\left(\frac{1}{\sqrt{3}}\right)^2 + C\left(\frac{1}{\sqrt{3}}\right)^2 = (A + B + C)\frac{1}{3} = (A + A + A)\frac{1}{3} = A$. The figures suggest that $C \equiv$ (M.I. of the solid cube about the z-axis) $= \frac{2}{3}Ma^2$. Since $A = C$, we get (M.I. of the solid cube about the diagonal) $= \frac{2}{3}Ma^2$. So, (angular momentum of the cube just before sudden change)

$= \frac{2}{3}Ma^2\left(\omega\left(\frac{1}{\sqrt{3}},\frac{1}{\sqrt{3}},\frac{1}{\sqrt{3}}\right)\right) = \frac{2Ma^2\omega}{3\sqrt{3}}(1,1,1)$. The following figures suggest how moment of inertia has been calculated

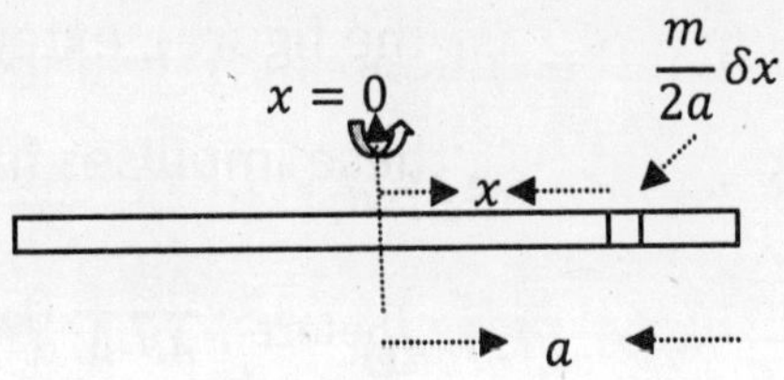

$$M.I. = 2\int_{x=0}^{x=a}\left(\frac{m}{2a}\delta x\right)x^2 = \frac{m}{a}\left[\frac{x^3}{3}\right]_0^a = \frac{1}{3}ma^2.$$

$$M.I. = \sum\left(\frac{1}{3}ma^2\right) = \left(\frac{1}{3}a^2\right)\sum m = \frac{1}{3}Ma^2$$

square plate

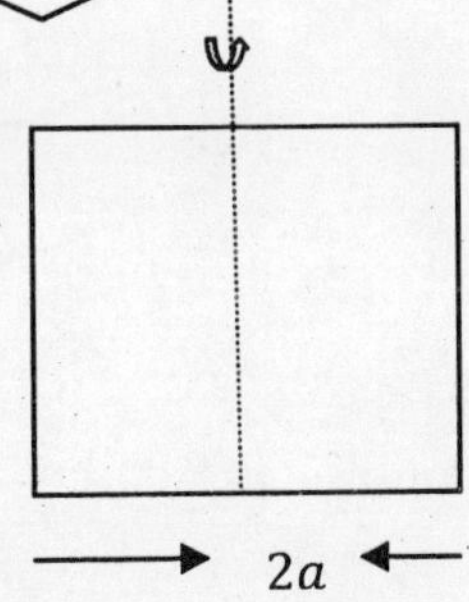

$$M.I. = \frac{1}{3}Ma^2 + \frac{1}{3}Ma^2 \text{ (using perpendicular axis theorem)}$$

So, $M.I. = \frac{2}{3}Ma^2$.

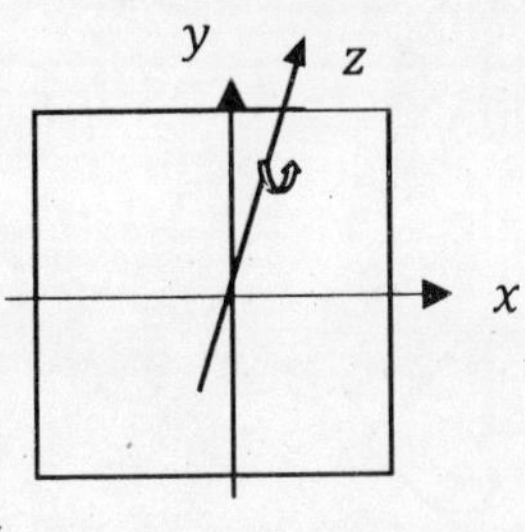

Similarly, for the solid cube $C = \sum\left(\frac{2}{3}ma^2\right) = \left(\frac{2}{3}a^2\right)\sum m = \frac{2}{3}Ma^2$. Further, by the parallel axis theorem, moment of inertia of a squared-shaped plate of side $2a$, about one edge, is $\frac{1}{3}Ma^2 + Ma^2 = \frac{4}{3}Ma^2$. Hence, by the perpendicular axis theorem, moment of inertia of a squared-shaped plate of side $2a$, about a line passing through one vertex and perpendicular to the plane of paper, is $\frac{4}{3}Ma^2 + \frac{4}{3}Ma^2 = \frac{8}{3}Ma^2$.

Hence, moment of inertia of a solid cube of side $2a$, about one edge is

$$\sum\frac{8}{3}ma^2 = \frac{8}{3}a^2\sum m = \frac{8}{3}a^2M = \frac{8}{3}Ma^2.$$

Also (angular momentum of the cube just after sudden change) $= \frac{8}{3}Ma^2 (\omega^*(0,0,1))$. Since edge AB is suddenly fixed, so, $\frac{2Ma^2\omega}{3\sqrt{3}} = \frac{2Ma^2\omega}{3\sqrt{3}}(1,1,1)\cdot(0,0,1) = (A.M.)_{\text{before}}\cdot\frac{\overrightarrow{AB}}{AB} = (A.M.)_{\text{after}}\cdot\frac{\overrightarrow{AB}}{AB} = \frac{8}{3}Ma^2\omega^*(0,0,1)\cdot(0,0,1) = \frac{8}{3}Ma^2\omega^*$. Hence $\omega^* = \frac{2Ma^2\omega}{3\sqrt{3}}\frac{3}{8Ma^2} = \frac{1}{4\sqrt{3}}\omega = \frac{\sqrt{3}}{12}\omega$. Proved.

(3):

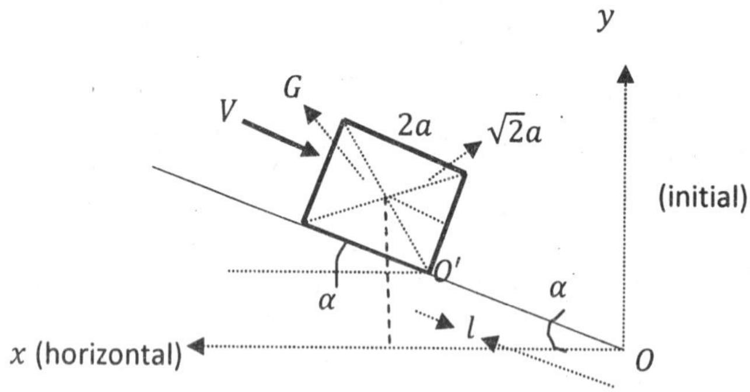

Let $OO' = l$. In our figure, O' is the small fixed nail about which inelastic solid cube tumbles. Let us take a point O of the inclined plane as the origin(see the figure). Let $OO' = l$. Here $O'(l\cos\alpha, l\sin\alpha)$ and $G((l\cos\alpha, l\sin\alpha) + (\sqrt{2}a)\left(\cos\left(\frac{\pi}{4}+\alpha\right), \sin\left(\frac{\pi}{4}+\alpha\right)\right)$ or $G(l\cos\alpha + a\cos\alpha - a\sin\alpha, l\sin\alpha + a\cos\alpha + a\sin\alpha)$.

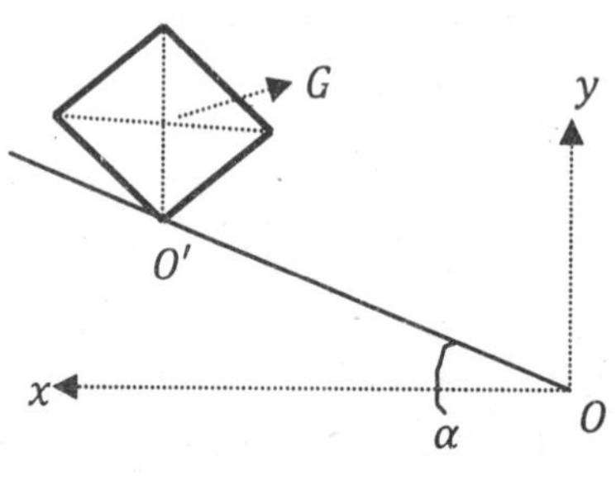

For minimum value of V, $(K.E.)_f = 0$. Just before tumbling, (angular velocity ω_i of the cube) is given by $\omega_i = \frac{(MV_{\min})(a)}{(M.I.)} = \frac{MV_{\min}\,a}{\frac{8}{3}Ma^2}$ (see the previous Ex.2). So, $\omega_i = \frac{3}{8}\frac{V_{\min}}{a}$. On using the work-energy theorem, we get

$0-\left\{\frac{1}{2}\left(\frac{8}{3}Ma^2\right)(\omega_i)^2\right\}=(K.E.)_f-(K.E.)_i=$ (work done by all forces) $=$ $-Mg\{(l\cos\alpha,\sqrt{2}a+l\sin\alpha)\cdot(0,1)-\ (l\cos\alpha+a\cos\alpha-a\sin\alpha,l\sin\alpha+a\cos\alpha+a\sin\alpha)\cdot(0,1)\}=-Mg\left(\sqrt{2}a+l\sin\alpha-(l\sin\alpha+a\cos\alpha+a\sin\alpha)\right)=-Mga(\sqrt{2}-\cos\alpha-\sin\alpha)$. Hence $-\frac{1}{2}\left(\frac{8}{3}Ma^2\right)\left(\frac{3}{8a}V_{\min}\right)^2=-Mga(\sqrt{2}-\cos\alpha-\sin\alpha)$ or, $(V_{\min})^2=\frac{16}{3}ga(\sqrt{2}-\cos\alpha-\sin\alpha)$. Proved.

(4):

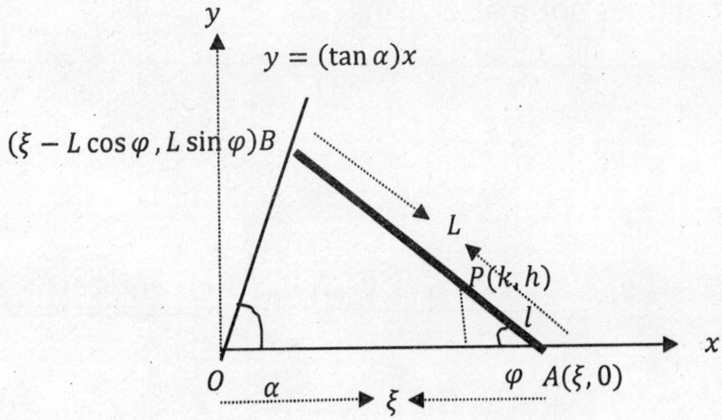

Let the two arbitrary lines be $y=0$ and $y=(\tan\alpha)x$, which intersect at the origin. Let L be the length of the rigid rod AB. Let one extremity of the rod be $A(\xi,0)$. Let φ be the angle which the rod makes with the first line. So, other extremity of the rod will be at $B(\xi-L\cos\varphi,L\sin\varphi)$ (see the figure). Let $P(k,h)$ be any point on the given rod. We have to find the direction of its motion.

From ΔOAB, $\frac{L}{\sin\alpha}=\frac{\xi}{\sin(\alpha+\varphi)}=\frac{\xi}{(\sin\alpha)\left(\frac{\xi-k}{l}\right)+(\cos\alpha)\left(\frac{h}{l}\right)}=\frac{\xi l}{(\sin\alpha)(\xi-k)+h(\cos\alpha)}$.

Hence $L(\xi-k)+Lh\cot\alpha=\xi l$. Now

$l^2=(k-\xi)^2+h^2=\left(k-\frac{Lh\cot\alpha-Lk}{l-L}\right)^2+h^2$ so, $l^2=\left(\frac{Lh\cot\alpha-kl}{l-L}\right)^2+h^2$.

Hence the locus of $P(k,h)$ is $l^2=\left(\frac{Ly\cot\alpha-lx}{l-L}\right)^2+y^2$ or,

$l^2x^2-2(lL\cot\alpha)xy+(l^2-2lL+L^2\csc^2\alpha)y^2=\{l(L-l)\}^2$, which is a central conic. Here $ab-h^2\equiv-4l^2L^2\cot^2\alpha+4l^2(l^2-2lL+L^2\csc^2\alpha)=4l^2L^2+4l^2(l^2-2lL)=4l^2(L-l)^2$, which is positive. So, the locus is an ellipse.

Here equation of path of point $P(k,h)$ in our figure is $l^2x^2 - 2(lL\cot\alpha)xy + (l^2 - 2lL + L^2\csc^2\alpha)y^2 = \{l(L-l)\}^2$. Now differentiating w. r. to x gives $l^2 2x - 2(lL\cot\alpha)(xy' + y) + (l^2 - 2lL + L^2\csc^2\alpha)2yy' = 0$ or, $y' = \frac{l(lx - L(\cot\alpha)y)}{lL(\cot\alpha)x - ((L-l)^2 + L^2\cot^2\alpha)y}$. This gives the direction of the path. If the two lines do not intersect, but are at right angle, then the motion cannot be represented by an angular velocity only. Reason: In pure rotation, path of each particle is circular one, but here paths of two extremes of the rod are straight lines, whether they intersect or not, it does not matter. Ans.

(5):

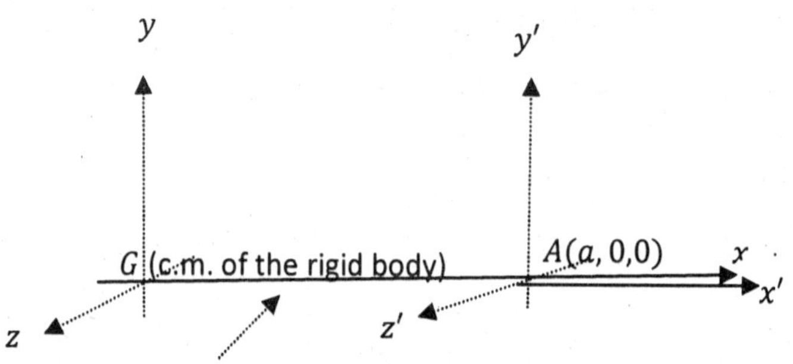

Let us choose the three principal axes through the centre of mass G to be the x-axis, y-axis and z-axis. Let us take $A(a,0,0)$ on the first principal axis. Let us draw two lines through A, one parallel to y-axis and another parallel to z-axis. It suffices to prove that (i) product of inertia relative to (Az', Ax')-axes is zero; (ii) product of inertia relative to (Ax', Ay')-axes is zero; (iii) product of inertia relative to (Ay', Az')-axes is zero. For (i): Here product of inertia relative to (Az', Ax')-axes is $\sum m(z'x') = \sum mz(x-a) = \sum(mxz - maz) = \sum mxz - a\sum mz = 0 - a(Mx_{\text{cm}}) = 0 - a(M(0)) = 0$. For (ii): product of inertia relative to (Ax', Ay')-axes is zero, as in (i). For (iii): (product of inertia relative to (Ay', Az')-axes) $= \sum my'z' = \sum myz = 0$. Proved.

(6): Here the rigid body is hinged to a fixed point O. Let us choose one principal axis through O as the z-axis. The force exerted by the hinge is acting at O, so, $(L_1, L_2, L_3) = \vec{0}$ and hence, by the Euler's dynamical equations,

$$\begin{cases} A\dot{\omega_1} = (B - C)\omega_2\omega_3 & \cdots (1) \\ B\dot{\omega_2} = (C - A)\omega_3\omega_1 & \cdots (2) \\ C\dot{\omega_3} = (A - B)\omega_1\omega_2 & \cdots (3) \end{cases}$$

where symbols have their usual meanings. Now $\omega_1(1) + \omega_2(2) + \omega_3(3)$ gives $A\omega_1\dot{\omega_1} + B\omega_2\dot{\omega_2} + C\omega_3\dot{\omega_3} = 0$. On integration, $A{\omega_1}^2 + B{\omega_2}^2 + C{\omega_3}^2 =$ constant, c_1. Given that $[\omega_1]_{t=0} = 0$, $[\omega_2]_{t=0} = 0, [\omega_3]_{t=0} = \omega_0$ that is, $(\omega_1, \omega_2, \omega_3) = (0,0, \omega_0)$ at $t = 0$. Hence $A(0)^2 + B(0)^2 + C(\omega_0)^2 = c_1$.

$\therefore\ c_1 = C{\omega_0}^2$. Hence $A{\omega_1}^2 + B{\omega_2}^2 + C{\omega_3}^2 = C{\omega_0}^2$ $\cdots (4)$

Now $A\omega_1(1) + B\omega_2(2) + C\omega_3(3)$ gives, $A^2\omega_1\dot{\omega_1} + B^2\omega_2\dot{\omega_2} + C^2\omega_3\dot{\omega_3} = 0$. On integrating, $A^2{\omega_1}^2 + B^2{\omega_2}^2 + C^2{\omega_3}^2 =$ constant, c_2. On using the initial conditions, we get $A^2{\omega_1}^2 + B^2{\omega_2}^2 + C^2{\omega_3}^2 = C^2{\omega_0}^2$ $\cdots (5)$

Here (4) can be written as $\dfrac{{\omega_1}^2}{\left(\frac{C{\omega_0}^2}{A}\right)} + \dfrac{{\omega_2}^2}{\left(\frac{C{\omega_0}^2}{B}\right)} + \dfrac{{\omega_3}^2}{{\omega_0}^2} = 1$ $\cdots (4')$

Also (5) can be written as $\dfrac{A{\omega_1}^2}{\left(\frac{C{\omega_0}^2}{A}\right)} + \dfrac{B{\omega_2}^2}{\left(\frac{C{\omega_0}^2}{B}\right)} + \dfrac{{\omega_3}^2}{{\omega_0}^2} = 1$ $\cdots (5')$

Clearly (4′) and (5′) both are satisfied by $(\omega_1, \omega_2, \omega_3) = (0,0, \omega_0)$ for all time t.

Proved.

(7): Here in place of θ of the question, we shall use α.

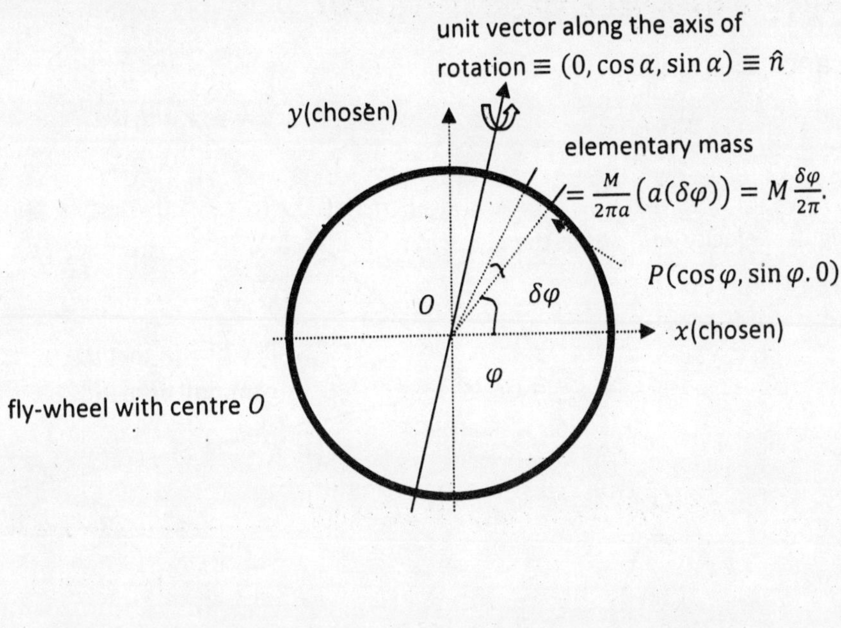

Here magnitude of force experienced by the elementary mass $= \left(M\frac{\delta\varphi}{2\pi}\right)\left|\overrightarrow{OP}\times(0,\cos\alpha,\sin\alpha)\right|\omega^2 = \left(M\frac{\delta\varphi}{2\pi}\right)|(\cos\varphi,\sin\varphi,0)\times(0,\cos\alpha,\sin\alpha)|\omega^2 =$
$\frac{M\omega^2}{2\pi}(\delta\varphi)|(\sin\alpha\sin\varphi,-\sin\alpha\cos\varphi,\cos\alpha\cos\varphi)| =$
$\left(\frac{M\omega^2}{2\pi}\delta\varphi\right)\sqrt{\sin^2\alpha+\cos^2\alpha\cos^2\varphi}$

Hence (the net torque experienced by the ring)
$= \int_{\varphi=0}^{\varphi=2\pi}\overrightarrow{OP}\times\left\{\left(\frac{M\omega^2}{2\pi}\delta\varphi\sqrt{\sin^2\alpha+\cos^2\alpha\cos^2\varphi}\right)\left\{\frac{(\overrightarrow{OP}\times\hat{n})}{|\overrightarrow{OP}\times\hat{n}|}\times\hat{n}\right\}\right\} =$
$\int_{\varphi=0}^{\varphi=2\pi}\frac{M\omega^2}{2\pi}(\delta\varphi)\frac{\sqrt{\sin^2\alpha+\cos^2\alpha\cos^2\varphi}}{|\overrightarrow{OP}\times\hat{n}|}\overrightarrow{OP}\times\left((\overrightarrow{OP}\cdot\hat{n})\hat{n}-\overrightarrow{OP}\right)$. Since $|\overrightarrow{OP}\times\hat{n}| =$
$|(\sin\alpha\sin\varphi,-\sin\alpha\cos\varphi,\cos\alpha\cos\varphi)| = \sqrt{\sin^2\alpha+\cos^2\alpha\cos^2\varphi}$, so, (the net torque experienced by the ring) $= \int_{\varphi=0}^{\varphi=2\pi}\frac{M\omega^2}{2\pi}(\delta\varphi)\left(\overrightarrow{OP}\cdot\hat{n}\right)\left(\overrightarrow{OP}\times\hat{n}\right) =$
$\int_{\varphi=0}^{\varphi=2\pi}\frac{M\omega^2}{2\pi}\delta\varphi(\cos\alpha\sin\varphi)(\sin\alpha\sin\varphi,-\sin\alpha\cos\varphi,\cos\alpha\cos\varphi) =$
$\frac{M\omega^2}{2\pi}\cos\alpha\left(\sin\alpha\int_0^{2\pi}\sin^2\varphi\,d\varphi,-\sin\alpha\int_0^{2\pi}\sin\varphi\cos\varphi\,d\varphi,\cos\alpha\int_0^{2\pi}\sin\varphi\cos\varphi\,d\varphi,\right) =$
$\frac{1}{2\pi}M\omega^2\cos\alpha\left(\frac{1}{2}\sin\alpha\int_0^{2\pi}(1-\cos2\varphi)d\varphi,-\right.$
$\left.\frac{1}{2}\sin\alpha\int_0^{2\pi}\sin2\varphi\,d\varphi,\frac{1}{2}\cos\alpha\int_0^{2\pi}\sin2\varphi\,d\varphi\right) = \frac{1}{2\pi}M\omega^2\cos\alpha$
$\left(\left(\frac{1}{2}\sin\alpha\right)2\pi,\left(-\frac{1}{2}\sin\alpha\right)(0),\left(\frac{1}{2}\cos\alpha\right)(0)\right) = \frac{1}{2}M\omega^2\sin\alpha\cos\alpha\,(1,0,0).$

- Its scalar moment is $\frac{1}{2}M\omega^2\sin\alpha\cos\alpha$ and the plane of this couple Is yz-plane. Here yz-plane contains $(0,0,1)$, the axis of the wheel and the axis of rotation along$(0,\cos\alpha,\sin\alpha)$. Proved.

(8): Let $\vec{A}\equiv\omega\hat{d}$ and $\vec{A'}\equiv\omega\hat{d}'$.

This is equivalent to velocity $v\hat{d}+\omega\hat{d}\times(\vec{a'}-\vec{a})$ at $(\vec{a'})$ and an angular velocity $\omega\hat{d}$ about the above line

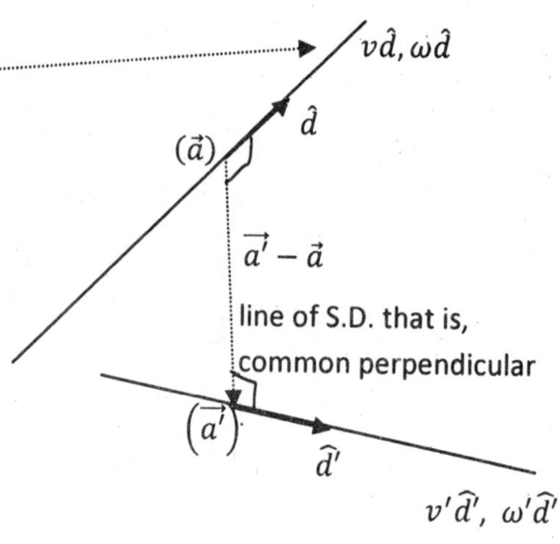

So, $\Gamma_{\text{net}} = \left(v'\hat{d}' + v\hat{d} + \omega\hat{d} \times \left(\vec{a'} - \vec{a}\right)\right) \cdot (\omega\hat{d} + \omega'\hat{d}') = \left(\vec{v'} + \vec{v} + \vec{\omega} \times \left(\vec{a'} - \vec{a}\right)\right) \cdot \left(\vec{\omega} + \vec{\omega'}\right) = \left(\vec{v'} + \vec{v}\right) \cdot \left(\vec{\omega} + \vec{\omega'}\right) + \left(\vec{\omega'} \times \vec{\omega}\right) \cdot \left(\vec{a'} - \vec{a}\right) = \left(\vec{v'} + \vec{v}\right) \cdot \left(\vec{\omega} + \vec{\omega'}\right) + \omega\omega'[\hat{d}' \ \hat{d} \ \vec{a'} - \vec{a}] = \left(\vec{v} + \vec{v'}\right) \cdot \left(\vec{\omega} + \vec{\omega'}\right) + \omega\omega' M$, where M denotes the mutual moments of the two lines.

Let us find λ such that the net velocity at the point $(1 - \lambda)\vec{a} + \lambda\vec{a'}$ is zero. Thus $v\hat{d} + \omega\hat{d} \times \left((1-\lambda)\vec{a} + \lambda\vec{a'} - \vec{a}\right) + \left\{v'\hat{d}' + \omega'\hat{d}' \times \left((1-\lambda)\vec{a} + \lambda\vec{a'} - \vec{a'}\right)\right\} = \vec{0}$ or, $\left(\vec{v'} + \vec{v}\right) + \omega\hat{d} \times (\lambda)\left(\vec{a'} - \vec{a}\right) + \omega'\hat{d}' \times (1-\lambda)\left(\vec{a} - \vec{a'}\right) = \vec{0}$ or, $\vec{v} + \vec{v'} + \left(-\lambda\vec{\omega} + (1-\lambda)\vec{\omega'}\right) \times \left(\vec{a} - \vec{a'}\right) = \vec{0}$ or,

$\underbrace{\vec{v} + \vec{v'}}_{\text{constant}} = \left\{(1-\lambda)\vec{\omega'} - \lambda\vec{\omega}\right\} \times \underbrace{\left(\vec{a} - \vec{a'}\right)}_{\text{constant}} \quad \cdots (*)$. Here $\left(\vec{v} + \vec{v'}\right) \cdot \left(\vec{a} - \vec{a'}\right) = \left(\vec{v} + \vec{v'}\right) \cdot (\text{constant})\left(\hat{d} \times \hat{d}'\right) = (\text{constant})\left(v\hat{d} + v'\hat{d}'\right) \cdot \left(\hat{d} \times \hat{d}'\right) = (\text{constant})(v(0) + v'(0)) = 0$. So, equation $(*)$ has a solution, say, $(1 - \lambda_0)\vec{\omega'} - \lambda_0\vec{\omega}$. It follows that the net velocity of the point $(1 - \lambda_0)\vec{a} + \lambda_0\vec{a'}$ lying on the common perpendicular is zero. So, this point must lie on the axis of the screw of resultant motion. So, axis of the resultant screw intersects the common perpendicular. Further,

since $\overrightarrow{\omega_{\text{net}}} \cdot \left(\hat{d} \times \hat{d}'\right) = \left(\omega\hat{d} + \omega'\hat{d}'\right) \cdot \left(\hat{d} \times \hat{d}'\right) = \omega(0) + \omega'(0) = 0$ so,

$\overrightarrow{\omega_{\text{net}}} \perp \left(\hat{d} \times \hat{d}'\right)$, that is, resultant screw motion is perpendicular to the common perpendicular. Proved.

(9): Let us take three points A, B, C with position vectors $\vec{r_1}, \vec{r_2}, \vec{r_3}$ respectfully and satisfying $\dot{\vec{r_1}} = \vec{v_0} + \vec{\omega} \times \vec{r_1}$, $\dot{\vec{r_2}} = \vec{v_0} + \vec{\omega} \times \vec{r_2}$, $\dot{\vec{r_3}} = \vec{v_0} + \vec{\omega} \times \vec{r_3}$. Let $D(\vec{r_4})$ be any point in the body satisfying $\dot{\vec{r_4}} = \vec{v_0} + \vec{\omega} \times \vec{r_4}$. It suffices to prove that D lies in the plane ABC. Here we have $\overrightarrow{BA} = \vec{\omega} \times (\vec{r_1} - \vec{r_2})$, $\overrightarrow{BC} = \vec{\omega} \times (\vec{r_3} - \vec{r_2})$ so, $\overrightarrow{BA} \times \overrightarrow{BC} = (\vec{\omega} \times (\vec{r_1} - \vec{r_2})) \times (\vec{\omega} \times (\vec{r_3} - \vec{r_2})) = \left(\left(\vec{\omega} \times (\vec{r_1} - \vec{r_2})\right) \cdot (\vec{r_3} - \vec{r_2})\right)\vec{\omega} - \left(\left(\vec{\omega} \times (\vec{r_1} - \vec{r_2})\right) \cdot \vec{\omega}\right)(\vec{r_3} - \vec{r_2}) = (\vec{\omega} \cdot (\vec{r_1} \times \vec{r_3} + \vec{r_2} \times \vec{r_1} + \vec{r_3} \times \vec{r_2}))\vec{\omega} - (0)(\vec{r_3} - \vec{r_2}) = ([\vec{\omega} \ \vec{r_1} \ \vec{r_3}] + [\vec{\omega} \ \vec{r_2} \ \vec{r_1}] + [\vec{\omega} \ \vec{r_3} \ \vec{r_2}])\vec{\omega}$.

Hence $(\overrightarrow{BA} \times \overrightarrow{BC}) \cdot \overrightarrow{BD} = ([\vec{\omega}\ \vec{r_1}\ \vec{r_3}] + [\vec{\omega}\ \vec{r_2}\ \vec{r_1}] + [\vec{\omega}\ \vec{r_3}\ \vec{r_2}])\vec{\omega} \cdot (\vec{\omega} \times (\vec{r_4} - \vec{r_2})) = 0$. So, D lies on the same plane in which A, B, C lie. Proved.

(10): In the following discussions, we shall use $\overrightarrow{\omega_1}, \overrightarrow{\omega_2}$ in place of $\vec{A}, \vec{A'}$. Let motion of a rigid body be given by $\overrightarrow{v_P} = \overrightarrow{v_A} + \vec{\omega} \times \overrightarrow{AP}$ where $\overrightarrow{v_P}$ denotes velocity of any point P (in the rigid body) relative to origin O, fixed in space, $\vec{\omega} \equiv$ (angular velocity of the body). Here $\overrightarrow{v_A}$ and $\vec{\omega}$ are known vectors. Here equation of axis of screw is $\vec{r} = \left(\vec{a} + \frac{\vec{\omega} \times \overrightarrow{v_A}}{\omega^2}\right) + \lambda\vec{\omega}$, where $\overrightarrow{OA} = \vec{a}$, $\overrightarrow{OP} = \vec{r}$ and λ is a parameter.

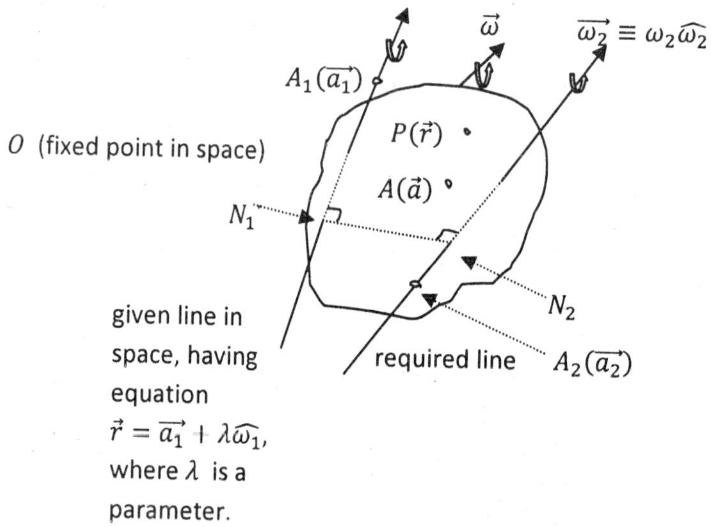

According to question, $\overrightarrow{v_A} + \vec{\omega} \times (\vec{r} - \vec{a}) = \overrightarrow{\omega_1} \times (\vec{r} - \overrightarrow{a_1}) + \overrightarrow{\omega_2} \times (\vec{r} - \overrightarrow{a_2})$ for every $\vec{r}$. That is, $(\vec{\omega} - \overrightarrow{\omega_1} - \overrightarrow{\omega_2}) \times \vec{r} + (\overrightarrow{v_A} - \vec{\omega} \times \vec{a} + \overrightarrow{\omega_1} \times \overrightarrow{a_1} + \overrightarrow{\omega_2} \times \overrightarrow{a_2}) \equiv \vec{0}$, for every $\vec{r}$. By using Lemma of Ex. 14, $\vec{\omega} - \overrightarrow{\omega_1} - \overrightarrow{\omega_2} = \vec{0}$ $\cdots(1)$

and $\overrightarrow{v_A} - \vec{\omega} \times \vec{a} + \overrightarrow{\omega_1} \times \overrightarrow{a_1} + \overrightarrow{\omega_2} \times \overrightarrow{a_2} = \vec{0}$. On using (1) in the last equation, we get $\overrightarrow{v_A} - \vec{\omega} \times \vec{a} + \overrightarrow{\omega_1} \times \overrightarrow{a_1} + (\vec{\omega} - \overrightarrow{\omega_1}) \times \overrightarrow{a_2} = \vec{0}$ or, $(\vec{\omega} - \overrightarrow{\omega_1}) \times \overrightarrow{a_2} = \vec{\omega} \times \vec{a} - \overrightarrow{\omega_1} \times \overrightarrow{a_1} - \overrightarrow{v_A}$ $\cdots(2)$

Hence $(\vec{\omega} - \overrightarrow{\omega_1}) \cdot (\vec{\omega} \times \vec{a} - \overrightarrow{\omega_1} \times \overrightarrow{a_1} - \overrightarrow{v_A}) = 0$ or, $(\vec{\omega} \times \overrightarrow{\omega_1}) \cdot \vec{a} - (\vec{\omega} \times \overrightarrow{\omega_1}) \cdot \overrightarrow{a_1} - (\vec{\omega} - \overrightarrow{\omega_1}) \cdot \overrightarrow{v_A} = 0$ or, $\overrightarrow{\omega_1} \cdot (\vec{a} \times \vec{\omega} - \overrightarrow{a_1} \times \vec{\omega} + \overrightarrow{v_A}) = \vec{\omega} \cdot \overrightarrow{v_A} = \Gamma$ or,

$\omega_1\,\widehat{\omega_1} \cdot ((\vec{a} - \overrightarrow{a_1}) \times \vec{\omega} + \overrightarrow{v_A}) = \Gamma$ or, $\omega_1 = \frac{\Gamma}{\widehat{\omega_1} \cdot (\overrightarrow{v_A} + \vec{\omega} \times (\overrightarrow{a_1} - \vec{a}))}$ $\cdots(*)$. Thus $\overrightarrow{\omega_1}$

is known, because the position of the first line is known. So, from (1), $\overrightarrow{\omega_2}$ is also known. Taking cross product with $\vec{\omega} - \overrightarrow{\omega_1}$ in (2), we get $|\vec{\omega} - \overrightarrow{\omega_1}|^2\overrightarrow{a_2} - (\overrightarrow{a_2} \cdot (\vec{\omega} - \overrightarrow{\omega_1}))(\vec{\omega} - \overrightarrow{\omega_1}) = (\vec{\omega} \times \vec{a} - \overrightarrow{\omega_1} \times \overrightarrow{a_1} - \overrightarrow{v_A}) \times (\vec{\omega} - \overrightarrow{\omega_1})$

so, $\overrightarrow{a_2} = \frac{\overrightarrow{a_2}\cdot(\vec{\omega}-\overrightarrow{\omega_1})}{|\vec{\omega}-\overrightarrow{\omega_1}|^2}(\vec{\omega} - \overrightarrow{\omega_1}) + \frac{(\vec{\omega}\times\vec{a}-\overrightarrow{\omega_1}\times\overrightarrow{a_1}-\overrightarrow{v_A})\times(\vec{\omega}-\overrightarrow{\omega_1})}{|\vec{\omega}-\overrightarrow{\omega_1}|^2}$ or,

$\overrightarrow{a_2} = \frac{(\vec{\omega}\times\vec{a}-\overrightarrow{\omega_1}\times\overrightarrow{a_1}-\overrightarrow{v_A})\times(\vec{\omega}-\overrightarrow{\omega_1})}{|\vec{\omega}-\overrightarrow{\omega_1}|^2} + \mu(\vec{\omega} - \overrightarrow{\omega_1})$ where μ is a parameter. This determines $\overrightarrow{a_2}$. This represents a line along $\vec{\omega} - \overrightarrow{\omega_1}$ ($\equiv \overrightarrow{\omega_2}$). One of its point is $\frac{(\vec{\omega}\times\vec{a}-\overrightarrow{\omega_1}\times\overrightarrow{a_1}-\overrightarrow{v_A})\times(\vec{\omega}-\overrightarrow{\omega_1})}{|\vec{\omega}-\overrightarrow{\omega_1}|^2}$. Thus equation of the second axis is

$\vec{r} = \frac{(\vec{\omega}\times\vec{a}-\overrightarrow{\omega_1}\times\overrightarrow{a_1}-\overrightarrow{v_A})\times(\vec{\omega}-\overrightarrow{\omega_1})}{|\vec{\omega}-\overrightarrow{\omega_1}|^2} + \mu(\vec{\omega} - \overrightarrow{\omega_1})$. This is the conjugate axis of $\vec{r} = \overrightarrow{a_1} + \lambda\widehat{\omega_1}$ and their resultant is the screw motion having axis of screw $\vec{r} = \left(\vec{a} + \frac{\vec{\omega}\times\overrightarrow{v_A}}{\omega^2}\right) + \lambda\vec{\omega}$.

Here N_1N_2 is the common perpendicular to the two axes. The direction of N_1N_2 is given by $\overrightarrow{\omega_1} \times \overrightarrow{\omega_2} = \overrightarrow{\omega_1} \times (\vec{\omega} - \overrightarrow{\omega_1}) = \overrightarrow{\omega_1} \times \vec{\omega}$. Also $(\overrightarrow{\omega_1} \times \overrightarrow{\omega_2}) \cdot \vec{\omega} = (\overrightarrow{\omega_1} \times \vec{\omega}) \cdot \vec{\omega} = 0$. So, $N_1N_2 \perp$ (axis of screw). It remains to prove that the resultant axis of screw intersects N_1N_2. Now it suffices to find a point K on the common perpendicular N_1N_2 where the net velocity due to $\overrightarrow{\omega_1}$ and $\overrightarrow{\omega_2}$ is parallel to $\vec{\omega}$ $(= \overrightarrow{\omega_1} + \overrightarrow{\omega_2})$. Let $\overrightarrow{N_1N_2} \equiv \vec{n}$. Let $N_1K: KN_2 = (1-\lambda):\lambda$. Here $\vec{n} \perp \overrightarrow{\omega_1}$ and $\vec{n} \perp \overrightarrow{\omega_2}$ so, $\vec{n} \cdot \overrightarrow{\omega_1} = 0$ and $\vec{n} \cdot \overrightarrow{\omega_2} = 0$ $\cdots (*)$. Now it suffices to find λ. Here $\left(\overrightarrow{\omega_1} \times (1-\lambda)\vec{n} + \overrightarrow{\omega_2} \times (-\lambda\vec{n})\right) \parallel (\overrightarrow{\omega_1} + \overrightarrow{\omega_2})$, so $(\overrightarrow{\omega_1} \times \vec{n} - \lambda(\overrightarrow{\omega_1} + \overrightarrow{\omega_2}) \times \vec{n}) \parallel (\overrightarrow{\omega_1} + \overrightarrow{\omega_2})$. This implies that $\left((\overrightarrow{\omega_1} - \lambda(\overrightarrow{\omega_1} + \overrightarrow{\omega_2})) \times \vec{n}\right) \times (\overrightarrow{\omega_1} + \overrightarrow{\omega_2}) = \vec{0}$ or, $\left\{\{\overrightarrow{\omega_1} - \lambda(\overrightarrow{\omega_1} + \overrightarrow{\omega_2})\} \cdot (\overrightarrow{\omega_1} + \overrightarrow{\omega_2})\right\}\vec{n} - (\vec{n} \cdot (\overrightarrow{\omega_1} + \overrightarrow{\omega_2}))(\overrightarrow{\omega_1} - \lambda(\overrightarrow{\omega_1} + \overrightarrow{\omega_2})) = 0$ or, $(\overrightarrow{\omega_1} \cdot (\overrightarrow{\omega_1} + \overrightarrow{\omega_2}) - \lambda|\overrightarrow{\omega_1} + \overrightarrow{\omega_2}|^2)\vec{n} = 0$, from $(*)$. Hence $\overrightarrow{\omega_1} \cdot (\overrightarrow{\omega_1} + \overrightarrow{\omega_2}) = \lambda|\overrightarrow{\omega_1} + \overrightarrow{\omega_2}|^2$ or $\lambda = \frac{\overrightarrow{\omega_1}\cdot(\overrightarrow{\omega_1}+\overrightarrow{\omega_2})}{|\overrightarrow{\omega_1}+\overrightarrow{\omega_2}|^2}$.

Hence common perpendicular to the axes intersects the axis of the resultant screw. Proved.

[**Note:** Here we further observe that $1-\lambda = 1-\frac{\overrightarrow{\omega_1}\cdot(\overrightarrow{\omega_1}+\overrightarrow{\omega_2})}{|\overrightarrow{\omega_1}+\overrightarrow{\omega_2}|^2} = \frac{\overrightarrow{\omega_1}\cdot(\overrightarrow{\omega_1}+\overrightarrow{\omega_2})+\overrightarrow{\omega_2}\cdot(\overrightarrow{\omega_1}+\overrightarrow{\omega_2})-\overrightarrow{\omega_1}\cdot(\overrightarrow{\omega_1}+\overrightarrow{\omega_2})}{|\overrightarrow{\omega_1}+\overrightarrow{\omega_2}|^2} = \frac{\overrightarrow{\omega_2}\cdot(\overrightarrow{\omega_1}+\overrightarrow{\omega_2})}{|\overrightarrow{\omega_1}+\overrightarrow{\omega_2}|^2}$. Hence $N_1K: KN_2 = \overrightarrow{\omega_2}\cdot(\overrightarrow{\omega_1}+\overrightarrow{\omega_2}):\overrightarrow{\omega_1}\cdot(\overrightarrow{\omega_1}+\overrightarrow{\omega_2})$.]

(11): Let one conjugate axis, corresponding to $\overrightarrow{\omega_1}$ be perpendicular to the resultant screw $\vec{\omega}$ $(=\overrightarrow{\omega_1}+\overrightarrow{\omega_2})$, so $\overrightarrow{\omega_1}\cdot(\overrightarrow{\omega_1}+\overrightarrow{\omega_2}) = 0$. So, on using the above note, we find that $\lambda = 0$. So, K and N_2 coincide. Hence the other axis meets with the axis of resultant screw. Conversely, let the other axis meets with the axis of resultant screw. So, $\lambda = 0$. Hence, from the above note,

$\overrightarrow{\omega_1}\cdot(\overrightarrow{\omega_1}+\overrightarrow{\omega_2}) = 0$ or, $\overrightarrow{\omega_1}\perp(\overrightarrow{\omega_1}+\overrightarrow{\omega_2})$. So, the first line is perpendicular to the axis of the resultant screw. Proved.

(12): Let $ABCD$ be any cube with centre O. Choose O as the origin. Choose x, y, z-axes as shown in the figure given below.

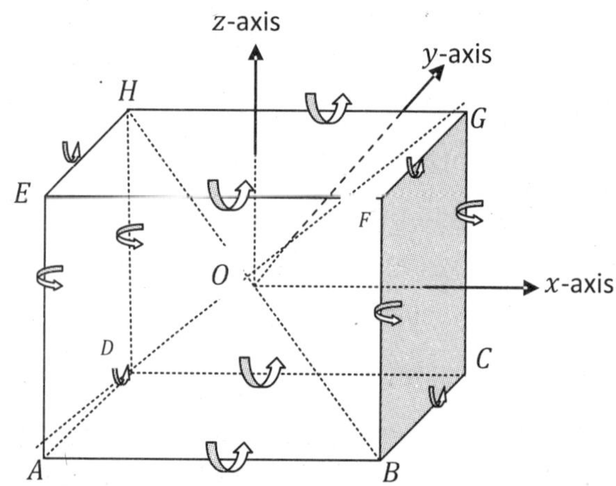

In our figure, let the 12 angular velocities to be compounded be $\omega(1,0,0)$ at A, $\omega(1,0,0)$ at D, $\omega(1,0,0)$ at E, $\omega(1,0,0)$ at H; $\omega(0,1,0)$ at A, $\omega(0,1,0)$ at B, $\omega(0,1,0)$ at E, $\omega(0,1,0)$ at F; $\omega(0,0,1)$ at A, $\omega(0,0,1)$ at B, $\omega(0,0,1)$ at C, $\omega(0,0,1)$ at D. By the above exercise 10, $\omega(1,0,0)$ at A and $\omega(1,0,0)$ at H, when compounded, gives an angular velocity $2\omega(1,0,0)$ and its screw motion passes through the origin O. Hence the resultant of the first four angular velocities is an angular velocity $4\omega(1,0,0)$ with the screw motion passing through the origin. Similarly we get two other angular velocities $4\omega(0,1,0)$ and $4\omega(0,0,1)$, both have screw motions passing through O. So the net screw motion will also pass through O. Further, the net angular velocity is $4\omega(1,0,0) + 4\omega(0,1,0) + 4\omega(0,0,1) = 4\omega(1,1,1)$. This shows that the resultant screw motion will be along the diagonal AOG. Proved.

(13): Let $A_1A_2A_3 \cdots A_n$ be any skew polygon. From question, the rigid body has simultaneously n angular velocities $k\overrightarrow{A_1A_2}$, $k\overrightarrow{A_2A_3}$, $k\overrightarrow{A_3A_4}, \cdots, k\overrightarrow{A_{n-1}A_n}$, $k\overrightarrow{A_nA_1}$, where k is the constant of proportionality. So, this is equivalent to the screw motion, whose (angular velocity part) = (sum of all angular velocities) = $k\overrightarrow{A_1A_2} + k\overrightarrow{A_2A_3} + k\overrightarrow{A_3A_4} + \cdots + k\overrightarrow{A_{n-1}A_n} + k\overrightarrow{A_nA_1} = k\overrightarrow{A_1A_3} + k\overrightarrow{A_3A_4} + \vec{0}$. So, there is only translational velocity. So, all particles of the rigid body will move with constant velocity. Proved.

(14): Before answering this question, let us prove the following

LEMMA : If $\vec{a} + \vec{b} \times \vec{r} = \vec{0}$ for every vector $\vec{r}$, then $\vec{a} = \vec{0}$ and $\vec{b} = \vec{0}$.

Proof of the Lemma: On taking $\vec{b}$ for $\vec{r}$ in the given condition, we get $\vec{a} = \vec{a} + \vec{0} = \vec{a} + \vec{b} \times \vec{b} = \vec{0}$. So, $\vec{a} = \vec{0}$. Now put $\vec{\imath}$ for $\vec{r}$ in the given condition. We get $-b_2\vec{k} + b_3\vec{\jmath} = \vec{b} \times \vec{\imath} = \vec{a} + \vec{b} \times \vec{\imath} = \vec{0}$, where $\vec{b} \equiv b_1\vec{\imath} + b_2\vec{\jmath} + b_3\vec{k}$. So, $b_2 = b_3 = 0$. Similarly $b_1 = 0$. Proved.

Proof of the main question: Let the four angular velocities be $(\overrightarrow{v_1}, \overrightarrow{\omega_1})$, $(\overrightarrow{v_2}, \overrightarrow{\omega_2})$, $(\overrightarrow{v_3}, \overrightarrow{\omega_3})$, $(\overrightarrow{v_4}, \overrightarrow{\omega_4})$, where $\overrightarrow{v_1} = \vec{0}$, $\overrightarrow{v_2} = \vec{0}$, $\overrightarrow{v_3} = \vec{0}$, $\overrightarrow{v_4} = \vec{0}$. Let the axis of screw of the i th angular velocity be passing through a point $A_i(\overrightarrow{a_i})$, where $i = 1, 2, 3, 4$. Let $P(\vec{r})$ be any point of the rigid body. From question, the net velocity of P is zero, that is, $\sum_{i=1}^{4}(\overrightarrow{v_i} + \overrightarrow{\omega_i} \times (\vec{r} - \overrightarrow{a_i})) = \vec{0}$ or, $\sum_{i=1}^{4}((\overrightarrow{v_i} + \overrightarrow{a_i} \times \overrightarrow{\omega_i}) + \overrightarrow{\omega_i} \times \vec{r}) = \vec{0}$ or, $\sum_{i=1}^{4}(\overrightarrow{v_i} + \overrightarrow{a_i} \times \overrightarrow{\omega_i}) + (\sum_{i=1}^{4}\overrightarrow{\omega_i}) \times \vec{r} = \vec{0}$ for every $\vec{r}$.

So, by the Lemma, $\sum_{i=1}^{4}(\overrightarrow{v_i} + \overrightarrow{a_i} \times \overrightarrow{\omega_i}) = \vec{0}$ $\quad\cdots(1)$

and $\sum_{i=1}^{4}\overrightarrow{\omega_i} = \vec{0}$ $\quad\cdots(2)$

We have to prove: **(1)** (the invariant of the resultant screw of $(\overrightarrow{v_1}, \overrightarrow{\omega_1})$ and $(\overrightarrow{v_2}, \overrightarrow{\omega_2})$) = (the invariant of the resultant screw of $(\overrightarrow{v_3}, \overrightarrow{\omega_3})$ and $(\overrightarrow{v_4}, \overrightarrow{\omega_4})$), etc; that is, $\left(\sum_{i=1}^{2}(\vec{v_i}+\vec{a_i}\times\vec{\omega_i})\right)\cdot\left(\sum_{i=1}^{2}\vec{\omega_i}\right)=\left(\sum_{i=3}^{4}(\vec{v_i}+\vec{a_i}\times\vec{\omega_i})\right)\left(\sum_{i=3}^{4}\vec{\omega_i}\right)$. **(2)** $\left(\sum_{i=1}^{3}(\vec{v_i}+\vec{a_i}\times\vec{\omega_i})\right)\cdot\left(\sum_{i=1}^{3}\vec{\omega_i}\right)=0$, etc. For (1): $LHS=\left(\sum_{i=1}^{2}(\vec{v_i}+\vec{a_i}\times\vec{\omega_i})\right)\cdot\left(\sum_{i=1}^{2}\vec{\omega_i}\right)=\left(-\sum_{i=3}^{4}(\vec{v_i}+\vec{a_i}\times\vec{\omega_i})\right)\cdot\left(-\sum_{i=3}^{4}\vec{\omega_i}\right)$, on using equations (1) and (2). Hence $LHS=RHS$. For (2): $LHS=\left(\sum_{i=1}^{3}(\vec{v_i}+\vec{a_i}\times\vec{\omega_i})\right)\cdot\left(\sum_{i=1}^{3}\vec{\omega_i}\right)=\{-(\vec{v_4}+\vec{a_4}\times\vec{\omega_4})\}\cdot\{-\vec{\omega_4}\}=\vec{v_4}\cdot\vec{\omega_4}=0$, because $\vec{v_4}=\vec{0}$. Proved.

(15): Let the plane lamina be rotating about OZ-axis, where OZ-axis is perpendicular to the plane of the lamina and OZ-axis intersects the lamina at O. Let us draw two rectangular axes OX and OY, fixed in space such that OX, OY, OZ constitute a rectangular coordinate system fixed in space. Let θ be the angle which Ox-axis makes with OX-axis at time t.

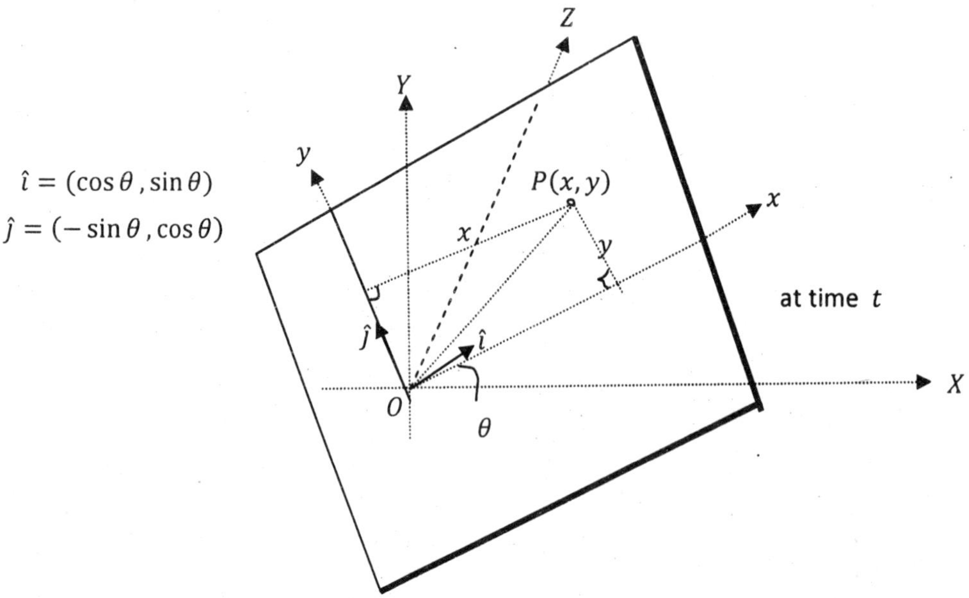

Given that $\dot{\theta}=\omega$, where ω is a function of time t. Let $\overrightarrow{OP}\equiv\vec{r}$. So, $\vec{r}=x(\cos\theta,\sin\theta)+y(-\sin\theta,\cos\theta)$. Hence $\dot{\vec{r}}=\dot{x}(\cos\theta,\sin\theta)+x\left((-\sin\theta)\dot{\theta},(\cos\theta)\dot{\theta}\right)+\dot{y}(-\sin\theta,\cos\theta)+y\left((-\cos\theta)\dot{\theta},(-\sin\theta)\dot{\theta}\right)$ or, $\dot{\vec{r}}=(\dot{x}\cos\theta-x\dot{\theta}\sin\theta-\dot{y}\sin\theta-y\dot{\theta}\cos\theta,\dot{x}\sin\theta+x\dot{\theta}\cos\theta+\dot{y}\cos\theta-y\dot{\theta}\sin\theta)$. Hence $\ddot{\vec{r}}=\left(\ddot{x}\cos\theta-\dot{x}\dot{\theta}\sin\theta-(\dot{x}\dot{\theta}\sin\theta+x\ddot{\theta}\sin\theta+x\dot{\theta}^2\cos\theta)-\ddot{y}\sin\theta-\dot{y}\dot{\theta}\cos\theta-(\dot{y}\dot{\theta}\cos\theta+y\ddot{\theta}\cos\theta-y\dot{\theta}^2\sin\theta),\ \ddot{x}\sin\theta+\dot{x}\dot{\theta}\cos\theta+\right.$

$\dot{x}\dot{\theta}\cos\theta + x\ddot{\theta}\cos\theta - x\dot{\theta}^2\sin\theta + \ddot{y}\cos\theta - \dot{y}\dot{\theta}\sin\theta - \left(\dot{y}\dot{\theta}\sin\theta + y\ddot{\theta}\sin\theta + y\dot{\theta}^2\cos\theta\right)\Big)$ or, $\ddot{\vec{r}} = \Big((\cos\theta)\left(\ddot{x} - x\dot{\theta}^2 - 2\dot{y}\dot{\theta} - y\ddot{\theta}\right) + (\sin\theta)\left(-2\dot{x}\dot{\theta} - x\ddot{\theta} - \ddot{y} + y\dot{\theta}^2\right),\ (\cos\theta)\left(2\dot{x}\dot{\theta} + x\ddot{\theta} + \ddot{y} - y\dot{\theta}^2\right) + (\sin\theta)\left(\ddot{x} - x\dot{\theta}^2 - 2\dot{y}\dot{\theta} - y\ddot{\theta}\right)\Big)$.

Hence the resolute of $\ddot{\vec{r}}$ along moving Ox-axis is $\ddot{\vec{r}}\cdot(\cos\theta, \sin\theta) = \cos^2\theta\left(\ddot{x} - x\dot{\theta}^2 - 2\dot{y}\dot{\theta} - y\ddot{\theta}\right) + \sin\theta\cos\theta\left(-2\dot{x}\dot{\theta} - x\ddot{\theta} - \ddot{y} + y\dot{\theta}^2\right) + \sin\theta\cos\theta\left(2\dot{x}\dot{\theta} + x\ddot{\theta} + \ddot{y} - y\dot{\theta}^2\right) + \sin^2\theta\left(\ddot{x} - x\dot{\theta}^2 - 2\dot{y}\dot{\theta} - y\ddot{\theta}\right) = \left(\ddot{x} - x\dot{\theta}^2 - 2\dot{y}\dot{\theta} - y\ddot{\theta}\right) = \ddot{x} - x\omega^2 - 2\dot{y}\omega - y\dot{\omega}$. Similarly the resolute of $\ddot{\vec{r}}$ along moving Oy-axis is $\ddot{\vec{r}}\cdot(-\sin\theta, \cos\theta) = -\sin\theta\cos\theta\left(\ddot{x} - x\dot{\theta}^2 - 2\dot{y}\dot{\theta} - y\ddot{\theta}\right) + \sin^2\theta\left(2\dot{x}\dot{\theta} + x\ddot{\theta} + \ddot{y} - y\dot{\theta}^2\right) + \cos^2\theta\left(2\dot{x}\dot{\theta} + x\ddot{\theta} + \ddot{y} - y\dot{\theta}^2\right) + \sin\theta\cos\theta\left(\ddot{x} - x\dot{\theta}^2 - 2\dot{y}\dot{\theta} - y\ddot{\theta}\right) = 2\dot{x}\dot{\theta} + x\ddot{\theta} + \ddot{y} - y\dot{\theta}^2 = 2\dot{x}\omega + x\dot{\omega} + \ddot{y} - y\omega^2$. Proved.

(16): Let A be a fixed point on the given curve. Let $B(\vec{r})$ be the position of the particle at time t, where $\overrightarrow{OB} = \vec{r}$. So, velocity of the particle relative to the fixed frame in space is given by $\overrightarrow{v_P} = \dfrac{\partial\vec{r}}{\partial t} + \vec{\omega}\times\vec{r}$ ... (1)

where $\dfrac{\partial}{\partial t}$ represents rate of change in the rotating frame and $\vec{\omega} \equiv \omega\hat{k}$ (in our figure). Hence acceleration of the particle B at time t relative to fixed frame in space is given by $\overrightarrow{a_P} = \dfrac{\partial\,\overrightarrow{v_P}}{\partial t} + \vec{\omega}\times\overrightarrow{v_P} = \dfrac{\partial}{\partial t}\left(\dfrac{\partial\vec{r}}{\partial t} + \vec{\omega}\times\vec{r}\right) + \vec{\omega}\times\left(\dfrac{\partial\vec{r}}{\partial t} + \vec{\omega}\times\vec{r}\right) = \dfrac{\partial^2\vec{r}}{\partial t^2} + \dfrac{\partial\vec{\omega}}{\partial t}\times\vec{r} + \vec{\omega}\times\dfrac{\partial\vec{r}}{\partial t} + \left(\vec{\omega}\times\dfrac{\partial\vec{r}}{\partial t} + \vec{\omega}\times(\vec{\omega}\times\vec{r})\right)$. Since $\dfrac{\partial\vec{\omega}}{\partial t} = \dfrac{d\vec{\omega}}{dt} - (\vec{\omega}\times\vec{\omega}) = \dfrac{d(\omega\hat{k})}{dt} - \vec{0} = \dfrac{d(\omega\hat{k})}{dt} = \vec{0}$ for uniform rotation, so in this case,

$$\boxed{\overrightarrow{a_P} = \frac{\partial^2\vec{r}}{\partial t^2} + 2\vec{\omega}\times\frac{\partial\vec{r}}{\partial t} + \vec{\omega}\times(\vec{\omega}\times\vec{r})}$$

. This is known as the *Coriolis formula*.

We shall need it later also.

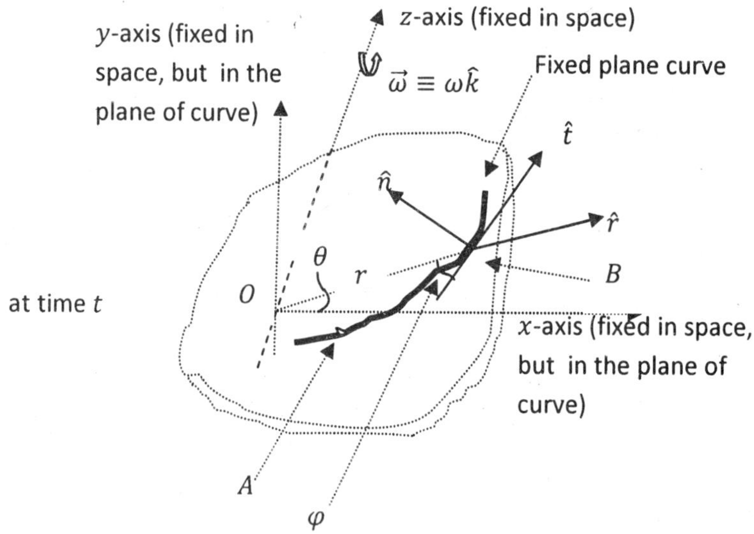

Hence $\overrightarrow{a_P} = (\ddot{s}\hat{t} + \kappa(\dot{s})^2\hat{n}) + 2\omega\hat{k} \times (\dot{s}\hat{t}) + \omega\hat{k} \times \left(\omega\hat{k} \times r\hat{r}\right)$, where s denotes arc-length AB. Now $\hat{k} \times \hat{t} = (1)(1)\sin 90°\,\hat{n} = \hat{n}$ and $\hat{k} \times \left(\hat{k} \times \hat{r}\right) = \left(\hat{k} \cdot \hat{r}\right)\hat{k} - 1\hat{r} = 0\hat{k} - \hat{r} = -\hat{r} = -\left((\hat{r} \cdot \hat{t})\hat{t} + (\hat{r} \cdot \hat{n})\hat{n}\right) = -\left((\cos\varphi)\hat{t} + (-\sin\varphi)\hat{n}\right) = -\hat{t}\cos\varphi + \hat{n}\sin\varphi$

so, $\overrightarrow{a_P} = \ddot{s}\hat{t} + \kappa(\dot{s})^2\hat{n} + 2\omega\dot{s}\hat{n} + \omega^2 r(-\hat{t}\cos\varphi + \hat{n}\sin\varphi) = (\ddot{s} - \omega^2 r\cos\varphi)\hat{t} + (\kappa\dot{s}^2 + 2\omega\dot{s} + \omega^2 r\sin\varphi)\hat{n}$. Hence, from question, $\frac{P}{m}\hat{t} + \frac{Q+R}{m}\hat{n} = \frac{1}{m}(P\hat{t} + Q\hat{n} + R\hat{n}) = \frac{1}{m}(\text{net force on the particle of mass } m) = \overrightarrow{a_P} = (\ddot{s} - \omega^2 r\cos\varphi)\hat{t} + (\kappa\dot{s}^2 + 2\omega\dot{s} + \omega^2 r\sin\varphi)\hat{n}$ or ($\ddot{s} - \omega^2 r\cos\varphi = \frac{P}{m}$ and $\kappa\dot{s}^2 + 2\omega\dot{s} + \omega^2 r\sin\varphi = \frac{Q+R}{m}$). So, $m(\ddot{s} - \omega^2 r\cos\varphi) = P$ and $m(\kappa\dot{s}^2 + 2\omega\dot{s} + \omega^2 r\sin\varphi) = Q + R$. Since $\cos\varphi = \frac{dr}{ds}$ and $\frac{d}{ds}\left(\frac{1}{2}v^2\right) = v\frac{dv}{ds} = \frac{ds}{dt}\frac{dv}{ds} = \frac{dv}{dt} = \ddot{s}$

so, $m\left(\frac{d}{ds}\left(\frac{1}{2}v^2\right) - \omega^2 r\frac{dr}{ds}\right) = P$ and $m(\kappa v^2 + 2\omega v + \omega^2 r\sin\varphi) = R + Q$.

Proved.

(17):

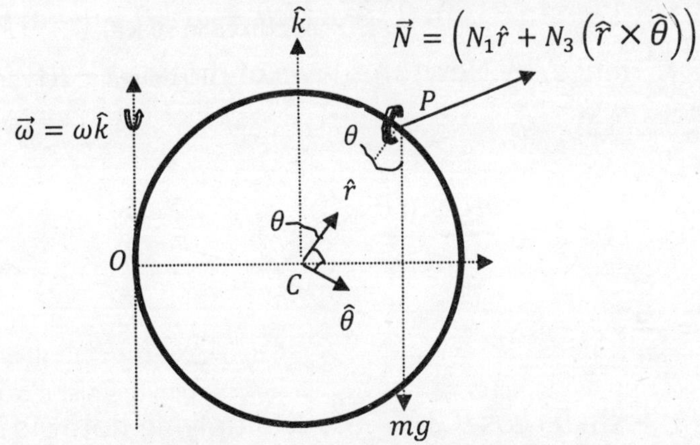

Let P be the position of the bead at time t. Let R be the radius of the circular wire and let C be its centre. Let the unit vector along $\overrightarrow{CP}$ be $\hat{r}$ and $\hat{\theta}$ be the unit vector perpendicular to $\overrightarrow{CP}$ and in the plane of wire. Let θ be the angle which CP makes with the upward vertical through C. Let $\hat{k}$ be the unit vector along upward vertical. So, $\vec{\omega} \equiv$ (angular velocity of the rotating wire) $=$ $\omega\hat{k}$. Given that ω is a constant. Clearly $\dot{\hat{r}} = \dot{\theta}\hat{\theta}$ and $\dot{\hat{\theta}} = -\dot{\theta}\hat{r}$. Also $\hat{k} = (\hat{k}\cdot\hat{r})\hat{r} + (\hat{k}\cdot\hat{\theta})\hat{\theta} = (\cos\theta)\hat{r} - (\sin\theta)\hat{\theta}$ and $\overrightarrow{OC} = (\overrightarrow{OC}\cdot\hat{r})\hat{r} + (\overrightarrow{OC}\cdot\hat{\theta})\hat{\theta} = (R\sin\theta)\hat{r} + (R\cos\theta)\hat{\theta}$. Hence velocity of P at time t relative to fixed frame in space is given by $\overrightarrow{v_P} = \left(\dot{R}\hat{r} + (R\dot{\theta})\hat{\theta}\right) + \omega\hat{k} \times \left(\overrightarrow{OC} + R\hat{r}\right) = \vec{0} + R\dot{\theta}\hat{\theta} + \omega\left((\cos\theta)\hat{r} - (\sin\theta)\hat{\theta}\right) \times \left((R\sin\theta)\hat{r} + (R\cos\theta)\hat{\theta} + R\hat{r}\right) = R\dot{\theta}\hat{\theta} + \omega R(\cos\theta\,\hat{r} - \sin\theta\,\hat{\theta}) \times \left((1+\sin\theta)\hat{r} + \cos\theta\,\hat{\theta}\right) = R\dot{\theta}\hat{\theta} + \omega R(\cos^2\theta + \sin\theta + \sin^2\theta)\hat{r} \times \hat{\theta}$ or, $\overrightarrow{v_P} = R\dot{\theta}\hat{\theta} + \omega R(1+\sin\theta)(\hat{r}\times\hat{\theta})$ $\cdots(*)$. On using Coriolis formula, $\boxed{\overrightarrow{a_P} = \frac{\partial^2\vec{r}}{\partial t^2} + 2\vec{\omega} \times \frac{\partial\vec{r}}{\partial t} + \vec{\omega} \times (\vec{\omega} \times \vec{r})}$, the acceleration of P at time t relative to fixed frame in space is given by $\overrightarrow{a_P} = R\left(\ddot{\theta}\hat{\theta} + \dot{\theta}(-\dot{\theta}\hat{r})\right) + 2\omega(\cos\theta\,\hat{r} - \sin\theta\,\hat{\theta}) \times (R\dot{\theta}\hat{\theta}) + \omega(\cos\theta\,\hat{r} - \sin\theta\,\hat{\theta}) \times \left(\omega(\cos\theta\,\hat{r}\theta - \sin\theta\,\hat{\theta}) \times (R\hat{r} + R\sin\theta\,\hat{r} + R\cos\theta\,\hat{\theta})\right) = -R\dot{\theta}^2\hat{r} + R\ddot{\theta}\hat{\theta} + 2\omega(R\cos\theta)(\dot{\theta})\hat{r} \times \hat{\theta} - \omega^2 R(1+\sin\theta)(\sin\theta\,\hat{r} + \cos\theta\,\hat{\theta})$ (obtained after calculation). Hence

$\overrightarrow{a_P} = \hat{r}\left(-R\dot{\theta}^2 - \omega^2 R(1+\sin\theta)\sin\theta\right) + \hat{\theta}\left(R\ddot{\theta} - \omega^2 R(1+\sin\theta)\cos\theta\right) + 2\omega R\cos\theta\,\dot{\theta}\left(\hat{r}\times\hat{\theta}\right)$. Hence, by Newton's laws of motion, $-R\dot{\theta}^2 - R\omega^2(1+\sin\theta)\sin\theta = \frac{-mg\cos\theta + N_1}{m}$ $\cdots(1)$

$$R\ddot{\theta} - \omega^2 R(1+\sin\theta)\cos\theta = \frac{mg\sin\theta}{m} \qquad \cdots(2)$$

and $2\omega R\cos\theta\,\dot{\theta} = \frac{N_3}{m}$ $\cdots(3)$

From (2), $\ddot{\theta} = \omega^2(1+\sin\theta)\cos\theta + \frac{g\sin\theta}{R}$. Multiplying both sides by $2\dot{\theta}$ and then integrating, we get $\dot{\theta}^2 = 2\omega^2\left(\sin\theta + \frac{1}{2}\sin^2\theta\right) - \frac{2g}{R}\cos\theta + C$, where $(0)^2 = 2\omega^2\left(\sin 0 + \frac{1}{2}\sin^2 0\right) - \frac{2g}{R}\cos 0 + C$ (on using the initial condition). So, $\dot{\theta}^2 = 2\omega^2\left(\sin\theta + \frac{1}{2}\sin^2\theta\right) + \frac{2g}{R}(1-\cos\theta)$. Now, from $(*)$, $v_P{}^2 = \left(R\dot{\theta}\right)^2 + \left(\omega R(1+\sin\theta)\right)^2 = R^2\left(2\omega^2\left(\sin\theta + \frac{1}{2}\sin^2\theta\right) + \frac{2g}{R}(1-\cos\theta)\right) + \omega^2 R^2(1 + 2\sin\theta + \sin^2\theta) = R^2\omega^2(1 + 4\sin\theta + 2\sin^2\theta) + 2gR(1-\cos\theta)$. Hence required magnitude of velocity is $\sqrt{2gR(1-\cos\theta) + R^2\omega^2(1 + 4\sin\theta + 2\sin^2\theta)}$. Ans. On putting the value of $\dot{\theta}^2$ in (1), we get $-R\left(2\omega^2\left(\sin\theta + \frac{1}{2}\sin^2\theta\right) + \frac{2g}{R}(1-\cos\theta)\right) - R\omega^2(1+\sin\theta)\sin\theta = \frac{-mg\cos\theta + N_1}{m}$ or, $-R\omega^2\left((1+\sin\theta)\sin\theta + 2\sin\theta + \sin^2\theta\right) - 2g(1-\cos\theta) = -g\cos\theta + \frac{N_1}{m}$. So, $N_1 = m\{g(3\cos\theta - 2) - R\omega^2(3\sin\theta + 2\sin^2\theta)\} = -m\{g(2 - 3\cos\theta) + R\omega^2\sin\theta\,(3 + 2\sin\theta)\}$. Also, from (3), $N_3{}^2 = (2m\omega R\cos\theta)^2\dot{\theta}^2 = (2m\omega R\cos\theta)^2\left\{2\omega^2\left(\sin\theta + \frac{1}{2}\sin^2\theta\right) + \frac{2g}{R}(1-\cos\theta)\right\}$. Hence magnitude of reaction is $\{N_1{}^2 + N_3{}^2\}^{\frac{1}{2}} = \left\{m^2\{g(2 - 3\cos\theta) + R\omega^2\sin\theta\,(3 + 2\sin\theta)\}^2 + (2m\omega R\cos\theta)^2\left\{2\omega^2\left(\sin\theta + \frac{1}{2}\sin^2\theta\right) + \frac{2g}{R}(1-\cos\theta)\right\}\right\}^{\frac{1}{2}}$. Ans.

(18): Let us take the vertically upward direction of the axis of the helical wire as the positive direction of the z-axis. Let equation of the helix be $\vec{r} = a(\cos\theta, \sin\theta, -b\theta)$ where a, b are positive constants. So, $\frac{ds}{d\theta}\hat{t} = \frac{d\vec{r}}{d\theta} =$ $a(-\sin\theta, \cos\theta, -b)$. $\therefore \frac{ds}{d\theta} = a\sqrt{1+b^2}$. Hence $\frac{d^2s}{d\theta^2}\hat{t} + \left(\frac{ds}{d\theta}\right)^2 \kappa\hat{n} = \frac{d^2\vec{r}}{d\theta^2} =$ $-a(\cos\theta, \sin\theta, 0)$. Here $\hat{t}$ is the unit vector along $(-\sin\theta, \cos\theta, -b)$ and hence $\hat{t} = \frac{1}{\sqrt{1+b^2}}(-\sin\theta, \cos\theta, -b)$. So, $\left(\frac{ds}{d\theta}\right)^3 \kappa\hat{b} =$ $\frac{d\vec{r}}{d\theta} \times \frac{d^2\vec{r}}{d\theta^2} = -a^2(b\sin\theta, -b\cos\theta, -1) = a^2(-b\sin\theta, b\cos\theta, 1)$. So, $\hat{b} = \frac{1}{\sqrt{1+b^2}}(-b\sin\theta, b\cos\theta, 1)$ and $a^3(1+b^2)^{\frac{3}{2}}\kappa = \left(\frac{ds}{d\theta}\right)^3 \kappa = a^2\sqrt{1+b^2}$. Hence $\kappa = \frac{1}{a(1+b^2)}$, which is a constant. Also $\hat{n} = \hat{b} \times \hat{t} = \frac{-1}{1+b^2}(\cos\theta +$ $b^2\cos\theta, b^2\sin\theta + \sin\theta, 0)$. So, $\hat{n} = -(\cos\theta, \sin\theta, 0)$. Further $(1,0,0) =$ $((1,0,0)\cdot\hat{t})\hat{t} + ((1,0,0)\cdot\hat{n})\hat{n} + \left((1,0,0)\cdot\hat{b}\right)\hat{b} = \frac{-\sin\theta}{\sqrt{1+b^2}}\hat{t} - \cos\theta\,\hat{n} + \frac{-b\sin\theta}{\sqrt{1+b^2}}\hat{b}$. Similarly $(0,1,0) = \frac{\cos\theta}{\sqrt{1+b^2}}\hat{t} - \sin\theta\,\hat{n} + \frac{b\cos\theta}{\sqrt{1+b^2}}\hat{b}$ and $(0,0,1) = \frac{-b}{\sqrt{1+b^2}}\hat{t} + \frac{1}{\sqrt{1+b^2}}\hat{b}$.

Here velocity at point $P(\vec{r})$ is given by $\vec{v_P} = \dot{s}\hat{t} + \omega\hat{k} \times a(\cos\theta, \sin\theta, -b\theta) =$ $\dot{s}\hat{t} + a\omega(-\sin\theta, \cos\theta, 0)$. So, on using the Coriolis formula

$$\boxed{\vec{a_P} = \frac{\partial^2\vec{r}}{\partial t^2} + 2\vec{\omega} \times \frac{\partial\vec{r}}{\partial t} + \vec{\omega} \times (\vec{\omega} \times \vec{r})}$$, acceleration at point $P(\vec{r})$

is given by
$\vec{a_P} = \{\ddot{s}\hat{t} + \kappa\dot{s}^2\hat{n}\} + 2\omega\hat{k} \times (\dot{s}\hat{t}) + \omega\hat{k} \times \left(\omega\hat{k} \times a(\cos\theta, \sin\theta, -b\theta)\right) = \ddot{s}\hat{t} +$ $\kappa\dot{s}^2\hat{n} + 2\omega\dot{s}\frac{\hat{n}}{\sqrt{1+b^2}} + \omega^2 a\{-b\theta\hat{k} - (\cos\theta, \sin\theta, -b\theta)\} = \ddot{s}\hat{t} + \kappa\dot{s}^2\hat{n} +$ $2\omega\dot{s}\frac{\hat{n}}{\sqrt{1+b^2}} + \omega^2 a\left[(-\cos\theta)\left\{\frac{-\sin\theta}{\sqrt{1+b^2}}\hat{t} - \cos\theta\,\hat{n} + \frac{-b\sin\theta}{\sqrt{1+b^2}}\hat{b}\right\} - \sin\theta\left\{\frac{\cos\theta}{\sqrt{1+b^2}}\hat{t} -\right.\right.$ $\left.\left.\sin\theta\,\hat{n} + \frac{b\cos\theta}{\sqrt{1+b^2}}\hat{b}\right\}\right] = \ddot{s}\hat{t} + \left[\dot{s}^2\kappa + \frac{2\dot{s}\omega}{\sqrt{1+b^2}} + \omega^2 a\right]\hat{n}$. Further the net force experienced by the particle is given by $\vec{F} = -mg\hat{k} + N_2\hat{n} + N_3\hat{b} = -mg\left(\frac{-b}{\sqrt{1+b^2}}\hat{t} + \frac{1}{\sqrt{1+b^2}}\hat{b}\right) + N_2\hat{n} + N_3\hat{b}$. So, on using the Newton's law, acceleration $= \frac{\text{net force}}{\text{mass}}$, we get $\ddot{s} = \frac{bg}{\sqrt{1+b^2}}$ $\cdots(1)$

$$\dot{s}^2\kappa + \frac{2\dot{s}\omega}{\sqrt{1+b^2}} + \omega^2 a = \frac{N_2}{m} \qquad \cdots(2)$$

and $$0 = \frac{-g}{\sqrt{1+b^2}} + \frac{N_3}{m} \qquad \cdots(3)$$

From (1), we find that $s = \frac{1}{2}\frac{bg}{\sqrt{1+b^2}}t^2$. Ans.

(19):

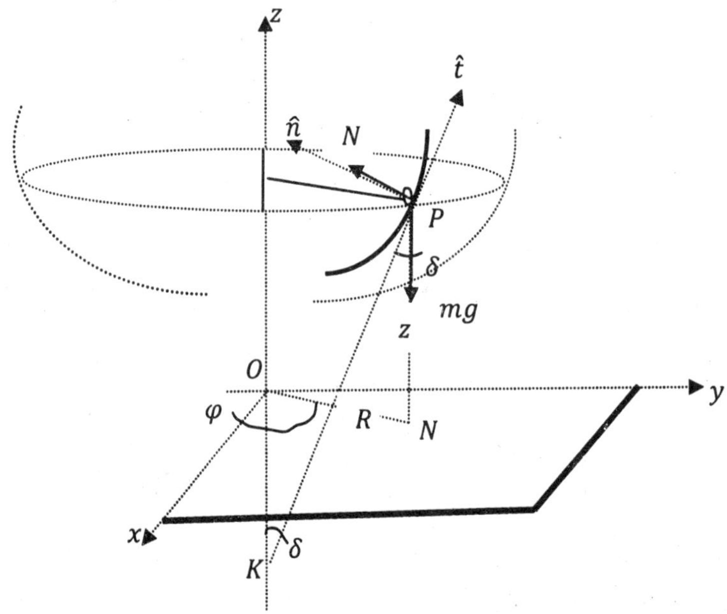

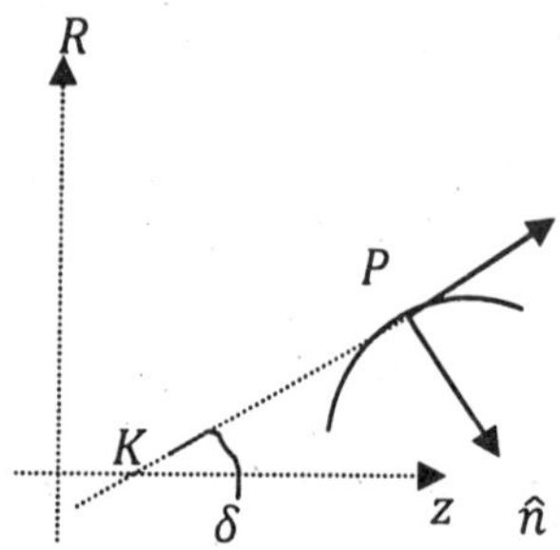

$$\hat{t} = (\cos\delta)\hat{k} + (\sin\delta)\hat{R} = \frac{dz}{ds}\hat{k} + \frac{dR}{ds}\hat{R}$$

$$\hat{n} = \frac{dR}{ds}\hat{k} - \frac{dz}{ds}\hat{R}$$

Let m be the mass of the particle moving on the inner surface of the surface of revolution (see the 1st figure). Let P, having cylindrical coordinates z, R, φ is the position of the particle at time t. Let $\hat{z}, \hat{R}, \hat{\varphi}$ be three unit vectors in the directions of only z increasing, only R increasing and only φ increasing respectively. Clearly $(1,0,0) = \cos\varphi\,\hat{R} + (-\sin\varphi)\hat{\varphi}$

and $(0,1,0) = (\sin\varphi)\hat{R} + (\cos\varphi)\hat{\varphi}$. Here $\vec{r} = \overrightarrow{OP} = R\hat{R} + z\hat{k} = (R\cos\varphi, R\sin\varphi, z)$, so, $\dot{\vec{r}} = \left(\dot{R}\cos\varphi + R(-\sin\varphi)\dot{\varphi}, \dot{R}\sin\varphi + R(\cos\varphi)\dot{\varphi}, \dot{z}\right)$ and hence $\ddot{\vec{r}} = \left(\ddot{R}\cos\varphi - \dot{R}\dot{\varphi}\sin\varphi - (\dot{R}\sin\varphi\,\dot{\varphi} + R\cos\varphi\,\dot{\varphi}^2 + R\sin\varphi\,\ddot{\varphi}), \ddot{R}\sin\varphi + \dot{R}\dot{\varphi}\cos\varphi + (\dot{R}\cos\varphi\,\dot{\varphi} - R\sin\varphi\,\dot{\varphi}^2 + R\cos\varphi\,\ddot{\varphi}), \ddot{z}\right) = \left(\ddot{R}\cos\varphi - 2\dot{R}\dot{\varphi}\sin\varphi - R\cos\varphi\,\dot{\varphi}^2 - R\sin\varphi\,\ddot{\varphi}\right)\left(\cos\varphi\,\hat{R} - \sin\varphi\,\hat{\varphi}\right) + \left(\ddot{R}\sin\varphi + 2\dot{R}\dot{\varphi}\cos\varphi - R\sin\varphi\,\dot{\varphi}^2 + R\cos\varphi\,\ddot{\varphi}\right)\left(\sin\varphi\,\hat{R} + \cos\varphi\,\hat{\varphi}\right) + \ddot{z}\hat{k} = \hat{R}\{\ddot{R} - R\dot{\varphi}^2\} + \hat{\varphi}\{2\dot{R}\dot{\varphi} + R\ddot{\varphi}\} + \ddot{z}\hat{k}$. So, the component of acceleration along $\hat{t}$ is $\ddot{\vec{r}}\cdot\hat{t} = \left\{\hat{R}\{\ddot{R} - R\dot{\varphi}^2\} + \hat{\varphi}\{2\dot{R}\dot{\varphi} + R\ddot{\varphi}\} + \ddot{z}\hat{k}\right\}\cdot\left\{\frac{dR}{ds}\hat{R} + \frac{dz}{ds}\hat{k}\right\} = \left(\ddot{R} - R\dot{\varphi}^2\right)\frac{dR}{ds} + \ddot{z}\frac{dz}{ds} = \left\{\ddot{R}\frac{\dot{R}}{\dot{s}} + \ddot{z}\frac{\dot{z}}{\dot{s}}\right\} - R\dot{\varphi}^2\frac{dR}{ds} = \ddot{s} - R\omega^2\frac{dR}{ds}$. (**Reason:** Here $(ds)^2 = (dz)^2 + (dR)^2$ so, $\dot{s}^2 = \dot{z}^2 + \dot{R}^2$. On differentiation, $2\dot{s}\ddot{s} = 2\dot{z}\ddot{z} + 2\dot{R}\ddot{R}$ so, $\ddot{s} = \ddot{R}\frac{\dot{R}}{\dot{s}} + \ddot{z}\frac{\dot{z}}{\dot{s}}$). Now, by Newton's laws of motion, we have $\ddot{s} - R\omega^2\frac{dR}{ds} = -mg\cos\delta = -mg\frac{dz}{ds}$. This establishes the first equation of motion. Further we see that $\kappa = -\frac{d\delta}{ds}$, in our figure, and $\tan\delta = \frac{dR}{dz}$ so $\kappa = \frac{\ddot{z}\dot{R} - \ddot{R}\dot{z}}{(\dot{z}^2 + \dot{R}^2)^{\frac{3}{2}}} = \frac{\ddot{z}\dot{R} - \ddot{R}\dot{z}}{(\dot{s})^3}$ or $\kappa v^2 = \kappa\dot{s}^2 = \frac{\ddot{z}\dot{R} - \ddot{R}\dot{z}}{\dot{s}} = \ddot{z}\frac{dR}{ds} - \ddot{R}\frac{dz}{ds}$ $\cdots(*)$. Next the component of acceleration along $\hat{n}$ is $\left\{\hat{R}\{\ddot{R} - R\dot{\varphi}^2\} + \hat{\varphi}\{2\dot{R}\dot{\varphi} + R\ddot{\varphi}\} + \ddot{z}\hat{k}\right\}\cdot\left\{\frac{dR}{ds}\hat{k} - \frac{dz}{ds}\hat{R}\right\} = -\frac{dz}{ds}\left(\ddot{R} - R\dot{\varphi}^2\right) + \ddot{z}\frac{dR}{ds} = -g\sin\delta + \frac{N}{m} = -g\frac{dR}{ds} + \frac{N}{m}$. It remains to prove that $\kappa v^2 + \omega^2 R\frac{dz}{ds} = \ddot{z}\frac{dR}{ds} - \frac{dz}{ds}\left(\ddot{R} - R\omega^2\right)$. This is true from $(*)$. This establishes the second equation of motion. Finally the component of acceleration along normal to the plane ONP is $\left\{\hat{R}\{\ddot{R} - R\dot{\varphi}^2\} + \hat{\varphi}\{2\dot{R}\dot{\varphi} + R\ddot{\varphi}\} + \ddot{z}\hat{k}\right\}\cdot\hat{\varphi} = 2\dot{R}\dot{\varphi} + R\ddot{\varphi} = \frac{1}{R}\frac{d}{dt}(R^2\dot{\varphi}) = 0$, by the Newton's laws of motion. Hence $\frac{d}{dt}(R^2\dot{\varphi}) = 0$ or, $\frac{d}{dt}(R^2\omega) = 0$. Proved.

(20): Let O be the origin of the rectangular Cartesian coordinate system of x, y, z-axes. Let P be the position of the particle at time t. Let N be the foot of perpendicular drawn from P on the xy-plane. Put $OP = r$, $\angle xON = \varphi$ and $\angle zOP = \theta$. Here r, θ, φ are called the *spherical polar coordinates* of point P.

So, $\hat{r} = \sin\theta \dfrac{\overrightarrow{ON}}{ON} + \cos\theta\,\hat{k}$, where $\hat{r}$ is the unit vector in the direction of $\overrightarrow{OP}$ and $\hat{k}$ is the unit vector in the direction of positive z-axis.

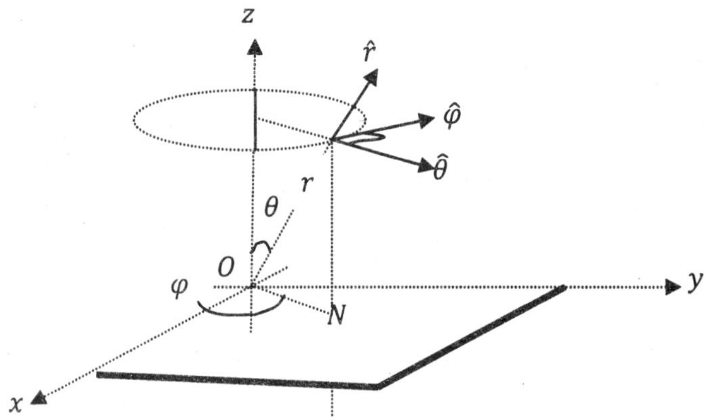

Since $\dfrac{\overrightarrow{ON}}{ON} = (\cos\varphi, \sin\varphi, 0)$ so, $\hat{r} = (\sin\theta\cos\varphi, \sin\theta\sin\varphi, \cos\theta)$. So the unit vector in the direction of only θ increasing is given by $\hat{\theta} = \left(\sin\left(\frac{\pi}{2} + \theta\right)\cos\varphi, \sin\left(\frac{\pi}{2} + \theta\right)\sin\varphi, \cos\left(\frac{\pi}{2} + \theta\right)\right) = (\cos\theta\cos\varphi, \cos\theta\sin\varphi, -\sin\theta)$. Hence the unit vector in the direction of only φ increasing is given by $\hat{\varphi} = \hat{r} \times \hat{\theta} = (\sin\theta\cos\varphi, \sin\theta\sin\varphi, \cos\theta) \times (\cos\theta\cos\varphi, \cos\theta\sin\varphi, -\sin\theta) = (-\sin^2\theta\sin\varphi - \cos^2\theta\sin\varphi, \cos^2\theta\cos\varphi + \sin^2\theta\cos\varphi, 0) = (-\sin\varphi, \cos\varphi, 0)$.

Now let us calculate $\dot{\hat{r}}, \dot{\hat{\theta}}, \dot{\hat{\varphi}}$. Here $\dot{\hat{r}} = \frac{d}{dt}(\sin\theta\cos\varphi, \sin\theta\sin\varphi, \cos\theta) = \left(\dot{\theta}\cos\theta\cos\varphi - \dot{\varphi}\sin\theta\sin\varphi, \dot{\theta}\cos\theta\sin\varphi + \dot{\varphi}\sin\theta\cos\varphi, -\dot{\theta}\sin\theta\right) = \dot{\theta}(\cos\theta\cos\varphi, \cos\theta\sin\varphi, -\sin\theta) + \dot{\varphi}(-\sin\theta\sin\varphi, \sin\theta\cos\varphi, 0) = \dot{\theta}\hat{\theta} + \dot{\varphi}\sin\theta\,\hat{\varphi}$. $\dot{\hat{\theta}} = \frac{d}{dt}(\cos\theta\cos\varphi, \cos\theta\sin\varphi, -\sin\theta) = \left(-\dot{\theta}\sin\theta\cos\varphi - \dot{\varphi}\cos\theta\sin\varphi, -\dot{\theta}\sin\theta\sin\varphi + \dot{\varphi}\cos\theta\cos\varphi, -\dot{\theta}\cos\theta\right) = -\dot{\theta}\hat{r} + \dot{\varphi}\cos\theta\,\hat{\varphi}$ and $\dot{\hat{\varphi}} = \frac{d}{dt}\left(\hat{r} \times \hat{\theta}\right) = \dot{\hat{r}} \times \hat{\theta} + \hat{r} \times \dot{\hat{\theta}} = \left(\dot{\theta}\hat{\theta} + \dot{\varphi}\sin\theta\,\hat{\varphi}\right) \times \hat{\theta} + \hat{r} \times \left(-\dot{\theta}\hat{r} + \dot{\varphi}\cos\theta\,\hat{\varphi}\right) = -\dot{\varphi}\sin\theta\,\hat{r} - \dot{\varphi}\cos\theta\,\hat{\theta} = -\dot{\varphi}\left(\sin\theta\,\hat{r} + \cos\theta\,\hat{\theta}\right)$. So, velocity at point P is given by $\vec{v} = \frac{d}{dt}\overrightarrow{OP} = \frac{d}{dt}\vec{r} = \frac{d}{dt}(r\hat{r}) = \dot{r}\hat{r} + r\dot{\hat{r}} = \dot{r}\hat{r} + r\left(\dot{\theta}\hat{\theta} + \dot{\varphi}\sin\theta\,\hat{\varphi}\right) = \dot{r}\hat{r} + r\dot{\theta}\hat{\theta} + r\sin\theta\,\dot{\varphi}\hat{\varphi}$. Hence acceleration at point P is given by

$a_r\hat{r} + a_\theta\hat{\theta} + a_\varphi\hat{\varphi} = \frac{d}{dt}(\dot{r}\hat{r} + r\dot{\theta}\hat{\theta} + r\sin\theta\,\dot{\varphi}\hat{\varphi}) = \ddot{r}\hat{r} + (\dot{r}\dot{\theta} + r\ddot{\theta})\hat{\theta} +$ $(\dot{r}\sin\theta\,\dot{\varphi} + r\dot{\theta}\dot{\varphi}\cos\theta + r\sin\theta\,\ddot{\varphi})\hat{\varphi} + \dot{r}(\dot{\theta}\hat{\theta} + \dot{\varphi}\sin\theta\,\hat{\varphi}) + r\dot{\theta}(-\dot{\theta}\hat{r} +$ $\dot{\varphi}\cos\theta\,\hat{\varphi}) + r\sin\theta\,\dot{\varphi}\left(-\dot{\varphi}(\sin\theta\,\hat{r} + \cos\theta\,\hat{\theta})\right) = (\ddot{r} - r\dot{\theta}^2 - r\dot{\varphi}^2\sin^2\theta)\hat{r} +$ $(\dot{r}\dot{\theta} + r\ddot{\theta} + \dot{r}\dot{\theta} - r\dot{\varphi}^2\sin\theta\cos\theta)\hat{\theta} + (\dot{r}\dot{\varphi}\sin\theta + r\dot{\theta}\dot{\varphi}\cos\theta + r\sin\theta\,\ddot{\varphi} +$ $\dot{r}\dot{\varphi}\sin\theta + r\dot{\theta}\dot{\varphi}\cos\theta)\hat{\varphi}$. Hence

$$\begin{cases} a_r = \ddot{r} - r\dot{\theta}^2 - r\dot{\varphi}^2\sin^2\theta \\ a_\theta = 2\dot{r}\dot{\theta} + r\ddot{\theta} - r\dot{\varphi}^2\sin\theta\cos\theta \\ a_\varphi = 2\dot{r}\dot{\varphi}\sin\theta + 2r\dot{\theta}\dot{\varphi}\cos\theta + r\sin\theta\,\ddot{\varphi} \end{cases}$$

Proved.

(21):

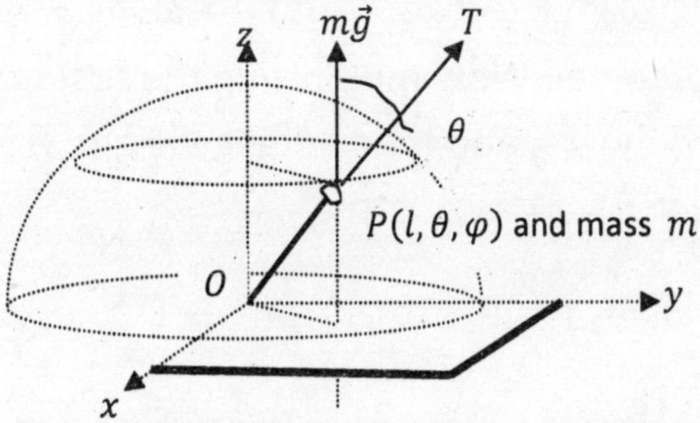

Here $r = l$, which is a constant in the above Ex. 20, so $\dot{r} = \ddot{r} = 0$. So, by the Newton's laws of motion and the results of Ex. 20, we get

$$\begin{cases} -l\dot{\theta}^2 - l\dot{\varphi}^2\sin^2\theta = a_r = \dfrac{mg\cos\theta - T}{m} \\ l\ddot{\theta} - l\dot{\varphi}^2\sin\theta\cos\theta = a_\theta = \dfrac{-mg\sin\theta}{m} \\ 2l\dot{\theta}\dot{\varphi}\cos\theta + l\sin\theta\,\ddot{\varphi} = a_\varphi = 0 \end{cases} \text{ or, }$$

$$\begin{cases} l\dot{\theta}^2 + l\sin^2\theta\,\dot{\varphi}^2 = \dfrac{T - mg\cos\theta}{m} \\ l\ddot{\theta} - l\sin\theta\cos\theta\,\dot{\varphi}^2 = -g\sin\theta \\ \dfrac{d}{dt}(\sin^2\theta\,\dot{\varphi}) = \sin\theta\,(2\dot{\theta}\dot{\varphi}\cos\theta + \sin\theta\,\ddot{\varphi}) = 0 \end{cases}$$

Proved.

(22): Here $A = B = C$ and $D = E = F$(see Ex. 2). So the formulae

$$\begin{cases} A\omega_1 - F\omega_2 - E\omega_3 = h_1 \\ B\omega_2 - D\omega_3 - F\omega_1 = h_2 \\ C\omega_3 - E\omega_1 - D\omega_2 = h_3 \end{cases} \text{ becomes } \begin{cases} A\omega_1 - D(\omega_2 + \omega_3) = h_1 & \cdots(1) \\ A\omega_2 - D(\omega_3 + \omega_1) = h_2 & \cdots(2) \\ A\omega_3 - D(\omega_1 + \omega_2) = h_3 & \cdots(3) \end{cases}$$

Since

no external force is acting, so h_1, h_2, h_3 are constants. Now $(1) + (2) + (3)$ gives $(\omega_1 + \omega_2 + \omega_3)(A - 2D) = (h_1 + h_2 + h_3)$ so, $\omega_1 + \omega_2 + \omega_3 = \frac{h_1+h_2+h_3}{A-2D}$, which is a constant. Proved. Differentiating $(1), (2), (3)$ we get

$$\begin{cases} A\dot{\omega_1} - D(\dot{\omega_2} + \dot{\omega_3}) = 0 & \cdots(4) \\ A\dot{\omega_2} - D(\dot{\omega_3} + \dot{\omega_1}) = 0 & \cdots(5) \\ A\dot{\omega_3} - D(\dot{\omega_1} + \dot{\omega_2}) = 0 & \cdots(6) \end{cases}$$

Now $(4) + \omega_2(3) - \omega_3(2)$ gives $\left(A\dot{\omega_1} - D(\dot{\omega_2} + \dot{\omega_3})\right) + \omega_2\left(A\omega_3 - D(\omega_1 + \omega_2)\right) - \omega_3\left(A\omega_2 - D(\omega_3 + \omega_1)\right) = \omega_2 h_3 - \omega_3 h_2$ or, $A\dot{\omega_1} - D(\dot{\omega_2} + \dot{\omega_3}) + D\left({\omega_3}^2 - {\omega_2}^2 + \omega_1(\omega_3 - \omega_2)\right) = \omega_2 h_3 - \omega_3 h_2$. Two similar equations can also be obtained. Now multiplying the first equation by ω_1, second by ω_2 and third by ω_3, and then adding we get $A\sum \omega_1\dot{\omega_1} - D\frac{d}{dt}\sum \omega_2\omega_3 = 0$ so, $\frac{A}{2}\sum\frac{d}{dt}{\omega_1}^2 - D\frac{d}{dt}\sum\omega_2\omega_3 = 0$

or, $\frac{d}{dt}\left(\frac{A}{2}\sum{\omega_1}^2 - D\sum\omega_2\omega_3\right) = 0$. So $\frac{A}{2}\sum{\omega_1}^2 - D\sum\omega_2\omega_3 = \text{constant, say}, c$.

Hence $\frac{A}{2}\sum{\omega_1}^2 - D\frac{1}{2}\{(\sum\omega_1)^2 - \sum{\omega_1}^2\} = c$ or, $\sum{\omega_1}^2 = \frac{c+\frac{D}{2}(\sum\omega_1)^2}{\frac{1}{2}(A+D)}$, which is a constant. Proved.

(23):

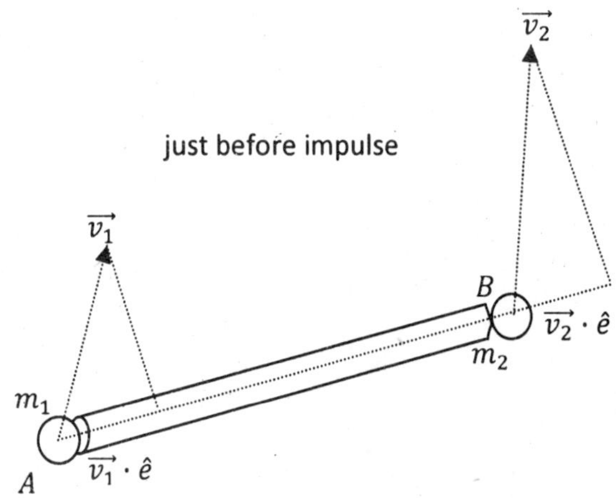

Since the rod is light, the net force experienced by the rod is zero. So, the impulsive reactions of the rod on masses m_1, m_2, just after external impulses, are $-P\hat{e}, P\hat{e}$ respectively. Since the rod is rigid, $\overrightarrow{v_1}\cdot\hat{e} = \overrightarrow{v_2}\cdot\hat{e}$ $\cdots(1)$

Just after external impulse, velocity of mass m_2 is $\frac{m_2\vec{v_2}+\vec{P_2}+P\hat{e}}{m_2}$ and velocity of mass m_1 is $\frac{m_1\vec{v_1}+\vec{P_1}-P\hat{e}}{m_1}$. Hence, due to rigid rod, $\frac{m_1\vec{v_1}+\vec{P_1}-P\hat{e}}{m_1}\cdot\hat{e}=\frac{m_2\vec{v_2}+\vec{P_2}+P\hat{e}}{m_2}\cdot\hat{e}$ or, $\vec{v_1}\cdot\hat{e}+\frac{\vec{P_1}\cdot\hat{e}-P}{m_1}=\vec{v_2}\cdot\hat{e}+\frac{\vec{P_2}\cdot\hat{e}+P}{m_2}$. Hence, from (1), $\frac{\vec{P_1}\cdot\hat{e}-P}{m_1}=\frac{\vec{P_2}\cdot\hat{e}+P}{m_2}$ or, $\left(\frac{\vec{P_2}}{m_2}-\frac{\vec{P_1}}{m_1}\right)\cdot\hat{e}=-P\left(\frac{1}{m_2}+\frac{1}{m_1}\right)$ or, $|P|=\frac{m_1m_2}{m_1+m_2}\hat{e}\cdot\left(\frac{\vec{P_2}}{m_2}-\frac{\vec{P_1}}{m_1}\right)$. Proved.

Also, required increase in $K.E.$ is $\frac{1}{2}m_1\left(\frac{m_1\vec{v_1}+\vec{P_1}-P\hat{e}}{m_1}\right)^2+\frac{1}{2}m_2\left(\frac{m_2\vec{v_2}+\vec{P_2}+P\hat{e}}{m_2}\right)^2-\left(\frac{1}{2}m_1{v_1}^2+\frac{1}{2}m_2{v_2}^2\right)=\frac{1}{2m_1}\left({P_1}^2+P^2+2m_1\vec{P_1}\cdot\vec{v_1}-2m_1P\vec{v_1}\cdot\hat{e}-2P\vec{P_1}\cdot\hat{e}\right)+\frac{1}{2m_2}\left({P_2}^2+P^2+2m_2\vec{P_2}\cdot\vec{v_2}+2m_2P\vec{v_2}\cdot\hat{e}+2P\vec{P_2}\cdot\hat{e}\right)=\frac{1}{2}\left(\frac{{P_1}^2}{m_1}+\frac{{P_2}^2}{m_2}\right)+\frac{1}{2}\left(\frac{1}{m_1}+\frac{1}{m_2}\right)P^2-P(\vec{v_1}\cdot\hat{e}-\vec{v_2}\cdot\hat{e})-P\left(\frac{\vec{P_1}}{m_1}-\frac{\vec{P_2}}{m_2}\right)\cdot\hat{e}+\vec{P_1}\cdot\vec{v_1}+\vec{P_2}\cdot\vec{v_2}=\frac{1}{2}\left(\frac{{P_1}^2}{m_1}+\frac{{P_2}^2}{m_2}\right)+\frac{1}{2}\left(\frac{1}{m_1}+\frac{1}{m_2}\right)P^2-P\left(P\left(\frac{1}{m_2}+\frac{1}{m_1}\right)\right)+\vec{P_1}\cdot\vec{v_1}+\vec{P_2}\cdot\vec{v_2}$, on using the above results. So, required increase in $K.E.$ is $\frac{1}{2}\left(\frac{{P_1}^2}{m_1}+\frac{{P_2}^2}{m_2}\right)-\frac{1}{2}\left(\frac{1}{m_1}+\frac{1}{m_2}\right)P^2+\vec{P_1}\cdot\vec{v_1}+\vec{P_2}\cdot\vec{v_2}$. Proved.

(24): By Newton's laws of motion for a particle,
$\frac{d}{dt}\overrightarrow{v_{\text{cm}}}=(\text{acceleration of the c. m.})=\frac{d^2}{dt^2}(\text{position vector of the c. m.})=\frac{d^2}{dt^2}\left(\frac{\int\vec{r}(\mu\,dv)}{\int\mu\,dv}\right)=\frac{\int\left(\frac{d^2}{dt^2}\vec{r}\right)(\mu\,dv)}{\int\mu\,dv}=\frac{\int\left(\frac{d^2}{dt^2}\vec{r}\right)(\mu\,dv)}{M}=\frac{\int\vec{F}(\mu\,dv)}{M}\Rightarrow M\frac{d}{dt}\overrightarrow{v_{\text{cm}}}=\int\vec{F}(\mu\,dv)=\int\mu\vec{F}\,dv$. Proved. Next since $\vec{H}=\int\vec{r}\times\{(\mu\,dv)\vec{v}\}$ so, $\frac{d}{dt}\vec{H}=\int\left\{\frac{d}{dt}(\vec{r}\times\vec{v})\right\}(\mu\,dv)=\int\left(\vec{v}\times\vec{v}+\vec{r}\times\frac{d\vec{v}}{dt}\right)(\mu\,dv)=\int\vec{r}\times\frac{d\vec{v}}{dt}(\mu\,dv)=\int\vec{r}\times\{\vec{F}(\mu\,dv)\}=\int\mu\vec{r}\times\vec{F}\,dv$. Proved.

8

STATICS OF A RIGID BODY

Examples (page 164): (1): This is solved in the text.

(2): This is solved in the text.

(3): This is solved in the text.

(4): The equation of the central axis of system of forces is $\vec{r} = \frac{1}{R^2}\vec{R} \times \vec{G} + \lambda\vec{R}$. Its Plücker's vectors are the unit vector in the direction of line, that is, $\frac{1}{R}\vec{R}$ and the vector $\left(\frac{1}{R^2}\vec{R} \times \vec{G}\right) \times \vec{R} = \frac{1}{R^2}\left(R^2\vec{G} - (\vec{R}\cdot\vec{G})\vec{R}\right) = \vec{G} - \frac{\vec{R}\cdot\vec{G}}{R^2}\vec{R} = \vec{G} - \frac{\Gamma}{R^2}\vec{R} = \frac{R^2\vec{G} - \Gamma\vec{R}}{R^2}$. Proved.

Examples (page 166): (1): Let forces $F_1\widehat{d_1}$ and $F_2\widehat{d_2}$ be acting along two skew lines. Let $ON(= p)$ be the common perpendicular to the two skew lines. Let us take O as the origin. Let $A(\lambda_0\widehat{d_1})$ be any fixed point on the first line and $K\left(\frac{p(\widehat{d_1}\times\widehat{d_2})}{\sin\theta} + \mu\widehat{d_2}\right)$ be any variable point on the second line, where θ is the angle between $\widehat{d_1}$ and $\widehat{d_2}$. In the figure given below, we have $\vec{R} = F_1\widehat{d_1} + F_2\widehat{d_2}$ and $\vec{G} \equiv (\text{net torque about } O) = \left(\frac{p\widehat{d_1}\times\widehat{d_2}}{\sin\theta}\right) \times F_2\widehat{d_2} == \frac{pF_2}{\sin\theta}\left(\cos\theta\,\widehat{d_2} - \widehat{d_1}\right)$. Hence equation of the null plane of A is $\left(\vec{r} - \lambda_0\widehat{d_1}\right)\cdot\left\{\left(\frac{p(\widehat{d_1}\times\widehat{d_2})}{\sin\theta} - \lambda_0\widehat{d_1}\right) \times F_2\widehat{d_2}\right\} = 0 \quad \cdots (*)$. Clearly it passes through A and N. Putting the position vector of K for $\vec{r}$ in $(*)$, we get $\left[\frac{p(\widehat{d_1}\times\widehat{d_2})}{\sin\theta} + \mu\widehat{d_2} - \lambda_0\widehat{d_1} \;\; \frac{p(\widehat{d_1}\times\widehat{d_2})}{\sin\theta} - \lambda_0\widehat{d_1} \;\; F_2\widehat{d_2}\right] = 0$ or, $\left[-\mu\widehat{d_2} \;\; \frac{p(\widehat{d_1}\times\widehat{d_2})}{\sin\theta} - \lambda_0\widehat{d_1} \;\; F_2\widehat{d_2}\right] = 0$, which is clearly true. This proves the first part of the question. Proved.

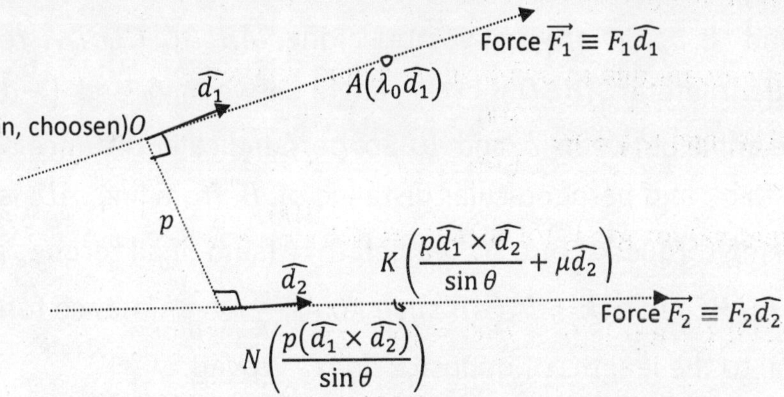

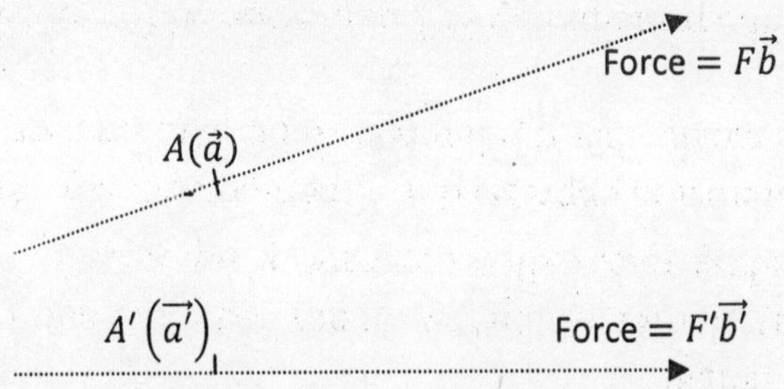

Hence the null plane of $A(\vec{a})$ is the plane containing the line of action of force $F'\vec{b'}$ and the point A. Its equation is $(\vec{r}-\vec{a})\cdot\left(\left(\vec{a}-\vec{a'}\right)\times F'\vec{b'}\right)=0$ or, $\left[\vec{r}-\vec{a}\ \ \vec{a}-\vec{a'}\ \ F'\vec{b'}\right]=0$ or, $\left[\vec{r}-\vec{a}\ \ \vec{a}-\vec{a'}\ \ \vec{b'}\right]=0$. Proved.

(2): It is given that $\vec{e}\times\left(\vec{G}-\vec{s}\times\vec{R}\right)=\vec{0}$ or, $\vec{e}\times\vec{G}-\left((\vec{e}\cdot\vec{R})\vec{s}-(\vec{e}\cdot\vec{s})\vec{R}\right)=0$ or, $\vec{s}=\frac{\vec{e}\times\vec{G}}{\vec{e}\cdot\vec{R}}+\lambda\vec{R}$. **Verification:** $LHS=\vec{e}\times\left(\vec{G}-\left(\frac{\vec{e}\times\vec{G}}{\vec{e}\cdot\vec{R}}+\lambda\vec{R}\right)\times\vec{R}\right)=\vec{e}\times\left(\vec{G}-\frac{(\vec{e}\cdot\vec{R})\vec{G}-(\vec{G}\cdot\vec{R})\vec{e}}{\vec{e}\cdot\vec{R}}\right)=\vec{e}\times\left(\vec{G}-\vec{G}+\frac{\vec{G}\cdot\vec{R}}{\vec{e}\cdot\vec{R}}\vec{e}\right)=\vec{0}=RHS$. Verified. Hence the locus of $\vec{s}$ is $\vec{s}=\frac{\vec{e}\times\vec{G}}{\vec{e}\cdot\vec{R}}+\lambda\vec{R}$, where λ is a parameter. This is a straight line through the point $\frac{\vec{e}\times\vec{G}}{\vec{e}\cdot\vec{R}}$ and parallel to $\vec{R}$. Proved.

Exercises on Chapter 8

(1): Let $ABCD$ be any cyclic quadrilateral. Let forces $F_1\hat{a}, F_2\hat{b}, F_3\hat{c}, F_4\hat{d}$ be acting on the lines AB, BC, CD, DA respectively, where F_1, F_2, F_3, F_4 are real numbers, and $\hat{a}, \hat{b}, \hat{c}, \hat{d}$ are unit vectors along $\overrightarrow{AB}, \overrightarrow{BC}, \overrightarrow{CD}, \overrightarrow{DA}$ respectively. Since the quadrilateral $ABCD$ is cyclic, angle between $\hat{a}$ and $(-\hat{d})$, say α, is the same as the angle between $\hat{c}$ and $\hat{b}$. So, perpendicular distance of B from line CD is $BC \sin\alpha$, and perpendicular distance of B from line AD is $AB \sin\alpha$. Now, on taking torques about B, we get $\vec{0} =$ (sum of all torques about B) $= \vec{0} + \vec{0} + F_3(BC \sin\alpha)\hat{k} + F_4(AB \sin\alpha)\hat{k} \Rightarrow \frac{F_3}{AB} = \frac{-F_4}{BC}$. Hence forces are proportional to the lengths of opposite sides. Proved.

(2):

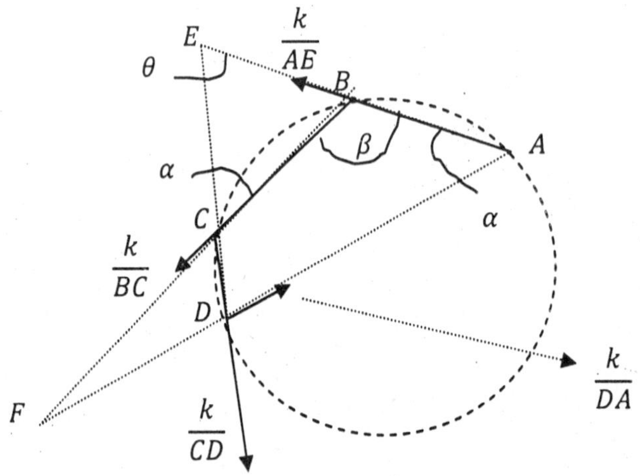

In our figure $ABCD$ is a cyclic quadrilateral. Here sides AB and CD intersect at E; sides AD and BC intersect at F. Let the angle DAB be α and angle ABC be β. Let the four coplanar forces be $\frac{k}{AB}, \frac{k}{BC}, \frac{k}{CD}$ and $\frac{k}{DA}$ as shown in the figure. Since these forces are coplanar, the system is equivalent to a single force, say $\vec{R}$, in the same plane. This shows that the net torque about any point is zero implies the line of action of $\vec{R}$ must pass through that point. Hence it is enough to prove that net torques about E and F both are zero. From figure, (net torque of the system about E) $= 0 + \left(-\frac{k}{BC}\right)(BE \sin\beta) + 0 + \frac{k}{DA}(DE \sin\beta) =$ $(k \sin\beta)\left(\frac{DE}{DA} - \frac{BE}{BC}\right) = (k \sin\beta)\left(\frac{\sin\alpha}{\sin\theta} - \frac{\sin\alpha}{\sin\theta}\right)$, from Δ's ADE and BCE. So, net torque of the system about E is zero. Similarly, net torque of the system about F is zero. Proved.

(3):

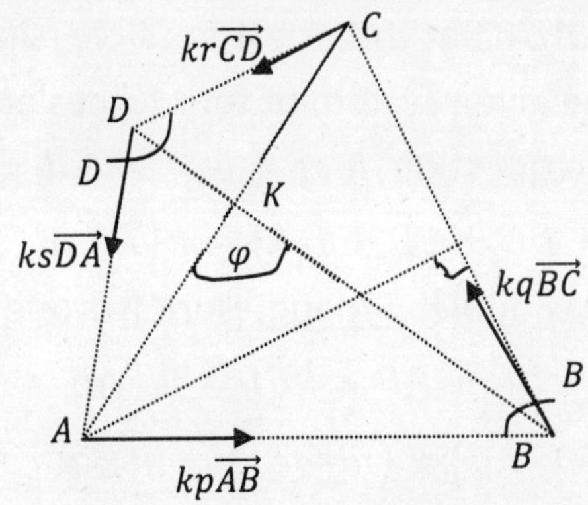

Let four forces acting at A, B, C, D be $kp\overrightarrow{AB}, kq\overrightarrow{BC}, kr\overrightarrow{CD}, ks\overrightarrow{DA}$ respectively. On taking torques (or moments) about A, we get $(kq\ BC)(AB\ \sin B) + (kr\ CD)(AD\sin(\pi - D)) = 0$ or,

$q(BC)(AB)\sin B = -r(CD)(AD)\sin D$ $\cdots(1)$

Similarly, by changing the symbols cyclically, we get $r(CD)(BC)\sin C = -s(DA)(BA)\sin A$ $\cdots(2)$

$s(DA)(CD)\sin D = -p(AB)(CB)\sin B$ $\cdots(3)$

and $p(AB)(DA)\sin A = -q(BC)(DC)\sin C$ $\cdots(4)$

Multiplying (2) and (4), we get $pr(AB)(BC)(CD)(DA)\sin C\sin A = qs(AB)(BC)(CD)(DA)\sin C\sin A$ or, $pr = qs$ $\cdots(*)$. Proved. Next, on taking torques about K (see the figure), we get $(kp\ AB)(AK\sin KAB)\hat{k} + \cdots + \cdots + \cdots = \vec{0}$ or, $p(AB)(AK)\left(\frac{\sin\varphi}{AB}KB\right)\hat{k} + \cdots + \cdots + \cdots = \vec{0}$ (using sine formula in triangle KAB). Hence $p(AK)(KB) + q(BK)(KC) + r(CK)(KD) + s(DK)(KA) = 0$ or, $\big(p(AK) + q(KC)\big)(KB) + r(KD)\left(CK + \frac{p}{q}KA\right) = 0$, from $(*)$. Hence $\big(p(AK) + q(KC)\big)\big(q(BK) + r(KD)\big) = 0$ $\cdots(5)$

So, $\frac{AK}{KC} = -\frac{q}{p}$, that is, K divides the diagonal AC in the ratio $p : q$. Similarly, from (5), K divides diagonal BD in the ratio $q : r$. Proved.

(4): If not, let they be in equilibrium. Here $ABCD$ is a skew quadrilateral, so let plane ACD and plane ACB be different. Now $\vec{0} =$ (the sum of all torques about C)$= \overrightarrow{CD} \times \left(d(\overrightarrow{DA})\right) + \overrightarrow{CB} \times \left(a(\overrightarrow{AB})\right) = d(\overrightarrow{CD} \times \overrightarrow{DA}) + a(\overrightarrow{CB} \times \overrightarrow{AB})$, and hence $\overrightarrow{CD} \times \overrightarrow{DA} \parallel \overrightarrow{CB} \times \overrightarrow{AB}$. But this is not possible, since planes ACD and ACB are different. So, the given system of four forces are not in equilibrium. This proves the first part of the question. If $a = b = c = d = k$ (say) then the resultant force is $k(\overrightarrow{AB} + \overrightarrow{BC} + \overrightarrow{CD} + \overrightarrow{DA}) = k(\overrightarrow{AC} + \overrightarrow{CA}) = \vec{0}$. So, the system of forces is equivalent to a couple. Proved. Here (torque of the couple) = (net torque about A) $= \overrightarrow{AB} \times k\overrightarrow{BC} + \overrightarrow{AD} \times k\overrightarrow{CD} = k(\overrightarrow{AB} \times \overrightarrow{BC} + \overrightarrow{AD} \times \overrightarrow{CD}) = k\left((\vec{b} - \vec{a}) \times (\vec{c} - \vec{b}) + (\vec{d} - \vec{a}) \times (\vec{d} - \vec{c})\right) = k(\vec{b} \times \vec{c} + \vec{c} \times \vec{a} + \vec{a} \times \vec{b} + \vec{c} \times \vec{d} + \vec{d} \times a + \vec{a} \times \vec{c}) = k(\vec{b} - \vec{d}) \times (\vec{c} - \vec{a}) = k(\overrightarrow{DB} \times \overrightarrow{AC}) \parallel$ (common perpendicular to AC and BD). So, the plane of the couple is parallel to AC and BD. Proved. Next let $ac = bd$ or, $\frac{b}{a} = \frac{c}{d} \equiv t$ (suppose). So, the four forces are $a\overrightarrow{AB}, ta\overrightarrow{BC}, td\overrightarrow{CD}, d\overrightarrow{DA}$. The resultant of the first two forces acts at B and is equal to $a(\overrightarrow{AB} + t\overrightarrow{BC}) = a(t-1)\left(\frac{t\overrightarrow{BC} - \overrightarrow{BA}}{t-1}\right) = a(t-1)\overrightarrow{BK_1}$, where K_1 divides AC in the ratio $AK_1 : K_1C = t : (-1)$ $\cdots (*)$. Again the resultant of last two forces acts at D and is equal to $d(t\overrightarrow{CD} + \overrightarrow{DA}) = -d(t-1)\left(\frac{t\overrightarrow{DC} - \overrightarrow{DA}}{t-1}\right) = -d(t-1)\overrightarrow{DK_2}$ where K_2 divides AC in the ratio $AK_2 : K_2C = t : (-1)$. Now, from $(*)$, K_1 and K_2 are the same points, say K, which lies on the diagonal AC. So, the resultant force of the four forces will act at K. Thus the line of action of resultant force intersects the diagonal AC. Similarly the line of action of resultant force intersects the diagonal BD. Proved.

(5): Let n forces $k\overrightarrow{A_1A_2}, k\overrightarrow{A_2A_3}, \cdots, k\overrightarrow{A_{n-1}A_n}, k\overrightarrow{A_nA_1}$ be acting along the sides $A_1A_2, A_2A_3, \cdots, A_{n-1}A_n, A_nA_1$ respectively, of a skew polygon. Here resultant force is $(k\overrightarrow{A_1A_2} + k\overrightarrow{A_2A_3} + \cdots + k\overrightarrow{A_{n-1}A_n} + k\overrightarrow{A_nA_1}) = k(\overrightarrow{A_1A_2} + \overrightarrow{A_2A_3} + \cdots + \overrightarrow{A_{n-1}A_n} + \overrightarrow{A_nA_1}) = k(\overrightarrow{A_1A_3} + \overrightarrow{A_3A_4} + \cdots + \overrightarrow{A_{n-1}A_n} + \overrightarrow{A_nA_1}) = \cdots = k(\overrightarrow{A_1A_n} + \overrightarrow{A_nA_1}) = \vec{0}$. So, the given system is equivalent to a couple. Proved.

(6): Let $OABC$ be tetrahedron. Let six forces $a\overrightarrow{OA}, b\overrightarrow{OB}, c\overrightarrow{OC}, p\overrightarrow{AB}, q\overrightarrow{BC}, r\overrightarrow{CA}$ be acting along the lines OA, OB, OC, AB, BC, CA respectively, where a, b, c, p, q, r are any nonzero real numbers (either positive or negative). We have to prove that the body under these forces will not be in equilibrium. If not, otherwise, let the body be in equilibrium. Hence torque about any point is $\vec{0}$. Here three forces $p\overrightarrow{AB}, q\overrightarrow{BC}, r\overrightarrow{CA}$ are coplanar, so it is equivalent to a single force, say $\vec{F}$, acting in the plane of ABC, or a couple having torque, say $\vec{L}$, perpendicular to the plane of ABC. Further forces $a\overrightarrow{OA}, b\overrightarrow{OB}$ and $c\overrightarrow{OC}$ have no torque about O. Hence (the net torque about O) = (torque of $\vec{F}$ about O) $= \vec{L}$, which is nonzero. This contradicts the assumption. Proved.

(7): For convenience, let us denote D by O, and let O be the origin. Let $A(\vec{a}), B(\vec{b}), C(\vec{c})$. Suppose the six forces in the directions of $\overrightarrow{AB}, \overrightarrow{BC}, \overrightarrow{CA}, \overrightarrow{OA}, \overrightarrow{OB}, \overrightarrow{OC}$ be $k\overrightarrow{AB}, k\overrightarrow{BC}, k\overrightarrow{CA}, k\overrightarrow{OA}, k\overrightarrow{OB}, k\overrightarrow{OC}$. These forces have equal magnitudes, because the tetrahedron is regular. Here resultant of $k\overrightarrow{AB}, k\overrightarrow{BC}, k\overrightarrow{CA}$ is $\vec{O}$, so the system of six forces acting on a body is equivalent to three concurrent forces $k\overrightarrow{OA}, k\overrightarrow{OB}, k\overrightarrow{OC}$ acting at O. This is equivalent to a single force $k\overrightarrow{OA} + k\overrightarrow{OB} + k\overrightarrow{OC}$ acting at O. It remains to prove that force $k\overrightarrow{OA} + k\overrightarrow{OB} + k\overrightarrow{OC}$ is perpendicular to the plane ABC; that is, $\vec{a} + \vec{b} + \vec{c}$ is perpendicular to the plane ABC. So, we must prove $(\vec{a} + \vec{b} + \vec{c}) \times \{(\vec{b} - \vec{a}) \times (\vec{c} - \vec{a})\} = \vec{0}$. Here $LHS = (\vec{a} + \vec{b} + \vec{c}) \times (\vec{b} \times \vec{c} + \vec{c} \times \vec{a} + \vec{a} \times \vec{b}) = \vec{a} \times (\vec{b} \times \vec{c}) + \vec{a} \times (\vec{c} \times \vec{a}) + \vec{a} \times (\vec{a} \times b) + \vec{b} \times (\vec{b} \times \vec{c}) + \vec{b} \times (\vec{c} \times a) + \vec{b} \times (\vec{a} \times \vec{b}) + \vec{c} \times (\vec{b} \times \vec{c}) + \vec{c} \times (\vec{c} \times \vec{a}) + \vec{c} \times (\vec{a} \times \vec{b}) = \{(\vec{a} \cdot \vec{c})\vec{b} - (\vec{a} \cdot \vec{b})\vec{c}\} + \cdots + \cdots + \cdots + \cdots + \cdots + \cdots + \cdots + \cdots = \vec{0}$ (after cancellation and using the facts that $a^2 = b^2 = c^2$, $\vec{b} \cdot \vec{c} = \vec{c} \cdot \vec{a} = \vec{a} \cdot \vec{b}$ in a regular tetrahedron). Hence central axis is the line from D perpendicular to plane ABC. Proved.

(8): Let $D(\vec{0})$. Here resultant force is given by $\vec{R} = k(\vec{b} - \vec{a}) + k(\vec{c} - \vec{b}) + k(\vec{a} - \vec{c}) + k(\vec{a} - \vec{d}) + k(\vec{b} - \vec{d}) + k(\vec{c} - \vec{d})$ so, $\vec{R} = k(\vec{a} + \vec{b} + \vec{c} - 3\vec{d}) = k(\vec{a} + \vec{b} + \vec{c} - 3\vec{0}) = k(\vec{a} + \vec{b} + \vec{c}) = 3k\overrightarrow{OG}$. Also, the net torque about O is given by $\vec{G} = \vec{a} \times k(\vec{b} - \vec{a}) + \vec{b} \times k(\vec{c} - \vec{b}) + \vec{c} \times k(\vec{a} - \vec{c}) + \vec{d} \times \left(k(\vec{a} - \vec{d}) + k(\vec{b} - \vec{d}) + k(\vec{c} - \vec{d})\right)$ so, $\vec{G} = k(\vec{a} \times \vec{b} + \vec{b} \times \vec{c} + \vec{c} \times \vec{a}) + k\vec{d} \times (\vec{a} + \vec{b} + \vec{c})$ or, $\vec{G} = k(\vec{a} \times \vec{b} + \vec{b} \times \vec{c} + \vec{c} \times \vec{a}) + k\vec{0} \times (\vec{a} + \vec{b} + \vec{c})$ or, $\vec{G} = k(\vec{a} \times \vec{b} + \vec{b} \times \vec{c} + \vec{c} \times \vec{a}) \perp$ (the plane ABC). So, equation of the central axis is $\vec{r} = \frac{3k\overrightarrow{OG} \times k(\vec{a}\times\vec{b}+\vec{b}\times\vec{c}+\vec{c}\times\vec{a})}{3k\overrightarrow{OG}} + \lambda(3k\overrightarrow{OG})$, where λ is a parameter. Clearly this line is parallel to OG (that is, DG).

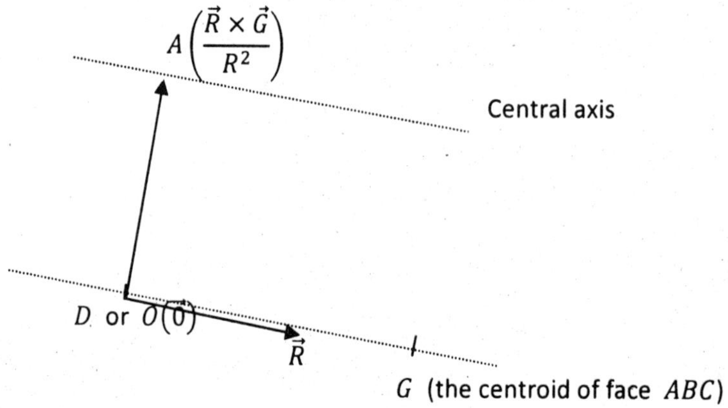

Here $\overrightarrow{AD} \cdot \vec{R} = \left(\frac{\vec{R}\times\vec{G}}{R^2} - \vec{0}\right) \cdot \vec{R} = \frac{1}{R^2}[\vec{R}\ \vec{G}\ \vec{R}] = 0$ so, $AD \perp \vec{R}$. Hence

perpendicular distance of the central axis from the line DG is $\left|\frac{\vec{R}\times\vec{G}}{R^2}\right|$ $\cdots(1)$

Further $\Delta =$ (area of triangle ABC) $= \left|\frac{1}{2}(\vec{a} - \vec{b}) \times (\vec{c} - \vec{b})\right| = \frac{1}{2}\left|-(\vec{a} \times \vec{b} + \vec{b} \times \vec{c} + \vec{c} \times \vec{a})\right| = \frac{1}{2}\left|\vec{a} \times \vec{b} + \vec{b} \times \vec{c} + \vec{c} \times \vec{a}\right|$ and hence $\frac{1}{DG}\left(\frac{2}{3}\Delta \cos\varphi\right) =$

$\frac{1}{\frac{1}{3}|\vec{a}+\vec{b}+\vec{c}|}\frac{2}{3}\left(\frac{1}{2}\left|\vec{a} \times \vec{b} + \vec{b} \times \vec{c} + \vec{c} \times \vec{a}\right|\right)\cos\varphi =$

$\frac{(k|\vec{a}+\vec{b}+\vec{c}|)(k|\vec{a}\times\vec{b}+\vec{b}\times\vec{c}+\vec{c}\times\vec{a}|)\sin(90°-\varphi)}{(k|\vec{a}+\vec{b}+\vec{c}|)^2} = \frac{|\vec{R}\times\vec{G}|}{R^2} =$ (perpendicular distance of the central axis from the line DG), from (1). Proved.

(9):

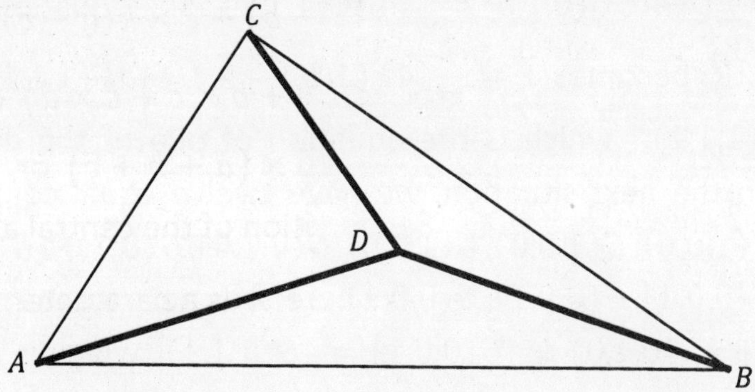

Let $A(\vec{a}), B(\vec{b}), C(\vec{c})\ D(\vec{d})$ be the vertices of a tetrahedron. The vectorial area of ΔABC is $\frac{1}{2}(\vec{b}-\vec{a}) \times (\vec{c}-\vec{a}) = \frac{1}{2}(\vec{b}\times\vec{c}+\vec{c}\times\vec{a}+\vec{a}\times\vec{b})$. From question, let the force acting at D be $k\left\{\frac{1}{2}(\vec{b}\times\vec{c}+\vec{c}\times\vec{a}+\vec{a}\times\vec{b})\right\}$, force acting at A be $\frac{k}{2}(\vec{d}\times\vec{c}+\vec{c}\times\vec{b}+\vec{b}\times\vec{d})$, force acting at C be $\frac{k}{2}(\vec{d}\times\vec{b}+\vec{b}\times\vec{a}+\vec{a}\times\vec{d})$ and force acting at B be $\frac{k}{2}(\vec{d}\times\vec{a}+\vec{a}\times\vec{c}+\vec{c}\times\vec{d})$. So, the resultant force is given by $\vec{R} = \frac{k}{2}\{(\vec{b}\times\vec{c}+\vec{c}\times\vec{a}+\vec{a}\times\vec{b}) + (\vec{d}\times\vec{c}+\vec{c}\times\vec{b}+\vec{b}\times\vec{d}) + (\vec{d}\times\vec{b}+\vec{b}\times\vec{a}+\vec{a}\times\vec{d}) + (\vec{d}\times\vec{a}+\vec{a}\times\vec{c}+\vec{c}\times\vec{d})\} = \frac{k}{2}\vec{0} = \vec{0}$ and the net torque about the origin O is given by $\vec{G} = \sum \vec{d}\times\frac{k}{2}(\vec{b}\times\vec{c}+\vec{c}\times\vec{a}+\vec{a}\times\vec{b}) = \frac{k}{2}\sum[\{(\vec{c}\cdot\vec{d})\vec{b} - (\vec{b}\cdot\vec{d})\vec{c}\} + \{(\vec{d}\cdot\vec{a})\vec{c} - (\vec{c}\cdot\vec{d})\vec{a}\} + \{(\vec{b}\cdot\vec{d})\vec{a} - (\vec{d}\cdot\vec{a})\vec{b}\}] = \frac{k}{2}\sum[\{(\vec{b}-\vec{c})\cdot\vec{d}\}\vec{a} + \{(\vec{c}-\vec{a})\cdot\vec{d}\}\vec{b} + \{(\vec{a}-\vec{b})\cdot\vec{d}\}\vec{c}] = \vec{0}$, after cancellation. So, the system of forces is in equilibrium. Proved.

(10): Let the twelve equal forces acting along the edges of a cube be $F(1,0,0), F(1,0,0), F(1,0,0), F(1,0,0), F(0,1,0), F(0,1,0), F(0,1,0), F(0,1,0), F(0,0,1), F(0,0,1), F(0,0,1), F(0,0,1)$. Let $2a$ be the length of each side of the cube. Here resultant force is given by $\vec{R} = F(1,0,0) + F(1,0,0) + F(1,0,0) + F(1,0,0) + F(0,1,0) + F(0,1,0) + F(0,1,0) + F(0,1,0) + F(0,0,1) + F(0,0,1) + F(0,0,1) + F(0,0,1) = F(4,4,4) = 4F(1,1,1)$, and the net torque about the origin O is given by $\vec{G} = \sum\{(a,-a,-a)\times F(1,0,0) + (a,a,-a)\times F(1,0,0) + (-a,a,-a)\times F(1,0,0) + (-a,-a,-a)\times F(1,0,0)\} = \sum(-4a)(0,0,1)\times$

$F(1,0,0) = \sum(-4aF)(0,1,0) = -4aF(1,1,1)$. So, equation of the central axis $\vec{r} = \frac{\vec{R}\times\vec{G}}{R^2} + \lambda\vec{R}$ becomes $\vec{r} = \frac{1}{R^2}\{4F(1,1,1)\} \times \{-4aF(1,1,1)\} + \lambda\vec{R} = \vec{0} + \lambda\vec{R}$ or, $\vec{r} = \lambda(4F)(1,1,1)$, which is the equation of one of the diagonals of the cube. Proved. In the next situation, we have twelve equal couples having moments $G(1,0,0), G(1,0,0), G(1,0,0), G(1,0,0), G(0,1,0), G(0,1,0), G(0,1,0), G(0,1,0)$, $G(0,0,1), G(0,0,1), G(0,0,1), G(0,0,1)$. Here net torque about any point is $4G(1,0,0) + 4G(0,1,0) + 4G(0,0,1) = 4G(1,1,1)$, which is parallel to one of the diagonals. Also net force is zero, because sum of forces constituting a couple is zero. Thus the system is equivalent to a couple having moment parallel to one of the diagonal of the cube. So, the central axis is parallel to one of the diagonal of the cube. Proved.

(11): The net torque about any point on the central axis is $\frac{\Gamma}{R^2}\vec{R}$. Let us take any point O on the central axis as the origin. Now, for any point $P(\vec{r})$, we have $|\text{net torque about } A| = \sqrt{\left(\frac{\Gamma}{R^2}\vec{R} + \vec{r}\times\vec{R}\right)^2} = \sqrt{\frac{\Gamma^2}{R^2} + \left(\vec{r}\times\vec{R}\right)^2} =$ $\sqrt{\frac{\Gamma^2}{R^2} + r^2R^2 - \left(\vec{r}\cdot\vec{R}\right)^2} = \sqrt{\frac{\Gamma^2}{R^2} + r^2R^2(1-\cos^2\theta)}$, where θ is the angle between $\overrightarrow{OP}$ and $\vec{R}$ (or the central axis). So, $|\text{net torque about } A| =$ $\sqrt{\frac{\Gamma^2}{R^2} + R^2(r\sin\theta)^2} =$ $\sqrt{\frac{\Gamma^2}{R^2} + R^2(\text{perpendicular distance of } P \text{ from the central axis})^2}$, which is minimum when P lies on the central axis. Proved.

(12): Let O be the point where the central axis intersects a plane Π perpendicular to it. Let us choose O as the origin. Let $A_i(\vec{a_i})$ and $A_i'(\vec{a_i'})$. Here resultant force is given by $\vec{R} = \sum(\vec{a_i'} - \vec{a_i}) = \sum\vec{a_i'} - \sum\vec{a_i} = n\overrightarrow{OG'} - n\overrightarrow{OG} =$ $n\overrightarrow{GG'}$, which is parallel to GG'. This proves the first part of the question. Also $\vec{G} = \sum\vec{a_i}\times(\vec{a_i'} - \vec{a_i}) = \sum(\vec{a_i}\times\vec{a_i'}) \quad \cdots(*)$. Here $\vec{R} \parallel \vec{G}$, because O lies on the central axis. Here resolute of the i th force in the direction of central axis is

$\left(\overrightarrow{a_\iota'}-\overrightarrow{a_\iota}\right)\cdot\frac{\vec{R}}{R}$. Also, equation of the plane Π is $\vec{r}\cdot\vec{R}=0$, and equation of the line of action of i th force is $\vec{r}=\overrightarrow{a_\iota}+\lambda\left(\overrightarrow{a_\iota'}-\overrightarrow{a_\iota}\right)$. Let $B_i\left(\overrightarrow{b_\iota}\right)$. So, $\overrightarrow{b_\iota}\cdot\vec{R}=0$ and $\overrightarrow{b_\iota}=\overrightarrow{a_\iota}+\lambda\left(\overrightarrow{a_\iota'}-\overrightarrow{a_\iota}\right)$. Hence $\lambda=\frac{-\overrightarrow{a_\iota}\cdot\vec{R}}{\left(\overrightarrow{a_\iota'}-\overrightarrow{a_\iota}\right)\cdot\vec{R}}$ and therefore $\overrightarrow{b_\iota}=\overrightarrow{a_\iota}+$ $\frac{-\overrightarrow{a_\iota}\cdot\vec{R}}{\left(\overrightarrow{a_\iota'}-\overrightarrow{a_\iota}\right)\cdot\vec{R}}\left(\overrightarrow{a_\iota'}-\overrightarrow{a_\iota}\right)=\frac{\left(\overrightarrow{a_\iota'}\cdot\vec{R}-\overrightarrow{a_\iota}\cdot\vec{R}\right)\overrightarrow{a_\iota}-\left(\overrightarrow{a_\iota}\cdot\vec{R}\right)\overrightarrow{a_\iota'}+\left(\overrightarrow{a_\iota}\cdot\vec{R}\right)\overrightarrow{a_\iota}}{\left(\overrightarrow{a_\iota'}-\overrightarrow{a_\iota}\right)\cdot\vec{R}}=\frac{\left(\overrightarrow{a_\iota'}\cdot\vec{R}\right)\overrightarrow{a_\iota}-\left(\overrightarrow{a_\iota}\cdot\vec{R}\right)\overrightarrow{a_\iota'}}{\left(\overrightarrow{a_\iota'}-\overrightarrow{a_\iota}\right)\cdot\vec{R}}=$ $\frac{\left(\overrightarrow{a_\iota'}\times\overrightarrow{a_\iota}\right)\times\vec{R}}{\left(\overrightarrow{a_\iota'}-\overrightarrow{a_\iota}\right)\cdot\vec{R}}=\frac{\left(\overrightarrow{a_\iota'}\times\overrightarrow{a_\iota}\right)\times\hat{e}}{\left(\overrightarrow{a_\iota'}-\overrightarrow{a_\iota}\right)\cdot\hat{e}}$ $\cdots(**)$ where $\hat{e}$ is the unit vector in the direction of $\vec{R}$.

So, the position vector of the required centroid is

$\frac{\Sigma\left\{\left(\overrightarrow{a_\iota'}-\overrightarrow{a_\iota}\right)\cdot\frac{\vec{R}}{R}\right\}\overrightarrow{b_\iota}}{\Sigma\left\{\left(\overrightarrow{a_\iota'}-\overrightarrow{a_\iota}\right)\cdot\frac{\vec{R}}{R}\right\}}=\frac{\Sigma\left\{\left(\overrightarrow{a_\iota'}-\overrightarrow{a_\iota}\right)\cdot\hat{e}\right\}\overrightarrow{b_\iota}}{\left\{\Sigma\left(\overrightarrow{a_\iota'}-\overrightarrow{a_\iota}\right)\right\}\cdot\hat{e}}=\frac{\Sigma\left\{\left(\overrightarrow{a_\iota'}\times\overrightarrow{a_\iota}\right)\times\hat{e}\right\}}{\vec{R}\cdot\hat{e}}$, from $(**)$. So, the position vector of the required centroid is $\frac{-\left(\Sigma\left(\overrightarrow{a_\iota}\times\overrightarrow{a_\iota'}\right)\right)\times\hat{e}}{R}=\frac{-\vec{G}\times\hat{e}}{R}=\vec{0}$, because $\vec{G}\parallel\hat{e}$. This was to be proved. Proved.

(13): Let $A_1A_2A_3\cdots A_{n-1}A_nA_1$ be a plane polygon. Let O be the origin in the plane of the polygon. Let $\hat{k}$ be the unit vector perpendicular to the plane of the polygon. So, we can suppose that force acting at the mid-point $M_1\left(\frac{\overrightarrow{a_1}+\overrightarrow{a_2}}{2}\right)$ of A_1A_2 is $\hat{k}\times\overrightarrow{A_1A_2}$. Similarly the force acting at the mid-point $M_2\left(\frac{\overrightarrow{a_2}+\overrightarrow{a_3}}{2}\right)$ of A_2A_3 is $\hat{k}\times\overrightarrow{A_2A_3}$, etc. So, the resultant is given by $\vec{R}=\hat{k}\times\overrightarrow{A_1A_2}+\hat{k}\times\overrightarrow{A_2A_3}+\cdots+\hat{k}\times\overrightarrow{A_nA_1}=\hat{k}\times\left(\overrightarrow{A_1A_2}+\overrightarrow{A_2A_3}+\cdots+\overrightarrow{A_nA_1}\right)=\hat{k}\times\vec{0}=\vec{0}$. Further the net torque about O is $\frac{\overrightarrow{a_1}+\overrightarrow{a_2}}{2}\times\{\hat{k}\times(\overrightarrow{a_2}-\overrightarrow{a_1})\}+\frac{\overrightarrow{a_2}+\overrightarrow{a_3}}{2}\times\{\hat{k}\times(\overrightarrow{a_3}-\overrightarrow{a_2})\}+\cdots+\frac{\overrightarrow{a_n}+\overrightarrow{a_1}'}{2}\times$ $\{\hat{k}\times(\overrightarrow{a_1}-\overrightarrow{a_n})\}=\frac{1}{2}[\{(a_2{}^2-a_1{}^2)\hat{k}+(a_3{}^2-a_2{}^2)\hat{k}+\cdots(a_1{}^2-a_n{}^2)\hat{k}\}+$ $\{\hat{k}\cdot(\overrightarrow{a_1}+\overrightarrow{a_2})\}(\overrightarrow{a_2}-\overrightarrow{a_1})+\{\hat{k}\cdot(\overrightarrow{a_2}+\overrightarrow{a_3})\}(\overrightarrow{a_3}-\overrightarrow{a_2})+\cdots]$. Since $\hat{k}\cdot\overrightarrow{a_\iota}=0$ for every $i=1,2,\cdots,n$ so, the net torque about O is zero. So, the body is in equilibrium. Proved.

(14): Let us cut the given polyhedron by parallel planes into very thin polygons. By the above Ex.13, forces corresponding to each polygon is in equilibrium. On taking aggregate, we find that the given system of forces is in equilibrium. Proved.

(15): Let us observe that every closed curved surface can be approximated by a polyhedron. Now, on applying the result of the above Ex.14 on polyhedron, the proof is clear. Proved.

(16): On using the usual sumbols, it is given that $\vec{F_1}+\vec{F_2}+\vec{F_3}+\vec{F_4}=\vec{0}$ $\cdots(1)$ and $\vec{r_1}\times\vec{F_1}+\vec{r_2}\times\vec{F_2}+\vec{r_3}\times\vec{F_3}+\vec{r_4}\times\vec{F_4}=\vec{0}$ $\cdots(2)$. To prove (i): $\frac{(\vec{F_1}+\vec{F_2})\cdot(\vec{r_1}\times\vec{F_1}+\vec{r_2}\times\vec{F_2})}{(\vec{F_1}+\vec{F_2})^2}=\frac{(\vec{F_3}+\vec{F_4})\cdot(\vec{r_3}\times\vec{F_3}+\vec{r_4}\times\vec{F_4})}{(\vec{F_3}+\vec{F_4})^2}$ (ii): $\frac{(\vec{F_1}+\vec{F_2}+\vec{F_3})\cdot(\vec{r_1}\times\vec{F_1}+\vec{r_2}\times\vec{F_2}+\vec{r_3}\times\vec{F_3})}{(\vec{F_1}+\vec{F_2}+\vec{F_3})^2}=0$. For (i), $LHS=\frac{\{-(\vec{F_3}+\vec{F_4})\}\cdot\{-(\vec{r_3}\times\vec{F_3}+\vec{r_4}\times\vec{F_4})\}}{\{-(\vec{F_3}+\vec{F_4})\}^2}$, on using (1) and (2). So, $LHS=RHS$. Proved. For (ii), $LHS=\frac{(-\vec{F_4})\cdot(-\vec{r_4}\times\vec{F_4})}{(-\vec{F_4})^2}=\frac{[\vec{F_4}\ \vec{r_4}\ \vec{F_4}]}{F_4{}^2}=0=RHS$. Proved.

(17): Here required equation of null plane of the point (pB,pA,C) is $(\vec{r}-(pB,pA,C))\cdot(G\hat{k}-(pB,pA,C)\times R\hat{k})=0$ where R denotes the magnitude of the resultant force and $G\equiv pR$. So, equation of null plane of the point (pB,pA,C) can be written as $(\vec{r}-(pB,pA,C))\cdot(-pAR,pBR,G)=0$ or, $(\vec{r}-(pB,pA,C))\cdot(-GA,GB,G)=0$ or, $(\vec{r}-(pB,pA,C))\cdot(A,-B,-1)=0$ or, $\vec{r}\cdot(A\hat{\imath}-B\hat{\jmath}-\hat{k})-pAB+pAB+C=0$ or, $\vec{r}\cdot(A\hat{\imath}-B\hat{\jmath}-\hat{k})+C=0$. Proved. This equation can also be written, in Cartesian form, as $Ax-By-z+C=0$. This is a null plane of point (pB,pA,C). For the second part of the question, let $ax-by-z+c=0$ be any plane. From the above result, we can suppose that this be the null plane of point (pb,pa,c), where p denotes the pitch of the system of forces. Clearly our plane intersects z-axis at $(0,0,c)$. So, $P(0,0,c)$ and $Q(pb,pa,c)$. Here $\overrightarrow{PQ}\cdot\hat{k}=(pb,pa,0)\cdot(0,0,1)=0$. So, $\overrightarrow{PQ}\perp\hat{k}$, that is, $\overrightarrow{PQ}$ is perpendicular to the central axis. Here $90^\circ-\varphi$ is the angle between $\hat{k}$ and normal to the plane, so $\sin\varphi=\cos(90^\circ-\varphi)=(0,0,1)\cdot\frac{(-a,b,1)}{|(-a,b,1)|}=\frac{1}{\sqrt{a^2+b^2+1}}$ and hence $\cot\varphi=\sqrt{\csc^2\varphi-1}=\sqrt{a^2+b^2}$. So, $PQ=\sqrt{p^2b^2+p^2a^2}=p\sqrt{a^2+b^2}=p\cot\varphi$. Proved.

(18): Let the two lines intersect at the origin O. Let $\hat{d}$ and $\widehat{d'}$ be two unit vectors in the direction of the two lines. Here equation of the conjugate line to the first line $\vec{r}=\lambda\hat{d}$ $\cdots(1)$

is given by the simultaneous equations

$$\begin{cases} \vec{G}\cdot\vec{r}=0 \\ \hat{d}\cdot(\vec{G}-\vec{r}\times\vec{R})=0 \text{ or } (\vec{R}\times\hat{d})\cdot\vec{r}=\hat{d}\cdot\vec{G} \end{cases}.$$

So, the line conjugate to (1) lies on the plane $\vec{r}\cdot\vec{G}=0$, which is free from $\hat{d}$. Similarly the line conjugate to the second line $\vec{r}=\mu\widehat{d'}$ ⋯ (2)

lie on the plane $\vec{r}\cdot\vec{G}=0$. In general, these lines will intersect at a point. Further we see that both conjugate lines of the given lines lie on the plane $\vec{r}\cdot\vec{G}=0$, which is the null plane of the point of intersection of the given lines. Proved.

(19): Let $A(\vec{a}), B(\vec{b}), C(\vec{c})$ be three collinear points. Here equations of null planes of A, B, C are

$$\begin{cases} (\vec{G}-\vec{a}\times\vec{R})\cdot\vec{r}=\vec{a}\cdot\vec{G} & \cdots(1) \\ (\vec{G}-\vec{b}\times\vec{R})\cdot\vec{r}=\vec{b}\cdot\vec{G} & \cdots(2) \\ (\vec{G}-\vec{c}\times\vec{R})\cdot\vec{r}=\vec{c}\cdot\vec{G} & \cdots(3) \end{cases}$$

respectively. Since $A(\vec{a}), B(\vec{b}), C(\vec{c})$ are three collinear points, so there exist three scalars λ, μ, ν such that $\lambda\vec{a}+\mu\vec{b}+\nu\vec{c}=\vec{0}$ ⋯ (4)

and $\lambda+\mu+\nu=0$ ⋯ (5)

Let us observe that $\lambda(\vec{G}-\vec{a}\times\vec{R})+\mu(\vec{G}-\vec{b}\times\vec{R})+\nu(\vec{G}-\vec{c}\times\vec{R})=$ $(\lambda+\mu+\nu)\vec{G}-(\lambda\vec{a}+\mu\vec{b}+\nu\vec{c})\times\vec{R}=0\vec{G}-\vec{0}\times\vec{R}=\vec{0}$, on using (4) and (5). So, $\vec{G}-\vec{a}\times\vec{R}, \vec{G}-\vec{b}\times\vec{R}, \vec{G}-\vec{c}\times\vec{R}$ are coplanar vectors. So, there exist a vector, say $\vec{d}$, which is perpendicular to each of the three vectors $\vec{G}-\vec{a}\times\vec{R}, \vec{G}-\vec{b}\times\vec{R}, \vec{G}-\vec{c}\times\vec{R}$ ⋯ $(*)$. Next let us observe that (1), (2), (3) are three simultaneous equations of 1st degree in catesian coordinates x, y, z, so, by Cramer's rule, we can find x, y, z which satisfy all the three equations; that is, there exists a vector, say $\vec{\alpha}$, which satisfy equations (1), (2), (3) ⋯ $(**)$. Now consider the equation of line: $\vec{r}=\vec{\alpha}+t\vec{d}$ where t is a parameter. Since $(\vec{G}-\vec{a}\times\vec{R})\cdot(\vec{\alpha}+t\vec{d})=(\vec{G}-\vec{a}\times\vec{R})\cdot\vec{\alpha}+t(\vec{G}-\vec{a}\times\vec{R})\cdot\vec{d}=\vec{a}\cdot\vec{G}+t(0)=$ $\vec{a}\cdot\vec{G}$, on using $(*)$ and $(**)$, so the line $\vec{r}=\vec{\alpha}+t\vec{d}$ lies on the plane (1). Similarly the same line lie on planes (2) and (3). Hence the three planes (1), (2), (3) intersect in a common line. Proved.

(20): Let the given system of forces be equivalent to a single force $R\hat{d}$ and a couple $(\vec{F}, -\vec{F})$ of torque $pR\hat{d}$ acting at a point $A(\vec{a})$ of the central axis $\vec{r} = \vec{a} + \lambda\hat{d}$, where p denotes pitch of the given system of forces, $\hat{d}$ denotes unit vector along the central axis and λ denotes parameter. Let us break the force $R\hat{d}$ acting at A into forces $\frac{1}{2}R\hat{d}$ acting at A and $\frac{1}{2}R\hat{d}$ acting at A. Here perpendicular distance between forces $\vec{F}$ and $-\vec{F}$ is $\frac{pR}{F}$ and $\vec{F}$ is perpendicular to $\hat{d}$. So, we can assume that $\vec{F}$ acts at a distance $\frac{1}{2}\left(\frac{pR}{F}\right)$ from A in the plane $(\vec{r} - \vec{a}) \cdot \hat{d} = 0$ and $-\vec{F}$ acts at a distance $\frac{1}{2}\left(\frac{pR}{F}\right)$ from A in the plane $(\vec{r} - \vec{a}) \cdot \hat{d} = 0$. Let A_1 be a point on the line of action of $\vec{F}$ and A_2 be a point on the line of action of $-\vec{F}$. Since $\vec{F} \cdot \hat{d} = 0$, so $\left|\vec{F} + \frac{1}{2}R\hat{d}\right| = \left|-\vec{F} + \frac{1}{2}R\hat{d}\right|$. Also $\hat{d} \cdot \left(\vec{F} + \frac{1}{2}R\hat{d}\right) = \frac{R}{2} = \hat{d} \cdot \left(-\vec{F} + \frac{1}{2}R\hat{d}\right)$. Hence $\vec{F} + \frac{1}{2}R\hat{d}$ and $-\vec{F} + \frac{1}{2}R\hat{d}$ are equally inclined to $\hat{d}$; that is, these two forces are equally inclined to the central axis. So, (the resultant of $\vec{F}$ at A_1 and $\frac{1}{2}R\hat{d}$ acting at A) and (the resultant of $-\vec{F}$ at A_2 and $\frac{1}{2}R\hat{d}$ acting at A) are the required two forces. Proved.

(21): Let the two conjugate lines be $\vec{r} = \vec{a} + \lambda\hat{d}$ $\cdots(1)$

and $\vec{r} = \vec{b} + \mu\widehat{d'}$ $\cdots(2)$

be two straight lines. Suppose the force $F_1\hat{d}$ acting at $A(\vec{a})$ and force $F_2\widehat{d'}$ acting at $B(\vec{b})$ be two forces equivalent to the given system of forces. Let AB be the transversal line. Here A is the point of intersection of the transversal line and line (1). Also, B is the point of intersection of the transversal line and line (2) Equation of the plane containing line (1) and line AB is $(\vec{r} - \vec{a}) \cdot \left(\hat{d} \times (\vec{a} - \vec{b})\right) = 0$ $\cdots(3)$

Also, equation of the null plane of $B(\vec{b})$ is $(\vec{r} - \vec{b}) \cdot (\text{net torque about } B) = 0$, that is, $(\vec{r} - \vec{b}) \cdot \left((\vec{a} - \vec{b}) \times F_1\hat{d} + (\vec{b} - \vec{b}) \times F_2\widehat{d'}\right) = 0$ or, $(\vec{r} - \vec{b}) \cdot \{(\vec{a} - \vec{b}) \times \hat{d}\} = 0$ or, $\left((\vec{r} - \vec{a}) + (\vec{a} - \vec{b})\right) \cdot \{(\vec{a} - \vec{b}) \times \hat{d}\} = 0$ or, $(\vec{r} - \vec{a}) \cdot$

$\{(\vec{a}-\vec{b})\times\hat{d}\}=0$ or, $(\vec{r}-\vec{a})\cdot\{\hat{d}\times(\vec{a}-\vec{b})\}=0$, which is the same as (3).

Proved.

(22): For convenience, we shall use $A(\vec{a})$ and $A'(\vec{a'})$ for P and Q. Let the two conjugate lines be $\vec{r}=\vec{a}+\lambda\hat{d}$ ... (1)

and $\vec{r}=\vec{a'}+\mu\hat{d'}$ ... (2)

Suppose force $F_1\hat{d}$ be acting along (1) and force $F_2\hat{d'}$ be acting along (2). Here these two forces are equivalent to the given system of forces. Let the given plane Π intersect the two lines (1) and (2) at A and A' respectively. Let $K(\vec{k})$ be the null point of the plane Π. Equation of the null plane Π of $K(\vec{k})$ is $(\vec{r}-\vec{k})\cdot\left\{(\vec{a}-\vec{k})\times F_1\hat{d}+(\vec{a'}-\vec{k})\times F_2\hat{d'}\right\}=0$. Since it passes through $A(\vec{a})$ so, $(\vec{a}-\vec{k})\cdot\left\{(\vec{a}-\vec{k})\times F_1\hat{d}+(\vec{a'}-\vec{k})\times F_2\hat{d'}\right\}=0$ or, $\left[\vec{a}-\vec{k}\ \ \vec{a'}-\vec{k}\ \ \vec{d'}\right]=0$ or, $\left[\overrightarrow{KA}\ \ \overrightarrow{KA'}\ \ \overrightarrow{A'P}\right]=0$ for every point P of the line (2). So, either $A'P$ lie in the plane KAA'(that is, Π) or $\overrightarrow{KA}\times\overrightarrow{KA'}=\vec{0}$. The first possibility is, in general, not true, so, $\overrightarrow{KA}\parallel\overrightarrow{KA'}$. So, K, A, A' are collinear points. Proved.

(23): Let us take A as the origin $O(\vec{0})$. Let $B(\vec{b})$ and $C(\vec{c})$. Let $\hat{d}$ be the unit vector along a fixed direction through A. Let the given system of forces is reduced into three forces $\vec{F_1}\equiv F_1\hat{d}$ acting at A, $\vec{F_2}$ acting at B and $\vec{F_3}$ acting at C. Let $\vec{R}$ be the resultant force of the given system of forces, and let $\vec{G}$ be the net torque of all forces about the origin O (or A). Here $\vec{R},\vec{G},\vec{b},\vec{c},\hat{d}$ are constant vectors. From question, on taking torque about O, we get

$$\vec{G}=\vec{b}\times\vec{F_2}+\vec{c}\times\vec{F_3} \quad \cdots (1)$$

Now, on taking torque about B, we get

$$\vec{G}-\vec{b}\times\vec{R}=-\vec{b}\times F_1\hat{d}+(\vec{c}-\vec{b})\times\vec{F_3} \quad \cdots (2)$$

and, on taking torque about C, we get

$$\vec{G}-\vec{c}\times\vec{R}=-\vec{c}\times F_1\hat{d}-(\vec{c}-\vec{b})\times\vec{F_2} \quad \cdots (3)$$

Taking dot product with $\vec{c}-\vec{b}$ in (2), we get $(\vec{c}-\vec{b})\cdot(\vec{G}-\vec{b}\times\vec{R}) = F_1(\hat{d}\times\vec{b})\cdot(\vec{c}-\vec{b})$. So, $F_1 = \frac{(\vec{c}-\vec{b})\cdot(\vec{G}-\vec{b}\times\vec{R})}{(\hat{d}\times\vec{b})\cdot(\vec{c}-\vec{b})}$, which is a constant. So, $\overrightarrow{F_1}(= F_1\hat{d})$ is a constant vector. Since $\vec{R} = \overrightarrow{F_1}+\overrightarrow{F_2}+\overrightarrow{F_3}$, so $\overrightarrow{F_2}+\overrightarrow{F_3} = \vec{R}-\overrightarrow{F_1}$, which is a constant vector. From (3), $(\vec{b}-\vec{c})\times\overrightarrow{F_2} = \vec{G}-\vec{c}\times\vec{R}+\vec{c}\times\overrightarrow{F_1}$ or, $\vec{\alpha}\times\overrightarrow{F_2} = \vec{\beta}$ $\cdots(*)$ where $\vec{\alpha}\equiv\vec{b}-\vec{c}$ and $\vec{\beta}\equiv\vec{G}-\vec{c}\times\vec{R}+\vec{c}\times\overrightarrow{F_1}$, both are constant vectors. On taking cross product with $\vec{\alpha}$ in $(*)$, we get $\vec{\alpha}\times(\vec{\alpha}\times\overrightarrow{F_2}) = \vec{\alpha}\times\vec{\beta}$ or, $(\vec{\alpha}\cdot\overrightarrow{F_2})\vec{\alpha} - \alpha^2\overrightarrow{F_2} = \vec{\alpha}\times\vec{\beta}$. So, the lines of action of $\overrightarrow{F_2}$ constitute a fixed plane through $B(\vec{b})$: $\left[\vec{r}-\vec{b}\ \ \frac{\vec{\beta}\times\vec{\alpha}}{\alpha^2}-\vec{\beta}\ \ \vec{\alpha}\right] = 0$. Since $\vec{\alpha} = \vec{b}-\vec{c}$ so, the above plane can be written as: $\left[\vec{r}-\vec{b}\ \ \frac{\vec{\beta}\times\vec{\alpha}}{\alpha^2}-\vec{\beta}\ \ \vec{b}-\vec{c}\right] = 0$. Clearly this plane is satisfied by $C(\vec{c})$, so BC lies on the fixed plane $\left[\vec{r}-\vec{b}\ \ \frac{\vec{\beta}\times\vec{\alpha}}{\alpha^2}-\vec{\beta}\ \ \vec{b}-\vec{c}\right] = 0$. Similarly the lines of action of $\overrightarrow{F_3}$ constitute a fixed plane through $C(\vec{c})$ and it contains the line BC. So, the two planes intersect in the line BC. Proved.

(24): Let $\mu(\delta v)$ be the elementary mass of the body situated at $P(\vec{r})$. So, the given system of forces is equivalent to a single force $\sum(\mu(\delta v))\vec{F}$ acting at O, together with a couple whose torque is $\sum\vec{r}\times(\mu(\delta v)\vec{F})$. For continuous distribution of mass in the rigid body $\sum(\mu(\delta v))\vec{F}$ can be written as $\int(\mu\vec{F})dv$ and $\sum\vec{r}\times(\mu(\delta v)\vec{F})$ can be written as $\int\{\mu(\vec{r}\times\vec{F})\}dv$. Proved.

(25): Let $P(\vec{r})$ be a point on the surface of the rigid body subjected to fluid pressure p. Here force exerted by the fluid on the elementary surface δS enclosing P is $p((-\vec{n})\delta S)$. So, the given system of forces on the rigid body is equivalent to a single force $\sum p((-\vec{n})\delta S)$ acting at origin O, together with a couple of torque $\sum\{\vec{r}\times(p(-\vec{n})\delta S)\}$. For continuous surface, $\sum p((-\vec{n})\delta S)$ can be written as $-\int(p\vec{n})dS$ and $\sum\{\vec{r}\times(p(-\vec{n})\delta S)\}$ can be written as $\int\{(p\vec{r})\times\vec{n}\}dS$. Proved.

(26): In the first case, $\vec{G}\equiv$ (net torque about O of the rigid body) $= \int c\vec{n}\ dS = c\int\vec{n}\ dS = c(\vec{0}) = \vec{0}$, by Ex. 12 of chapter V. Further, since sum of two forces of a couple is zero, the body will be in equilibrium. In the second case, the given system of forces is equivalent to a single force $\vec{R}$ acting at the origin O, together with a couple of torque $\vec{G}$ where $\vec{R} = \sum(k\vec{n})\delta S$ and $\vec{G} = \sum \vec{r} \times \{(k\vec{n})\delta S\}$. For continuous surface $\sum(k\vec{n})\delta S$ can be written as $k\int \vec{n}\, dS = k(\vec{0}) = \vec{0}$, by Ex. 12 of chapter V. Further $\sum \vec{r} \times \{(k\vec{n})\delta S\}$ can be written as $k\int(\vec{r} \times \vec{n})dS$. Now we shall show that $\int(\vec{r} \times \vec{n})dS = 0$ over the closed surface. Let us imagine that the space within the closed surface is filled with small-sized cubes with common walls. It is easy to see that $\sum(\vec{r} \times \vec{n})dS$ for each cube is zero. Further we observe that the sum of all $\sum(\vec{r} \times \vec{n})dS$ for all cubes is equal to $\sum(\vec{r} \times \vec{n})dS$ over the given closed surface, because sum of $\sum(\vec{r} \times \vec{n})dS$ of two cubes over their common surface area is zero. Hence $\int(\vec{r} \times \vec{n})dS = 0$. Hence the body is in equilibrium. Proved.

(27): Let $\hat{a}, \hat{b}, \hat{c}$ be the unit vectors along the positive directions of x, y, z-axes respectively, and they are oblique coordinate axes. Let $P(x\hat{a} + y\hat{b} + z\hat{c})$ be any point on the plane $\frac{L}{\sin\alpha}x + \frac{M}{\sin\beta}y + \frac{N}{\sin\gamma}z = 0$ which passes through the origin O. Here net torque about O is given by $\vec{G} = L\left(\frac{\hat{b}\times\hat{c}}{\sin\alpha}\right) + M\left(\frac{\hat{c}\times\hat{a}}{\sin\beta}\right) + N\left(\frac{\hat{a}\times\hat{b}}{\sin\gamma}\right)$.

We have to prove that $\vec{G} \perp \overrightarrow{OP}$, that is, $\vec{G} \cdot \overrightarrow{OP} = 0$.

$LHS = \vec{G} \cdot \overrightarrow{OP} = \left\{L\left(\frac{\hat{b}\times\hat{c}}{\sin\alpha}\right) + M\left(\frac{\hat{c}\times\hat{a}}{\sin\beta}\right) + N\left(\frac{\hat{a}\times\hat{b}}{\sin\gamma}\right)\right\} \cdot (x\hat{a} + y\hat{b} + z\hat{c}) =$
$\left(\frac{L}{\sin\alpha}x + \frac{M}{\sin\beta}y + \frac{N}{\sin\gamma}\right)[\hat{a}\ \hat{b}\ \hat{c}] = 0[\hat{a}\ \hat{b}\ \hat{c}] = 0 = RHS$. Proved.

(28): Here condition of equilibrium is $\vec{0} = T(\kappa\hat{n}) + \frac{dT}{ds}\hat{t} + \vec{F} = \left(T\frac{d\hat{t}}{ds} + \frac{dT}{ds}\hat{t}\right) +$
$\vec{F} = \frac{d(T\hat{t})}{ds} + \vec{F} = \frac{d}{ds}\left(T\frac{d}{ds}\vec{r}\right) + \vec{F} = \frac{d}{ds}\left(T\frac{d}{ds}(x, y, z)\right) + (X, Y, Z) =$
$\frac{d}{ds}\left(T\left(\frac{dx}{ds}, \frac{dy}{ds}, \frac{dz}{ds}\right)\right) + (X, Y, Z) = \frac{d}{ds}\left(T\frac{dx}{ds}, T\frac{dy}{ds}, T\frac{dz}{ds}\right) + (X, Y, Z) =$
$\left(\frac{d}{ds}\left(T\frac{dx}{ds}\right), \frac{d}{ds}\left(T\frac{dy}{ds}\right), \frac{d}{ds}\left(T\frac{dz}{ds}\right)\right) + (X, Y, Z) = \left(\frac{d}{ds}\left(T\frac{dx}{ds}\right) + X, \frac{d}{ds}\left(T\frac{dy}{ds}\right) +\right.$
$\left. Y, \frac{d}{ds}\left(T\frac{dz}{ds}\right) + Z\right)$. Hence $\frac{d}{ds}\left(T\frac{dx}{ds}\right) + X = 0$, etc. Proved.

(29): Let β be the constant angle which each tangent to the given helical string makes with fixed direction of $\hat{a}$ of the axis of helix. Since $\frac{dT}{ds}\hat{t} + T\kappa\hat{n} + \vec{F} = \vec{0}$ and $T =$ constant, from question, so $\frac{dT}{ds} = 0$ and $\vec{F} = -T\kappa\hat{n} = -T\frac{d\hat{t}}{ds}$. Hence $\vec{F}\cdot\hat{a} = -T\frac{d\hat{t}}{ds}\cdot\hat{a} = -T\frac{d}{ds}(\hat{t}\cdot\hat{a}) = -T\frac{d}{ds}(\cos\beta) = -T(0) = 0$. So, $\vec{F} \perp \hat{a}$. Further we know that $\vec{F}$ lies in the plane of $\hat{t}$ and $\hat{n}$ passing through elementary mass. So, force on any element of the string is directly from the axis of the helix. Proved.

(30):

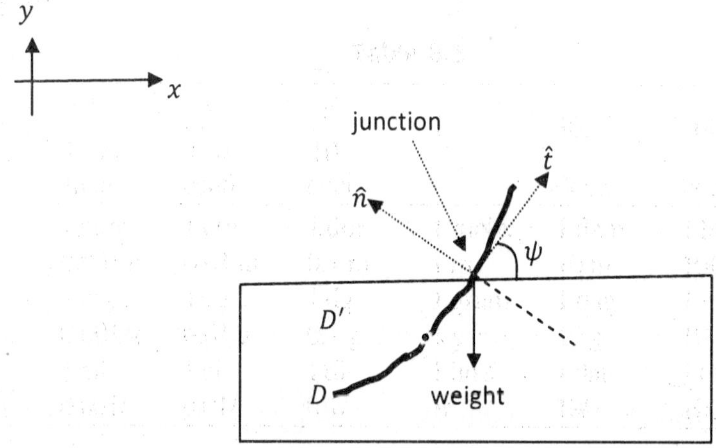

Let A be the cross-section of the wire. On using the equation $\kappa T + \vec{F}\cdot\hat{n} = 0$ we get $\kappa_{\text{out}}T_{\text{junction}} - \{(DA)g\}\cos\psi_{\text{junction}} = 0$ and, by Archimedes' principle, $\kappa_{\text{in}}T_{\text{junction}} - \{((D-D')A)g\}\cos\psi_{\text{junction}} = 0$. Hence $\frac{\kappa_{\text{in}}}{\kappa_{\text{out}}} = \frac{D-D'}{D}$. Proved.

(31): Let equation of the equiangular spiral in polar coordinates be $r = ke^{(\cot\alpha)\theta}$, where k, α are constants. Hence $\ln r = \ln k + (\cot\alpha)\theta$. On taking differential, we get $\frac{1}{r}\frac{dr}{d\theta} = \cot\alpha$ or, $\cot\varphi = \cot\alpha$ or, $\varphi = \alpha$. Hence $p = r\sin\varphi = r\sin\alpha$ and therefore $\frac{d\psi}{ds} = \kappa = \frac{1}{r}\frac{dp}{dr} = \frac{\sin\alpha}{r}$. Here equations of equilibrium are $\frac{dT}{ds} - \lambda g\sin\psi = 0$ and $T\kappa - \lambda g\cos\psi = 0$. From the second equation $T = \frac{\lambda g\cos\psi}{\kappa} = \frac{\lambda g\cos\psi}{\frac{\sin\alpha}{r}} = \frac{r\lambda g\cos\psi}{\sin\alpha}$. Hence $\lambda g\sin\psi = \frac{dT}{ds} =$

$\frac{d}{ds}\left(\frac{r\lambda g\cos\psi}{\sin\alpha}\right) = \frac{g}{\sin\alpha}\frac{d}{ds}(r\lambda\cos\psi) =$
$\frac{g}{\sin\alpha}\left(\frac{dr}{ds}\lambda\cos\psi + \frac{d\lambda}{ds}r\cos\psi - r\lambda\sin\psi\frac{d\psi}{ds}\right) = \frac{g}{\sin\alpha}\Big((\cos\varphi)\lambda\cos\psi +$
$\frac{d\lambda}{ds}r\cos\psi - r\lambda\sin\psi\frac{\sin\alpha}{r}\Big)$. Hence $\lambda\sin\psi = \frac{\lambda\cos\varphi\cos\psi}{\sin\alpha} + \frac{d\lambda}{ds}r\frac{\cos\psi}{\sin\alpha} - \lambda\sin\psi$ or,
$\lambda\cos\varphi\cos\psi + \frac{d\lambda}{ds}r\cos\psi - 2\lambda\sin\psi\sin\alpha = 0$. Now multiplying by $\cos\psi$, we get $\lambda\cos\varphi\cos^2\psi + \frac{d\lambda}{ds}r\cos^2\psi - 2\lambda\sin\psi\cos\psi\sin\alpha = 0$ or, $(\lambda\cos^2\psi)\frac{dr}{ds} +$
$\frac{d\lambda}{ds}r\cos^2\psi - 2\lambda\sin\psi\cos\psi\left(r\frac{d\psi}{ds}\right) = 0$ or,

$\cos^2\psi\left(\lambda\frac{dr}{ds} + r\frac{d\lambda}{ds}\right) + \left(\frac{d}{ds}\cos^2\psi\right)\lambda r = 0$ or,

$(\cos^2\psi)\frac{d}{ds}(\lambda r) + \left(\frac{d}{ds}\cos^2\psi\right)\lambda r = 0$ or, $\frac{d}{ds}(\lambda r\cos^2\psi) = 0$ or, $\lambda r\cos^2\psi =$ constant. Proved.

(32): Using the figure in the above Ex. 30, we have $\frac{dT}{ds} - (\sin\psi)\lambda g = 0$ and $\kappa T - (\lambda g)\cos\psi + N = 0$, where λ denotes mass per unit length of the wire (that is, $\lambda = DA$) and N denotes the normal force exerted by the smooth curve on the heavy string. From the first equation $dT = \lambda g\sin\psi\, ds = \lambda g\frac{dy}{ds}ds =$ $\lambda g\, dy$. On integration, $T_B - T_A = \lambda g(y_B - y_A)$. This proves the first part of the question. Further, $N = \lambda g\cos\psi - \kappa T = \lambda g\cos\psi - \kappa\big(T_A - \lambda g(y_A - y)\big)$. Ans. If the string is light then $\lambda \approx 0$ and hence $\frac{dT}{ds} = 0$ or, $T =$ constant. Proved. Also, $N = -\kappa T = (\text{constant})\kappa \propto \kappa$. Proved.

(33): Using the figure in the above Ex. 30, we have $\frac{dT}{ds} - (\sin\psi)\lambda g + \mu N = 0$ and $\kappa T - (\lambda g)\cos\psi + N = 0$, where λ denotes mass per unit length of the wire (that is, $\lambda = DA$), N denotes the normal force exerted by the smooth curve on the heavy string and μ is the coefficient of friction. Given that $\lambda \approx 0$, so $\frac{dT}{ds} + \mu N = 0$ and $\kappa T + N = 0$. Hence $\frac{dT}{ds} + \mu(-\kappa T) = 0$ or $\frac{dT}{T} = \mu k\, ds$. On integration, $\ln\frac{T_f}{T_i} = \mu(\psi_f - \psi_i)$, where suffix i, f indicate initial and final respectively. Hence $\frac{T_f}{T_i} = e^{\mu(\psi_f - \psi_i)} = e^{\mu(\text{angle between the tangents at the two points})}$.

Proved.

(34): Using the figure in the above Ex. 30, we have $\frac{dT}{ds} - (\sin\psi)\lambda g + \mu N = 0$ and $\kappa T - (\lambda g)\cos\psi + N = 0$, where λ denotes mass per unit length of the wire (that is, $\lambda = DA$), N denotes the normal force exerted by the smooth curve on the heavy string and μ denotes the coefficient of friction. Ans.

FIVE LAWS OF INERTIAL FRAME*

ABSTRACT: *In order to establish the celebrated time dilation and length contraction formulae, Einstein used only two axioms. On the other hand he tacitly used many other also. Further there are three objectionable things in Einstein's axiom: all laws of physics are the same in all inertial frames. Firstly, it is not a single axiom, but a bundle of axioms. Secondly, "physics" is not a defined term, but a vague term. One should avoid using this term in stating fundamental laws of nature. Thirdly, since the subject physics is going through evolution, no one can ever claim that the complete list of laws of physics has been found out. So, the phrase "all laws' looks awkward.*

In this paper, I have tried to rectify these anomalies. I have given minimum number of axioms, so that the above said two formulae can be derived. In the derivation, all the axioms are used up, and are well-indicated. These axioms have been named here as "the five laws of inertial frame". Further, Einstein established these formulae with the help of two hypothetical experiments, while our derivation uses only one. The five axioms, given below, are the exact foundation for the special theory of relativity. It naturally turns out that length and velocity are fundamental quantities, while time is a derived quantity – a contrast to the popular view.

LAWS OF INERTIAL FRAME

1. In all inertial frames, speed of light in vacuum is the same.
2. An event occurring in one inertial frame must occur in all other inertial frames.

* This paper of the author was published in the journal Acta Ciencia Indica. Vol. XXXV P. No. 4. 665 (2009).

3. In all inertial frames, light ray travels in a straight line.
4. In all inertial frames, laws of reflection of light are valid for mirror moving with constant velocity.
5. For all inertial frames, longitudinal length of an object is the same.

In the above laws, the terms inertial frame, speed of light, event are self-explanatory. We shall take distance s and speed v as more primitive concepts than time t.

For this we define duration as follows:

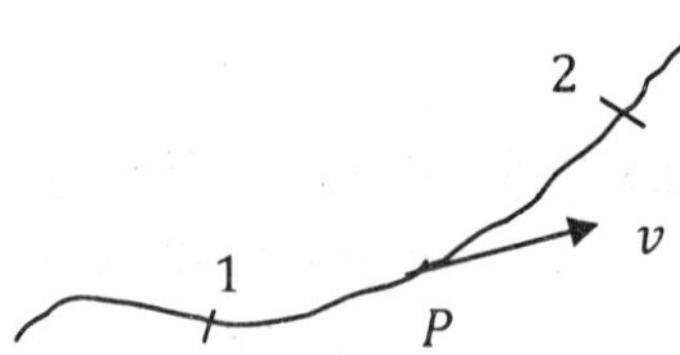

(duration in going from position $1(s = s_1)$ to position $2(s = s_2)$ with speed v at position P) $\equiv \int_{s_1}^{s_2} \frac{1}{v} ds$.

Using our five laws, we shall try to derive the following well-known formulae:

$$\Delta t = \gamma(\Delta t_0) \quad \text{(dilation of time)}$$

and

$$\Delta l = \frac{\Delta l_0}{\gamma} \quad \text{(length contraction)}$$

Let S be an inertial frame in which a lightclock with two mirrors at its ends, is fixed as shown in Figure (1):

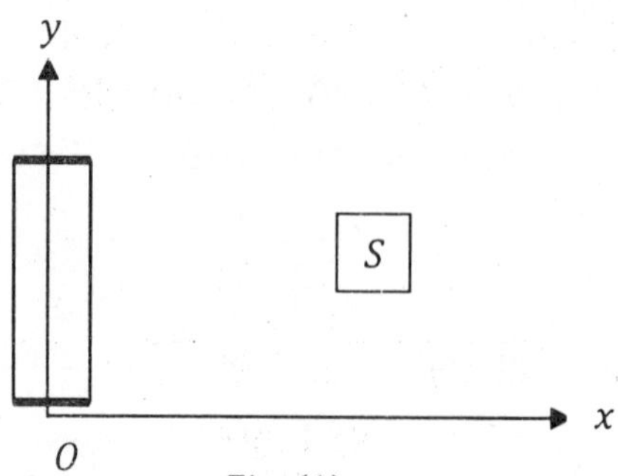

Fig. (1)

Let S' be an inertial frame with origin O' and coordinate axes x' and y'. Let O' be moving along the negative direction of x-axis with constant speed $c\beta$, where $0 < \beta < 1$, and c is the speed of light in vacuum. By **Law (1)**, c has the same value for all inertial frames. Let the positive directions of x'-axis and x-axis be the same, and let the positive directions of y'-axis and y-axis are the same. Let a rod AB be placed on the x'-axis of the frame S' such that its left end A is situated at O' (see Figure (2)):

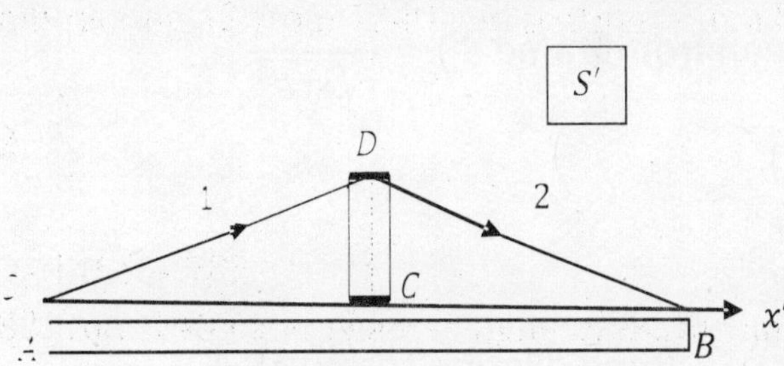

Fig. (2)

Suppose that, as O coincides with O', $t = 0 = t'$ and a light signal starts travelling along the positive y-axis in frame S. Let the length L of the lightclock CD be such that as the light returns back to O in frame S, it strikes the right end B of the rod AB in frame S'.

By the definition of duration,

$\Delta t_0 \equiv$ (duration in going from lower end C of the mirror to the upper end D of the mirror) $= \left(\frac{1}{c}\right) L$ ... (1)

where c denotes the constant speed of light in all frames.

Here "incidence of light with upper end mirror" is an event that occurred in frame S, so, by **law (2)**, this event will also occur in frame S'. Let D be the position of the upper mirror, when the light ray incident on it. By **law (5)**, the y'-coordinate of D is L. Also, by **law (3)**, the path of the incident light will be the straight line joining O' and D in frame S'

Here, in order to occur the incidence of light ray 1 with the upper mirror in frame S', the "horizontal" component of the velocity of light in S' must be equal to the speed, βc, of the frame S as seen from S'.

In frame S', upper end mirror of the lightclock is moving, so, by **law (4)**, incident light $O'D$ will reflect at D such that $O'C = CB$.

Now, by **law (1)**, vertical component of the velocity of light in frame S' is $\sqrt{c^2 - (\beta c)^2}$ $\left(= c\sqrt{1-\beta^2}\right)$. Hence

$\Delta t \equiv$ (duration in going from the lower end of the mirror to the upper end of the mirror, as seen from frame S') $= \frac{L}{c\sqrt{1-\beta^2}}$ $\cdots(2)$

From (1) and (2),

$$\Delta t = \frac{\Delta t_0}{\sqrt{1-\beta^2}}$$

or, $\Delta t = \gamma(\Delta t_0)$ (time dilation formula)

where $\gamma \equiv \frac{1}{\sqrt{1-\beta^2}}$.

Now observer in S and observer in S' want to measure the length of the rod in their respective frames. For observer in S', rod is static. They argue that light ray with horizontal velocity βc takes duration $2\big(\gamma(\Delta t_0)\big)$ in covering the same length as that of the horizontal rod, so the length l_0 of the static rod in frame S' is given by

$$l_0 = (\beta c)(2(\gamma(\Delta t_0))) \qquad \cdots(3)$$

Now, since rod is fixed in S' so, for observers in S, rod is moving towards the negative x-direction with the same speed $(= \beta c)$ as that of frame S'.

Hence, for observer in S, rod takes duration $2(\Delta t_0)$ in order to disappear from the lower end mirror. Hence for observer in S, length of moving rod is given by

$$l = (\beta c)\big(2(\Delta t_0)\big) \qquad \cdots(4)$$

So, from (3) and (4),

$$l = \frac{l_0}{\gamma} \qquad \text{(length contraction formula).}$$